LUNAR GRAVIMETRY

ASTROPHYSICS AND SPACE SCIENCE LIBRARY

VOLUME 273

LUNAR GRAVIMETRY

Revealing the Far-Side

by

RUNE FLOBERGHAGEN

ESA/ESTEC,
Noordwijk, The Netherlands

KLUWER ACADEMIC PUBLISHERS

DORDRECHT / BOSTON / LONDON

A C.I.P. Catalogue record for this book is available from the Library of Congress.

ISBN 978-94-015-7117-3   ISBN 978-90-481-9552-7 (eBook)
DOI 10.1007/978-90-481-9552-7

Published by Kluwer Academic Publishers,
P.O. Box 17, 3300 AA Dordrecht, The Netherlands.

Sold and distributed in North, Central and South America
by Kluwer Academic Publishers,
101 Philip Drive, Norwell, MA 02061, U.S.A.

In all other countries, sold and distributed
by Kluwer Academic Publishers,
P.O. Box 322, 3300 AH Dordrecht, The Netherlands.

Front Cover Picture:
This colour image of the Moon was taken by Galileo at 9:35 a.m. PST, December 9, 1990,
at a range of about 350,000 miles. The colour composite uses monochrome images taken through
violet, red, and near-infrared filters. The concentric, circular Orientale Basin, some 1000 km
across, is near the centre, the near-side is to the right, and the far-side to the left. The image
beautifully captures the dichotomy of the two lunar hemispheres. *JPL image P-37329.*

http://www.lpi.usra.edu/expmoon/galileo/galileo.html

Printed on acid-free paper

To Silvia,

for your endless love and tenderness

Preface

In today's specialised world, scientists and engineers, in an ever-increasing fashion, "feed and breed" on detailed issues of specific and narrow research fields. On the other side, the greater and often multi-disciplinary challenges to be tackled are steadily becoming more complex. Being a scientist therefore means more than being a specialist. Successful scientific endeavour equally depends on the ability to bridge gaps and create cross-links between isolated research niches and related fields. In most branches there is a fascinating story to tell that puts together the individual components of a greater whole and also succeeds in explaining the details and particularities of a problem to non-specialists. As so suitably told in an old Chinese proverb: "When the wise man points to the Moon, the ignorant look at his finger". These words lie at the heart of the book you are now about to read. In a complex reality, in a certain sense, we are all ignorant in one field or the other. Undeniably, the Leonardo da Vincis of the present time are rare and far apart.

Obviously, such considerations are not restricted to lunar science alone, but apply to the study of all "open systems" characterised by widespread interactions with neighbouring systems and subsystems. Most if not all geosciences and planetary sciences, for example, fall into this category. The discipline called *lunar gravimetry* is no different. It is a research field with a growing number of active players, in particular due to the new boost given to lunar science over the past decade. Nonetheless, in my opinion, the consequence of specialisation is that researchers in many ways increasingly depend on the work and words of other specialists from affiliated domains. In other words, the perspective on lunar gravity field models tends to be slightly different for model developers and for end users.

There is always a caleidoscope of scientific and technical issues at the foundation of a "science product". A satellite-based gravity field model is for example derived in an inverse problem formulation from the orbit perturbations due to all forces acting on satellites in low lunar orbit. Therefore, all the complex issues of orbit computation are related to the quality of the final product. Another example is the role of the under-sampling of the gravimetric problem, which demands for the use of *a priori* information, if the goal is to derive a global solution. The quality assessment of lunar gravity field models as well as the understanding of the trade-offs and choices made during the data reduction process are therefore issues that are largely hidden to non-specialists in lunar satellite geodesy, or selenodesy.

This book is my humble attempt, for what concerns the Moon's gravity field, to bridge the gap between the wise man's finger and the core of the matter (literally the Moon). Although the book certainly does not pretend to *be* the wise man himself, it is my hope that the reader upon reading it will recognise and better understand the issues involved in satellite-based gravity field model development in general, and all the peculiarities of the lunar gravimetric problem in particular. Then and only then may the book be viewed upon as successful in fulfilling its task.

Over the past years I have grown to believe that something I can only loosely describe as "participation and anticipation" is a fundamental driver for any science. The fact that someone, somewhere, actually cares about what one is doing, and is impatiently waiting to see the results, is probably the key factor that motivates to push for new achievement. It is also a crucial element of the overall relevance of a research effort. Without such involvement it is my personal belief that science would quite frequently be an unrewarding occupation.

In that sense, I have been fortunate to enjoy the support of numerous people throughout the research period, all of whom deserve credit for helping me make this book the product it has become. In the first place, Karel Wakker and Boudewijn Ambrosius are gratefully thanked for offering me a position within their organisation, where I enjoyed the freedom to individually and in cooperation with international partners, build up a (small, but fine) lunar gravity field research branch. It is my hope that I by this end product have paid respect to the faith they put in me. Obviously, no research can be done without funding. The Research Council of Norway is greatly acknowledged for generously granting me what I needed to enjoy a decent living while working on my favourite topic. In terms of on-topic discussions I have fiercely enjoyed the discussions with Pieter Visser. In the course of the past four to five years it appears to me that we have developed similar views on satellite-based gravity field modelling and the underlying orbit computation problem. I am sure that without his contribution, the book would have been a less attractive document on the current and near-future state-of-affairs in lunar gravimetry.

On the international side, the interest, suggestions and comments from several people and groups cannot be underestimated. I am much indebted to Oliver Montenbruck, Eberhard Gill and Frank Weischede of DLR/GSOC, not only for their keen interest in lunar and planetary orbit and gravity field modelling in general, but in particular for setting up (and paying) the tracking campaign of Lunar Prospector, using DLR antenna hardware. Many thanks for the free ride. Likewise, I owe my participation in the selenodetic experiments foreseen for SELENE to a number of people within NAO and ISAS, in particular Kosuke Heki of NAO. You definitely gave me an excellent opportunity to place parts of my research within an on-going project. Hopefully, this book is useful for you in your further preparations for the actual data reduction. *Arigato*. Alex Konopliv of JPL and Frank Lemoine of NASA/GSFC are both gratefully thanked for providing the full information matrices of their lunar gravity field models. Without such information, any attempt to assess the true quality of the models would be in vain. Dave Rowlands of NASA/GSFC is equally acknowledged for providing the GEODYN II software, as well as for technical assistance in connection with the use of the software. Massimiliano Vasile of Politecnico di Milano, finally, gave me some new impulses and perspectives on the low lunar orbit problem.

Other people also contributed, some even without knowing it. Specifically, Johannes Bouman brought in several ideas concerning regularisation and quality assessment. Radboud Koop was a most critical reader of draft versions of the manuscript. People close to me will know that that is something I really appreciate. Wencke van der Meulen reprocessed the Lunar Prospector measurements and also made a start with simulations for the SELENE mission. Sander Goossens has taken up the challenge to further develop the lunar gravity field work, with particular emphasis on local and regional gravity field estimation based on compactly supported basis functions. If I am not mistaken, that will be an excellent continuation of my own humble efforts, as well as a solid contribution to the advancement of selenodesy in general.

Delft, The Netherlands, October 2001
Rune Floberghagen

Summary

The synchronisation of the orbital and rotational motions of the Moon, driven by tidal dissipation, directly limits the spatial sampling of the lunar gravity field by means of conventional tracking techniques to slightly more than 50% of the lunar sphere. As a consequence, the gravimetric problem of dynamically solving for the global gravity field from the same satellite observations is severely ill-posed. The main goal of this book is to give a thorough analysis of present-day lunar gravimetry and, subsequently, discuss and analyse satellite measurement concepts that are capable of lifting the current knowledge to a new level. Beginning with the the scientific and technical motivations for lunar gravimetric research, the book describes the numerical and physical problems inherent to the conventional tracking concepts, predominantly Doppler tracking by Earth-based stations. In the second part of the book, *satellite-to-satellite tracking* (SST) measurement concepts are advocated as a well-performing and relatively cost-effective means to achieve a fully global sampling of the gravity field, and, hence, a much better determination of the gravity field. Although likely to be suitable for the task, a detailed discussion of *satellite gravity gradiometry* (SGG) as an alternative to SST, and presently under study for high-resolution terrestrial gravity field recovery, is outside the scope of this work. SGG principles and techniques are therefore only discussed in general terms.

A very logical companion to words like "determination" or "improvement" of a gravity field model is obviously the aspect of quality assessment. Since global lunar gravimetry is a case of an ill-posed problem taken to an extreme, a gravity field solution based on a given set of Earth-based tracking data set is not unique. Rather, a whole range of design parameters and design "philosophies" are found to influence the end product. In this book quality is therefore mainly measured on the basis of numerical information, extracted from the least squares adjustment process, and on the basis of satellite orbit information, such as precise orbit determination results. In the absence of adequate data sets for validation and verification, these two types of information are in fact the only ones directly available to the analyst. A fundamentally different approach would be to focus on the selenophysical interpretation of a gravity field model. Such methods are, however, only briefly discussed in the present work. Rather, it is chosen the emphasise the true information content of present and near-future gravity field solutions, essentially in honour of the fact that it is the physics of the Moon that one wants to derive from (among others) the gravity field, and not the other way around. This point of view does, of course, not argue against any kind of physical interpretation of a gravity field model, but simply states that such analysis in itself does not constitute an objective assessment technique with a clear relation to the data reduction process.

In terms of orbit-related information it is shown that the behaviour of recent lunar gravity field models, developed by different research groups over the past decade, is far from coherent. Similarly, when zooming in on the numerical properties of the normal equations and on the role played by *regularisation methods* and *regularisation parameters* intended to compensate for the incomplete sampling of the problem, it becomes clear that, in spite of a very high-quality sampling of the near-side, the uncertainty (or error) in global models is still significant. These facts are in essence the biggest motivating factors for a dedicated lunar gravity field experiment in the near future. Likewise, by means of an experiment aimed to improve an existing model by the inclusion of new data, it is equally striking that high-quality orbit solutions, on which the gravity field solutions are based, require uniform and near-continuous sampling of the force field.

Semi-analytical covariance analysis based on a linear perturbation theory as well as full-scale simulations of a *low–low* mode SST experiment clearly show the prospects of inter-satellite tracking for the lunar sciences. Even though the achievable accuracy and spatial resolution depends on a whole range of instrument parameters as well as the (improved) quality of the modelling of non-gravitational phenomena, the overall trend is that near-future lunar gravity field solutions can be expected to be self-contained, *i.e.* not needing numerical regularisation, up to a high degree and order. One experiment shows that the achievable maximum spherical harmonic degree and order approximates that of currently available regularised models. In other words, models of degree and order 90 – 100, corresponding to a half-wavelength resolution of $\sim$50 km, seem within reach. As a direct comparison, this number presently lies somewhere in the range 10 – 12, or $\sim$500 km. As such, SST techniques are likely to lead to an order of magnitude improvement in the knowledge of the global selenopotential.

Contents

Chapter *1*

Introduction

'*...a mission that includes either satellite-to-satellite tracking or a gravity gradiometer to permit direct gravity mapping of the farside should remain a high priority for lunar science.*'

F. G. Lemoine *et al.*, 1997

The study of the shape and gravitational attraction of celestial bodies has a longstanding history. Already more than 2000 years ago Greek astronomers realised that the Earth is basically spherical in shape, and efforts were made to determine its radius by means of geometric methods. Galilei was the first man who perceived that mathematics and physics were going to join forces, and he was able to unify celestial and terrestrial phenomena into one theory, destroying the traditional division between the world above and the world below the Moon. After Galilei articulated gravity as a uniform acceleration, Huygens suggested it was the natural quantity to define the unit of length, being the constant relating the period of oscillation with the length of a pendulum. This idea was, however, short-lived since oscillation periods were soon found to exhibit latitudinal variations. Following Newton's discovery of the law of gravitation, Huygens and Newton postulated a pole-flattened equilibrium figure for the Earth, an idea which even today is used for simple geometrical modelling of the Earth's figure. Another major breakthrough soon came with Clairaut's study of equilibrium shapes of rotating fluids, which led to the concept of a rotating ellipsoidal Earth. His observations advanced both a geodetic application (the shape of the Earth) and provided a possible geophysical interpretation (hydrostatic equilibrium), and, more importantly, marked the beginning of physical geodesy and potential theory.

The development of the least-squares estimation technique by Gauss and Legendre , as well as its application in geodesy, soon showed that the ellipsoidal Earth did not match reality. Finally, this led to the introduction of a shape based on the physics of the Earth: the equipotential surface of the gravity field. Originally named the *mathematical* shape of the Earth by Gauss, it was later renamed the *geoid* by Listing. Ever since, gravimetry has been a cornerstone of geodesy and geophysics, indispensable in determining the detailed shape and potential differences on the surface of the Earth and other planets, as well as constraining density models for their internal structure.

The launch of *Sputnik–1* on 4 October 1957 initiated the era of satellite geodesy. Measurements of time-variations of a satellite orbit yield a comprehensive knowledge of the satellite motion under influence of all acting forces, as well as the description of the position of satellites and (possibly) ground stations in a suitable reference frame. Hence, through satellite orbit perturbation analysis, it became possible to deduce parameters of the underlying force models, amongst them the gravitational potential. The roots of satellite geodesy can, however, be traced back as far as the early nineteenth century when Laplace in 1802 determined the dynamical flattening of the Earth from the motion of the lunar node, thereby essentially treating the Moon as an artificial satellite. In the years prior to Sputnik–1, further foundations to satellite geodesy were given, which in turn made it possible to obtain significant results very soon after the launch of the first satellite. One of the first outstanding results was the accurate determination of the Earth's flattening, or essentially the second spherical zonal harmonic coefficient of the gravity field, from observations of *Sputnik–2* and *Explorer–1* [*e.g., Merson and King-Hele*, 1958; *O'Keefe*, 1959]. In 1959 the third zonal harmonic (the pear shape of the Earth) was determined [*O'Keefe*, 1959], followed by the rapid development of adequate theories of artificial satellite motion [*e.g., Brouwer*, 1959; *Kozai*, 1959, 1962, 1966; *Kaula*, 1966; *Aksnes*, 1970]. A neat summary of all the major milestones in the development of theories of artificial satellite orbits is provided by *Wnuk* [1999].

Thereafter, progress was rapid. Orbital analysis of a range of satellites with different orbit characteristics evidently yielded much more accurate geopotential models over long wavelengths. Today, through continuous improvement of tracking systems and the choice of suitable measurement types exhibiting strong observability over a wide range of orbital frequencies, satellite methods almost exclusively define the coarse scales of state-of-the-art gravity field models. In the near future, techniques like *satellite gravity gradiometry* (SGG) and *satellite-to-satellite tracking* (SST) will dominate the determination of the geopotential down towards 50 km spatial resolution [*Watkins et al.*, 1995; *ESA*, 1996, 1999, 2000; *Balmino et al.*, 1998; *Woodworth et al.*, 1998; *Davies et al.*, 1999; *Visser*, 1999; *Mazanek et al.*, 2000]. The advent of *satellite altimetry* in 1975 [*Stanley*, 1975] moreover revolutionised the short-wavelength determination of the marine gravity field. Directly related to this, the required orbit determination accuracy of altimetric satellite missions has also motivated significant advances for the longer wavelengths [*Nerem et al.*, 1995]. Finally, temporal variations in the gravity field of the Earth are currently being investigated by means of satellite techniques, primarily by analysis of long time series of satellite laser ranging measurements, a method first demonstrated by *Yoder et al.* [1983]. A new elan in the field of time-variable gravity field estimation is expected with the realisation of the GRACE mission, which will employ SST techniques for pseudo-monthly solutions of the Earth's gravitational potential [*Wahr et al.*, 1998].

From the 1960s on, satellite geodesy also found its natural extension in *planetodesy*, meaning the investigation of the figure, topography, rotation parameters and gravitational potential of neighbouring planets and moons. Exploration and scientific studies of our neighbours in the Solar System began with the race to the Moon and the first pictures of the far-side of the Moon taken by *Luna–3* in 1959, and have since been extended to include two of the other terrestrial planets, Venus and Mars. The gravity fields of these planets have been determined from Earth-based tracking of spacecraft, for the larger part performed by the *Deep Space Network* (DSN), owned and operated by the U.S. National Aeronautics and Space Administration (NASA), with stations at Goldstone, Madrid and Canberra. Recent solutions include (for Mars) *Smith et al.* [1993, 1999b] and (for Venus) *Konopliv et al.* [1999]. Flybys have moreover allowed the determination of the gravity parameter (mass) and a few lower degree harmonic coefficients for the outer planets [*e.g.*, *Thomas*, 1991] and their natural satellites [*Haw et al.*, 2000], and the combination of radio tracking and camera observations has allowed for the determination of the global shape and gravitational potential of small satellites and asteroids [*e.g.*, *Thomas*, 1993; *Yeomans et al.*, 1997; *Rossi et al.*, 1999]. Generally speaking, the latter type of planetodetic parameter estimation is of a rather poor quality, and allows only limited geophysical interpretation; in any case, such interpretation is frequently based on additional assumptions, *e.g.* concerning density or state of load compensation.

The gravity field of the Moon, being our closest neighbour in the Solar System, was first investigated in 1966 when the Russian *Luna–10* was placed in orbit around the Moon and provided dynamical proof that the oblateness of the Moon's gravitational potential (the second zonal harmonic) was larger than the shape predicted from hydrostatic equilibrium [*Akim*, 1966]. The initial results also indicated the pear shape of the equipotential surface of the gravitational potential and the related offset of the centre-of-mass with respect to the centre-of-figure. A second milestone was the discovery of mass concentrations, or *mascons*, from the residual signature of *Lunar Orbiter line-of-sight* (LOS) Doppler data [*Muller and Sjogren*, 1968]. Mascons are significant positive gravity anomalies associated with areas of high density under the ringed lowland maria, mainly found on the near-side of the Moon. Such buried and largely uncompensated mascons contain information on the impact processes that led to mascon formation and possibly, through the strength of the lithosphere (the Moon's rigid outer shell), information on the early thermal history of the Moon. The mascon discovery was the first evidence of medium-wavelength gravity field variations on the Moon. Moreover, mascons proved to be of immense practical importance during the *Apollo* missions, where they pulled the *Lunar Module* spacecraft away from their prescribed trajectories and endangered the crew safety. One famous example is that of the *Apollo 11* landing, where it was only by manual control that astronaut Neil Armstrong was able to safely land, some five kilometres downrange of the target landing area and with only a few seconds left of propellant [*Dooling*, 1994]. This pure engineering problem triggered a significant effort in modelling the mascons as well as the global selenopotential in the late 1960s and early 1970s.

Despite a significant lobby for follow-on missions to the projects of the Apollo era, such a project was not to materialise until 1994, when the *Clementine* probe was inserted into lunar orbit [*Nozette et al.*, 1994]. Clementine was, however, originally not designed for mapping of the lunar surface [*Garret and Rustan*, 1995], and therefore suffered from unfortunate orbital characteristics, which were chosen to optimise the mission's primary military purpose. Truly significant gravity field improvements were, therefore, not achieved until *Lunar Prospector*, as part of NASA's *Discovery* program of *faster, better, cheaper* science missions [*Huntress*, 1999], entered lunar orbit in early 1998. A full year in a polar orbit at a mean altitude of 100 km, plus a six-months extended phase at a mean altitude of 25 km, with periapsis as low as 15 km, have delivered excellent data sets suitable for high-resolution nearside gravity modelling [*Konopliv et al.*, 1998; *Konopliv and Yuan*, 1999; *Konopliv et al.*, 2001; *Arkani-Hamed*, 1999b; *Hubbard et al.*, 1999].

However, as long as spherical harmonic models, which require globally defined basis functions, remain the preferred mathematical representation formula of the gravitational potential, the lack of direct far-side tracking contributes to aliasing in the models. This sampling problem is a consequence of Cassini's laws, which describe the phase-lock of the Moon's rotational and orbital motions due to tidal dissipation. Whereas the spherical harmonics form an orthonormal basis for functions uniformly sampled over the surface of the sphere, the actual tracking data distribution only allows unambiguous determination of certain linear combinations of harmonics. In lunar gravity field modelling from Earth-based Doppler measurements, this is taken to an extreme, as data collection beyond the limbs and the poles is only marginally possible, and the sampling basically takes place over one hemisphere only. As an aid to the estimation process, it is common practice to invoke prior constraints (mathematically one speaks of *regularisation*) based on the early observation [*Kaula*, 1966] that the variance spectrum of the gravitational potential is well approximated by a power law. Originally developed for the geopotential, this so-called *Kaula's rule of thumb* was soon extended to other planets, based on the assumption of equal stress, and in many cases later modified when more satellite tracking data became available, [*e.g., Tscherning and Rapp*, 1974]. The use of this *a priori* constraint helps partition the data variance between the spherical harmonic degrees, but gives no guidance about how the variance should be partitioned within each degree. As a result, lunar gravity field models, as do many other planetary gravity field models, exhibit features believed to represent artifacts resulting from a combination of an uneven data distribution and occasionally overzealous application of existing constraint schemes [*Bills and Lemoine*, 1995].

The merit of such constraints is basically found in improved tracking data fits of the models derived thereof. Moreover, the dynamic effect of far-side orbit perturbations is somewhat identified from the integral behaviour of the satellite orbit. It is therefore believed that some far-side features of the gravitational potential are for real. However, several problems still arise. The first problem concerns observability: the combination of observing satellite orbit changes due to the accumulated acceleration of the far-side gravity field as the satellite comes out of occultation and a general power law constraint is never going to replace a fully global data set. One is fundamentally interested in addressing the spatial variations of the gravitational potential and underlying mass and stress distributions of the Moon, and this requires adequately sampled satellite data. Second, there is a danger of premature reasoning: as constraints lead to improved measurement fits, they tend to increase the confidence in the solutions, and, hence, the derived gravity field model becomes the *true* gravity field on the basis of which a significant level of geophysical analysis may be performed. In reality, however, it is only the global satellite orbit measure which fits well with the data. A third and related problem concerns the size of the models. In order to exhaust the increasingly good near-side data, an extended spherical harmonic parameterisation is required. The global nature of the basis functions does, however, not allow for such a determination, a fact which is also generally reflected in the error measures. Therefore, there is a pitfall danger that the extended series of spherical harmonic coefficients serves as absorption parameters during the satellite orbit determination process. Fourth and finally, for the determination of the gravitational structure of the Moon at a range of scales it is required that the measurements contain information on all harmonic coefficients defining that same scale. Hence, for detailed gravity mapping one would like to have a measurement technique which works well for both the long wavelengths (global structure), medium wavelengths (basins and mascons) and relatively short-wavelengths (smaller craters, transition zones, etc.). Such a mapping requires an augmentation of the information which can be derived from Earth-based Doppler observations alone.

   Kaula [1969] realised these problems. One year after the discovery of the lunar mascons he wrote, [quote]: *'The problem of how to use the lunar satellites most effectively to determine the gravitational field on the back side of the Moon must be regarded as still unsolved. Possible solutions are: (i) a greater variety of orbital inclinations; (ii) a satellite-to-satellite tracking system; (iii) satellite-born laser altimetry; or (iv) satellite-born measurements of gravity gradients.'*, [unquote]. This was, most probably, the first proposal for extra-terrestrial use of non-conventional tracking techniques. Indeed, more than thirty years later, and despite the undisputable improvements achieved over the past years, the problem remains the same.

   The topic of this book is the transition phase from the present era of conventional satellite-based techniques into a decade of lunar gravimetric research, when global lunar satellite tracking data are expected to become available [*e.g., Namiki et al.*, 1999; *Matsumoto et al.*, 1999; *Heki et al.*, 1999; *Kaneko et al.*, 1999; *Sasaki et al.*, 1999]. Starting with a thorough review and analysis of present knowledge of the gravitational potential, including both orbit and orbit error behaviour as well as the quality assessment of selenopotential models, and regularisation, this work presents a study of SST-based gravity field mapping of the Moon. The overall objective is to illustrate the quality and limitations of currently available models as well as the prospective benefits of a truly global data set. The outline of the book is as follows. Chapter 2 discusses, in general terms, the rationale for lunar gravity mapping, *i.e.* the gravity-related science of the Moon and the overall benefits of a new mission providing global data. Moreover, the research objectives are outlined in further detail. Chapter 3 presents an analysis of current state-of-the-art models on the basis of satellite orbit information. Error studies based on calibrated covariances are complemented by orbit propagation results. Moreover, the models are compared in terms of their long-term behaviour, an aspect not unimportant for mission design, and in particular for SST configurations, for which one of the two (or more) spacecraft involved is likely to be a small, free-flying and non-propelled craft deployed from a larger lunar orbiter. Chapter 4 deals with the assessment of the true information content provided by the low lunar satellite tracking data sets, and the resulting ill-conditioning of the lunar gravimetric inverse problem. The general role of regularisation is discussed, with particular emphasis on the problem of the possibly overzealous application of constraint schemes. A crucial part of the analysis is a discussion on the perspective on the underlying estimator, which is typically understood as a case of unbiased collocation. However, collocation does not yield the optimal solution in the case that the regularisation scheme is incorrect, a fact which is frequently neglected in gravity field analysis. Furthermore, an alternative to empirical regularisation in the framework of both collocation and biased estimation is presented. Logically, error measures that include the effect of the regularisation error (or *bias*) are therefore advocated. In Chap. 5, the first non-U.S. or non-Russian efforts to model the gravitational field are presented. Through 3-way tracking of the Lunar Prospector spacecraft, it is attempted to improve the available models. The main goal of this exercise is to prepare for future missions, to demonstrate the technology of both hardware and software models and to test the tools discussed in the previous chapters. Results are presented in terms of both orbit and gravity field modelling. Chapter 6 finally details possible SST solution strategies and relates them to on-going satellite projects. Full-scale simulations of both self-contained solutions and solutions requiring regularisation are presented in an effort to determine the full potential of an SST experiment, with emphasis on aspects of quality assessment.

Chapter 2

Fundamentals of lunar gravity field recovery

'*It may be no surprise that human mind can deduce the laws of falling objects, because the brain has evolved to devise strategies for dodging them.*'

Paul Davies

2.1 Scientific rationale

There are four principal scientific reasons for a return to the Moon, popularly described as the science *of* the Moon, science *from* the Moon, science *on* the Moon and, fourth, the exploration and possible future utilisation of lunar resources. All of these contribute in their own way to the understanding of the solar system and in particular to the knowledge of the origin and development of the Earth–Moon system. In this book, focus is on the role of gravity analysis as an independent discipline as well as the scientific return through interaction with other selenophysical and selenochemical data sets; in other words gravity-related science *of* the Moon. Such analysis is, however, inherently tangled with satellite orbit analysis, and, hence, both precise orbit requirements for mapping purposes and more general mission design considerations (long-term orbit behaviour, frozen or periodic orbits) are also discussed.

A striking paradox in lunar science is the fact that, although the Moon is our closest neighbour in the solar system, there is still a large number of open questions on its origin, formation, evolution and present-day structure and chemistry. Many of these questions are tightly linked to the Earth and other planets and, indeed, important discoveries in Earth and planetary sciences have frequently found their origin in lunar studies. Through comparative planetology important lessons are learned (and will continue to be learned) on our own place and home in the universe, as well as on the structure and evolution of the solar system as a whole. For a further discussion on Earth-Moon relationships in the most widespread sense of the word, the reader is referred to *Barbieri and Rampazzi* [2001].

2.1.1   Science of the Moon

The Moon is anomalous in its mass compared to its primary; the Earth-Moon system brings together the largest most evolved and the smallest and least evolved of the terrestrial planets (disregarding the icy Pluto–Charon system because of the very low mass of the primary). Of the other terrestrial planets only Mars has satellites, but these are small and most likely captured asteroids [*Kaula et al.*, 1986]. The Moon is furthermore unusual in the facts that (i) its mean density is significantly lower than that of other terrestrial planets; (ii) lunar petrology and geochemistry indicate an extraordinary dryness (the only volatile elements appear to come from the solar wind); (iii) bulk geological analysis, although proving many similarities between the Moon and the Earth's mantle and crust, and thus providing evidence that the Earth and the Moon were created in the same part of the solar system (unlike *e.g.* most meteorites), also exhibits distinct differences, particularly in the abundance of heavier elements like iron [*Lucey et al.*, 1995]; (iv) the Moon has a very small core, a feature which is tightly linked to the formation and subsequent differentiation processes; and (v) it is quite distant from its primary: 60.39 mean Earth radii. The last of these anomalous properties is well understood as a consequence of the exceptional dissipative character of the Earth, because of the oceans undulating between continents in response to the Moon's tidal attraction. The dryness strongly suggests the Moon's low bulk density comes from a depletion in iron and possibly other heavy elements. These compelling facts altogether suggest that the formation of the Moon was an unusual event, and that an unusual hypothesis may be required to explain it.

The Moon is important to planetary science for two prime reasons. For one, the Moon is an ancient world, with more than 99% of the surfaces predating 2 Gyr (and more than 70% predates 4.2 Gyr) [*Spudis*, 1996]. The story of the Moon is therefore the story of the early solar system. The small size of the Moon caused large-scale internal activity to come to an end 3 Gyr ago, and the lunar highlands are the oldest geological units in the inner solar system; moreover, they have preserved the early stages of crustal formation. Consequently, for about 4 billion years, impacts have been the main shaper of the lunar surface. Impacts are even known to control the locales of lunar volcanism (another shaper of surface features): the basaltic maria are all situated in areas created by earlier impacts. As a consequence, the Moon contains a stratigraphic history record and is unique in providing information on the early history and evolution of the solar system. Furthermore, the unique combination of remote sensing data, manned reconnaissance, in situ geophysical stations and sample return makes the Moon a primary body for the study of a variety of geophysical problems for which data are inadequate or unavailable elsewhere in the solar system.

The second specific interest of the Moon for planetary science concerns its role as the companion of the Earth in a binary system. The well-preserved record of impacts and volcanism on the Moon gives reliable information on what has occurred on Earth, where plate tectonics and other sources of resurfacing have removed nearly all traces of such events. Finally, the inter-dependence of the Earth and the Moon is likely to have affected the emergence and evolution of life on Earth itself, as research over the last decade has suggested that the presence of the Moon in the vicinity of the Earth not only controlled the evolution of the spin rate of the Earth, but also stabilised the orientation of the spin axis [*Laskar and Robutel*, 1993; *Laskar et al.*, 1993; *Laskar*, 1996; *Néron de Surgy and Laskar*, 1997].

In large, the role of gravity field analysis in this conglomerate of scientific issues consists of determining the size of the core, jointly with other selenosciences, like seismology and lunar laser ranging and possibly very-long-baseline interferometry, and in determining the gravitational structure of craters, basins and highland areas, over a vast (*i.e.* as large as possible) range of selenophysical scales. Detection of dynamically supported structures constitute a direct link to lithospheric strength, and, in a similar fashion, isostatically compensated structures will provide information on lunar rheology. Furthermore, information deduced from gravity mapping is used to constrain models from related selenosciences, like seismology, topographic mapping, thermal analysis and geochemistry. Lunar science is therefore truly a multi-disciplinary occupation.

Origin and evolution of the Earth-Moon system

The origin of the Earth-Moon system is one of the major unsolved mysteries of the planetary sciences. At least four scenarios have been proposed for the creation of such a binary system [*Wood*, 1986], three of which were ruled out by the end of the 1980s: fission of a part of the Earth's mantle; capture in Earth orbit of a Moon formed elsewhere; accretion of planetesimals in Earth orbit. All origin models must necessarily comply with a range of geochemical as well as dynamical constraints. Geochemical constraints are provided by the bulk chemical composition resulting from the particular impact model, devised from surface chemistry and mineralogy, while the dynamical constraints mainly concern the angular momentum (amplitude and orientation) of the Earth-Moon system. The Earth and the Moon have the greatest amount of angular momentum of all terrestrial planet-satellite systems in the solar system [*Spudis*, 1996], and, moreover, the orbit and rotation of the Moon do not coincide with the Earth equator. Second, dynamical considerations in many cases are found to lead to a swarm of smaller moons rather than a single large partner of the Earth [*Kaula et al.*, 1986].

The surviving scenario is that of the *giant impact* model, [*Hartmann and Davies*, 1975; *Benz et al.*, 1986; *Wänke and Dreibus*, 1986; *Benz et al.*, 1987, 1989; *Cameron and Benz*, 1991], in which one or more giant impacts on the early Earth caused ejection of parts of both mantles into Earth orbit, while the core of the impactor would merge with the core of the Earth. The Moon then formed from the ejected mantle material. This model resolves many of the problems of previous scenarios, and indeed meets key constraints on the global chemical (*e.g.* depletion in iron) and mineralogical (has there been a global melting episode in the form of a magma ocean?) composition of the Moon as well as its internal structure [*Wänke and Dreibus*, 1986]. Despite the accordance with composition data, the possible explanation for the high angular momentum of the Earth-Moon system and the inclination of the lunar orbit, as well as the accommodation of an early magma ocean phase of the Moon, the details of the process are not yet fully understood, nor is the process finally proven. A disclaimer maintained by some is furthermore that the model might be too flexible and therefore stretchable to fit the available data too easily [*e.g.*, *Spudis*, 1996]. Therefore, the problem is still regarded as one of the unsolved mysteries of the Earth-Moon system.

The main property requiring a giant impact theory in the last phases of Earth accretion is the Moon's extraordinary depletion in iron, mainly determined from bulk density data. The iron-depletion may be confirmed in the case that the metallic core is small, such that the Fe/Mg ratio is low, like in the upper mantle of the Earth. Recent gravity field research [*Konopliv et al.*, 1998, 2001], supported by lunar laser ranging results [*Dickey et al.*, 1994], improves the core size estimation, but the internal consistency is still hampered by uncertainties in the satellite-derived second degree harmonics [*Bills*, 1995]. More work on the nature of the lunar core, involving several geophysical data sets and in particular improved gravity field modelling will directly address the amount and nature of metals (*e.g.* by palaeomagnetic research) as well as the bulk composition of the Moon, and as such directly impose constraints on the lunar origin.

Role of impact processes in planetary evolution

Early observations of the Moon revealed the widespread significance of cratering as a surface shaping process. Later, the Apollo exploration provided evidence for their impact origin, as was proposed already in 1949 by *Baldwin* [1949]. Craters range from microscopic pits on returned samples to huge impact basins > 2000 km. This enormous collection of craters, combined with low degradation in time, allows crater morphology to be studied over a vast range of crater types, and therefore complements and enhances the information derived from investigating partially (or completely) eroded terrestrial craters. Hence, apart from studying the nature and history of cratering on the Moon, one is concerned with the geophysics of impact processes in the solar system and its role in surface formation.

Related to the nature of impact processes one is also interested in the subsequent geophysical response, in particular for the larger craters and basins. The effects of the oldest large impacts are complex because of multiple (subsequent) events within the same area. Precise measurements of both geochemistry and topography/gravity may therefore reveal hints of ancient impact basins. Studying such large impacts is also much easier on the Moon than on the Earth. The only known multi-ring basin on the Earth is Chicxulub located at the tip of the Yucatan peninsula in the Gulf of Mexico. This crater is not directly accessible for observation and imaging since it is covered by a 300 m thick sedimentary layer, and its very existence was discovered only as a result of oil drills and subsequent gravimetric mapping. Given the relevance of such impacts to the history of our planet, for example through the association of Chicxulub with the Cretaceous-Tertiary epoch boundary and, hence, the extinction of major species [*e.g.*, *Smit*, 1994; *Alvarez*, 1997], it is only natural to resort to the Moon. Moreover, there is a growing awareness of the problem of potentially hazardous near-Earth objects.

Volcanism, maria and mass concentrations in lunar physics

The lunar maria represent the most obvious and well-documented example of extrusive volcanism. The different volcanic evolution of the Earth and the Moon can be explained when considering the heat sources of the two bodies, like accretional heat, heat from gravitational differentiation of the crust, mantle and core and radiogenic heat. The much smaller Moon obviously had weaker heat sources and cooled faster than the Earth; therefore, the Moon is a body which was active primarily in the early phase of its evolution, *i.e.* the first 600 Myr–1.2 Gyr. Lunar volcanism is believed to have consisted of several phases, beginning with the early magma ocean [*Toksöz et al.*, 1972; *Runcorn*, 1977; *Solomon and Longhi*, 1977; *Warren*, 1985] and subsequent crust formation [*Longhi*, 1977], continued as highland volcanism and finally ceased as a large scale resurfacing process when basaltic mare lavas extruded onto the surface [*Head III*, 1976; *Solomon and Head*, 1980]. While the emergence of volcanic formations is strictly related to the thermal history of the Moon, a careful combination of geological studies and gravity and other geophysical data will help to gain better understanding of the lunar interior, *e.g.* mantle depth, heterogeneities and crustal thickness variations, as volcanism reflects in a quite straightforward way the state and evolution of the interior.

A truly stunning and unexpected result of the Lunar Orbiter missions was the discovery of large positive gravity anomalies associated with near-side circular basins [*Muller and Sjogren*, 1968]. A circular basin is formed by a large impact that excavates near-surface strata and deposits ejecta on the surroundings as an encircling high rim. In general, for such impacts, because of differential pressure at the base of the excavated zone, the surrounding area would collapse and the lower strata would rebound to attain an isostatic equilibrium. Such a structure would, on the contrary of what is observed over the near-side circular basins, give rise to a negative gravity anomaly. This indicates excess mass concentrations (baptised *mascons* by Muller and Sjogren) in the basins and, furthermore, emphasises that processes other than simple impacts are responsible for their creation [*Arkani-Hamed*, 1998]. In terms of age determination of the mascons, crater counts and relative age determination, combined with absolute data from the returned lunar samples, have revealed that the circular basins were formed during catastrophic cratering events about 4.0–3.9 Gyr ago, and that mare flooding of the basins occurred within 100–800 Myr after basin formation. Mascons are therefore for the major part created at least 3.6 Gyr ago and have survived since [*Head III*, 1976].

A related discovery was the presence of *negative anomalies over the highland areas surrounding the basins*, implying significant mass *deficiencies* [*e.g., Sjogren et al.*, 1972*a*, 1972*b*] caused by a combination of impact ejecta and thickening of the crust by subsequent shock waves [*Neumann et al.*, 1996; *Von Frese et al.*, 1997]. Such negative anomaly rings have been verified as real features for many mascons, and are hence no artifacts of the gravity modelling process [*Lemoine et al.*, 1997; *Konopliv et al.*, 1998, 2001]. This close relationship between excess mass concentrations in the basins and mass deficiencies in the surroundings firmly suggests an intimate relationship between their formation processes. The detection of the detailed gravitational shape of the transition zone between these areas is a strong argument for high resolution gravity mapping.

Following the discovery of the mascons, new models were proposed for their formation and support mechanism. Typically two schools of formation theories have been proposed [*Spudis*, 1996; *Arkani-Hamed*, 1998], both of which describe mascons as near-surface features: (i) passive formation models which state that molten basalt was created in the deep interior, caused by global heating of the Moon, and flooded the basins [*e.g., Runcorn*, 1974]. The impacts thus had no direct control on the formation of the mascons, but rather produced favourable places for the mare basalt to extrude. Thick fills of high-density lava would constitute the excess mass; and (ii) active formation models that relate mascon formation directly to the effects of the giant impacts, *e.g.* suggesting that partial melting occurred beneath the surrounding highlands [*e.g., Wise and Yates*, 1970; *Arkani-Hamed*, 1973*a*]. This would involve uplift of dense rocks from the mantle by the unloading of the crust during excavation of an impact basin and possibly lava fill by lateral transformation of molten basalt.

The support mechanisms proposed for the survival of the mascons over 3.6 Gyr also fall into two distinct characters: elastic support models [*Kuckes*, 1977; *Solomon and Head*, 1980] and viscous decay models [*Arkani-Hamed*, 1973*b*, 1973*c*; *Meissner*, 1977]. An elastic support would suggest that the mascons are supported by an elastic layer on the surface of the Moon, while viscous decay models emphasise the role of viscous deformation of the lunar interior, and therefore suggest that the surface loads produced by mascons might decay over geological timescales. While the formation of mascons is generally a debated issue, it is likely that both the extrusive component as well as related mantle uplift are complementary causes of the large positive gravity anomalies [*Bowin et al.*, 1975; *Spudis*, 1996] and that mascons experienced viscous decay until the lithosphere reached the strength to elastically support the mascons [*Arkani-Hamed*, 1998]. Given the multitude of data sets and modelling techniques involved, the lunar mascons are probably the most compelling example of how the different scientific disciplines are entangled, and where improvements in any field are likely to enhance the overall picture of the Moon.

Additional rationales from detailed gravity mapping, in particular including the far-side, arise from the near-side/far-side dichotomy of mare fills. Until Lunar Prospector [*Binder*, 1998; *Konopliv et al.*, 1998, 2001] there was no direct evidence for far-side mascons, and, moreover, several of the far-side basins appear nearly compensated (*e.g.* the giant 2500 km diameter ancient South Pole Aitken basin). Even till this day it may be argued that the evidence for far-side mascons is circumstantial at best. Related to crustal strength and thickness on the one side, and crater age and size on the other side, there is therefore possibly a transition between mare fill and associated mascons, and craters and basins exhibiting negative gravity anomalies or full compensation. In the context of relaxation modelling for floors of large craters and impact basins, higher resolution gravimetry will provide much needed information about the rheology of the Moon.

Lunar tectonics

The present idea of a tectonically quiet, one-plate Moon, characterised by a single unsegmented global lithospheric plate, does not apply for the early Moon. Even in the absence of the still slightly debated magma ocean concept, *i.e.* if only partial global melting occurred, most of the older surface rocks have taken part in endogenic processes on a global scale. The tectonics of the Moon were, however, rather simple in comparison to the more active planets, like the Earth. Most lunar tectonic structures are connected to impact basins and large craters, creating an interesting link between internal processes and processes related to impacts and volcanism [*Spudis*, 1996]. The highest intensity tectonic features are linked to areas of highest volcanic activity, such as the lava-flooded basaltic basins. The link with gravimetry is found in the realisation that the most important aspect of lunar tectonics is how tectonic structures reflect the interior; this is most obvious in areas of intense tectonic activity (albeit in the earlier phases of the Moon's history). Major unresolved issues on lunar tectonics include the nature of early heterogeneities and temporal growth of the lithosphere, and the extent of viscous relaxation in the early Moon [*Solomon and Head*, 1979, 1980]. There is, hence, a clear link between impact processes, volcanism, tectonics and gravimetry in studies of large craters and basins.

Lunar crust and interior

Inference of the structure and composition of the crust and mantle mainly come from geophysical investigations: seismology for velocity profiling [*Toksöz et al.*, 1974; *Nakamura et al.*, 1977; *Nakamura*, 1983]; surface and orbit geochemistry for mineralogical mapping; and gravity and topography for the determination of lateral inhomogeneities [*Zuber et al.*, 1994; *Spudis et al.*, 1994]. Aside from mascons, one famous result is the near-side/far-side dichotomy of both topography and crustal thickness.

   While the topography is the product of altimetric data processing using the *selenoid* (an equipotential surface of the lunar gravitational potential) as a reference surface, inference of crustal thickness, a key parameter in the analysis of the structure of basins and large craters, is based on a combination of a selenopotential and a spherical harmonic expansion for the topography, with additional assumptions on the crustal density. A variety of analysis strategies exists, but the quality of end product of selenoid-to-topography ratios or crustal thicknesses is crucially related to the quality of both the topography and gravity estimates. The far-side crust is significantly thicker than its near-side counterpart, about 4–12 km on average [*Zuber et al.*, 1994; *Neumann et al.*, 1996; *Wieczorek and Phillips*, 1997, 1998] with overall variations of more than 100 km [*Neumann et al.*, 1996]. The net effect of crustal variations, with the additional effect of mantle topography gives rise to an offset of the centre of mass with respect to the centre of figure of about 1.6 km. The physical figure of the Moon may therefore be described as slightly pear-shaped, with the elongated half pointing away from Earth. Given a good crustal thickness estimate, estimation of the topography of the lunar Moho (crust/mantle boundary) is possible, which in turn are direct indications of deep density variations in the Moon [*Arkani-Hamed*, 1999b]. Estimations of crustal thickness and Moho topography are key applications of high-resolution and high-accuracy gravity field models both at present and in the future.

2.1.2   Flight dynamics and satellite orbit modelling

Lunar exploration is not limited to scientific research. Assuming there is general consensus that lunar exploration is a worthwhile endeavour, in other words that the scientific rationale along with technical, economical and possibly socio-cultural motivations, form legitimate reasons to explore and utilise the Moon to the benefit of mankind, navigation of manned and unmanned spacecraft in lunar orbit in itself becomes a driving force behind gravimetric research. A second motivation for improved gravimetric mapping of the Moon is therefore rooted in the practical aspects of satellite flight dynamics, *i.e.* orbit prediction and orbit determination. Uncertainties in the satellite force model directly affect both mission planning as well as the ability to compute orbits with an accuracy level adequate for the achievement of the mission goals, given the fact that several mapping instruments, not only instruments for gravity field studies, require accurate knowledge of the spacecraft orbit and orientation.

From a lunar satellite mission planning point of view, the most important factor concerns orbit maintenance. Given the relatively rough nature of the lunar gravity field – *e.g.*, the equipotential surface variations are approximately a factor four larger than for the terrestrial gravity field –, combined with the fact that the lunar far-side still has to be mapped, sizeable propellant margins are usually incorporated in order to ensure that the mission goals are met. It is therefore not surprising that lunar gravity models in the past have been known to predict widely different orbit behaviour [*e.g.*, *Konopliv et al.*, 1993; *Meyer et al.*, 1994; *d'Avanzo et al.*, 1995; *Milani and Knežević*, 1995; *Eckstein and Montenbruck*, 1995; *Vasile*, 1996; *Finzi and Vasile*, 1997; *Floberghagen and Visser*, 1997; *Goossens et al.*, 1999]. Improved knowledge of the gravity field translates into accurate propellant allocation for orbit control, and, hence, into lower mission cost. Along the same line, a better knowledge of the force model evidently also leads to better navigation of the spacecraft, which might be necessary for critical applications, like *e.g.* unguided pinpointed landings at the lunar surface [*Tuckness*, 1995*b*, 1995*a*; *Floberghagen and Visser*, 1997; *Goldstein et al.*, 1999]. For most landing problems, knowledge of the parking orbit prior to the descent phase is, however, more critical than the gravity field perturbations induced during the relatively short descent phase.

The long-term evolution of a lunar satellite orbit is evidently not independent from the previous considerations. Ability to understand the behaviour of satellite orbits over considerable periods of time is instrumental to a range of applications, for example future missions foreseeing global mapping of the gravity field through inter-satellite tracking experiments. In this case, it is likely that one of the space-craft will be unpropelled, and purely passively stabilised [*e.g., Milani et al.*, 1996; *Häusler*, 1998; *Namiki et al.*, 1999; *Sasaki et al.*, 1999]. In other words, after deploy-ment of the spacecraft in its designated orbit, it will not be possible to actively control it any longer. A crucial aspect of the experiment performance, and satellite lifetime, is therefore the understanding of the long-term evolution of the satellite orbit under the influence of the gravity field and possible also other forces, which obviously must be studied, prior to the mission, for all suitable and acceptable combinations of mission parameters.

Similar to the case of Earth-orbiting remote sensing missions, there are also orbit precision requirements, or at least precision goals, for lunar and planetary mapping missions. A range of remote sensing data are intrinsically depending on knowledge of the satellite orbit, and experiment statistics are accordingly de-pending on orbit accuracy figures. One such example is the application of laser altimetry for the mapping of the lunar topography. Here, the radial orbit accu-racy is directly related to the measurement quality, and, hence, the quality of the topographic end product.

All in all, it may therefore safely be concluded that there is a range of both purely scientific reasons for gravimetric mapping of the Moon, as well as a cate-gory of more engineering-oriented applications. The borderline between the two is far from clean-cut, and the resulting inter-disciplinary nature of the problem is, to many researchers, illustrative of the scientific and engineering appeal of lunar gravimetric research.

In connection with recent project proposals for new lunar satellite missions it has also been discussed whether studies of low lunar satellite orbit behaviour com-puted on the basis of a high-quality gravity field model might be helpful in the search for lunar resources. However unrealistic this may seem, it may or may not become an application of satellite tracking data in the future, depending on the quality of the measurements provided by future missions.

2.2   State-of-the-art of lunar gravity field modelling

Since the dawn of lunar exploration by spacecraft, a variety of selenopotential models has been produced using Earth-based tracking data of U.S. and Russian satellites in lunar orbit. Generally speaking, research into the lunar gravity field to date knows two eras. The first period roughly covers, on the U.S. side, the Apollo project, including the precursor *Lunar Orbiter* missions (1966-1972), while on the Russian side the main data source were the *Luna* missions (1959-1976). Although these initial efforts to determine the gravitational field of the Moon were primarily triggered by the need to bring man to the lunar surface and return him safely to Earth (navigation), much like the entire Apollo project was never science-driven, they nevertheless sparked off important scientific discoveries, like the discovery of the mascons and also the first signs of both radial (differentiation) and other, non-mascon related, lateral density variations inside the Moon. While the inter-pretation of the lunar gravity field models continued throughout the 1980s, the lack of new spacecraft tracking data to work on prevented new gravity field mod-elling efforts for about 15 years.

However, a background lobby for a lunar polar orbiter mission had never ceased [*Spudis*, 1996]. Triggered by, on the one side, the promising results deliv-ered by analysis of data from Apollo and contemporary projects, and, on the other side, the limitations of the data sets returned by the same missions, numerous studies were conducted into a purely science driven follow-on mission. A major conclusion of one such study, that of the *Lunar Observer* mission [*Cook et al.*, 1990; *Ridenoure*, 1991; *Konopliv*, 1991], was that both mission design (fuel budget calcula-tions), navigation and lunar science would all benefit from reprocessing of the his-torical data using modern (super)computers and geodetic modelling techniques. The second phase of lunar gravity field modelling activities, which commenced around 1991, is, hence, characterised by similar sets of tracking observables as those of the first era, but taking advantage of new developments in the fields of high-speed computers and geodetic data processing. Next to the re-use of the historical Apollo-era data, the data sources from this second period comprise the Clementine [*Spudis*, 1992; *Nozette et al.*, 1994; *Zuber et al.*, 1994; *Lemoine et al.*, 1997] and Lunar Prospector [*Binder*, 1998; *Konopliv et al.*, 1998, 2001] missions. While the Clementine mission suffered from orbit characteristics which were suboptimal for gravity modelling purposes, there is no doubt that Lunar Prospector has led to quality improvements in lunar gravity field modelling, in particular due to its polar, low-altitude orbit.

A third phase, and a major step forward in lunar gravimetry, is planned shortly into the first decade of the third millennium, with global data collection through inter-satellite tracking experiments [*SELENE*, 1996; *Namiki et al.*, 1999; *Matsumoto et al.*, 1999; *Heki et al.*, 1999; *Kaneko et al.*, 1999; *Sasaki et al.*, 1999]. This book should be seen as an explorative step towards self-contained solutions based on such data sets.

2.2.1   Representation formulae for the gravitational potential

There exist several representation formulae for the gravitational potential of celestial bodies. Since the major interest in the gravitational potential of the Moon, or *selenopotential*, concerns its deviations from sphericity, any such formula must 1) adequately describe the nature of the lateral variations of the potential, and possibly aid estimates of radial density variations and centre-of-mass determinations, and, 2) in the particular case where the potential model is derived directly from satellite orbit measurements, also be functionally related to the perturbed orbital motion. The latter requirement is valid for tracking techniques, in contrast to *in situ* measurements, which are obviously – by their nature – located in a specified reference frame.

The gravitational potential U in the exterior of a celestial body is known to satisfy Laplace's equation [*e.g., Kaula*, 1966]

$$\Delta U = \frac{\partial^2 U}{\partial x^2} + \frac{\partial^2 U}{\partial y^2} + \frac{\partial^2 U}{\partial z^2} = 0 \qquad (2.1)$$

where the Cartesian coordinates $\{x, y, z\}$ are body-fixed and rotating with the celestial body in inertial space. Since the choice of coordinate system is largely dependent on the geometry of the boundaries, for global gravity field modelling the coordinate system of choice is spherical and Moon-fixed, for which the solution reads, in a body-fixed reference frame rotating with the planetary rotation and the alignment of the axes according to international definition or some estimate of a principal axis system,

$$U = \frac{GM}{r} \sum_{l=0}^{\infty} \left(\frac{a_e}{r}\right)^l \sum_{m=0}^{l} K_{lm}^T Y_{lm}(\phi, \lambda) \qquad (2.2)$$

with

$$K_{lm} = \begin{bmatrix} \overline{C}_{lm} \\ \overline{S}_{lm} \end{bmatrix} \qquad (2.3)$$

being the normalised coefficients of the spherical harmonic expansion of the gravitational field and

$$Y_{lm} = \overline{P}_{lm}(\sin\phi) \begin{bmatrix} \cos m\lambda \\ \sin m\lambda \end{bmatrix} \qquad (2.4)$$

being the orthogonal harmonic basis functions of degree l and order m on the sphere. Furthermore,

$$GM = \text{the gravitational parameter of the Moon}$$
$$a_e = \text{the mean equatorial lunar radius}$$
$$r, \lambda, \phi = \text{the selenocentric spherical coordinates: radius, longitude and}$$
$$\text{latitude}$$
$$\overline{P}_{lm}(\sin\phi) = \text{the fully normalised associated Legendre functions}$$

In most practical applications, the origin of the body-fixed coordinate frame is chosen to coincide with the centre of mass, and the z-axis is oriented along the principal polar moment of inertia. Consequently, the degree one terms are all zero. The normalisation of the associated Legendre functions implies that

$$\int_0^{2\pi} \int_{-1}^{1} Y_{lm}^2 (\xi, \lambda) \, d\xi d\lambda = 4\pi \tag{2.5}$$

where $\xi = \sin \phi$, *i.e.* the integration of the square of a spherical harmonic basis function over the unit sphere evaluates to the surface area. Orthogonality relations furthermore ensure that products of non-coinciding degree and order harmonics integrate to zero over the same unit sphere. It is the same type of orthogonality relationships, combined with the quite direct applications in spacecraft dynamics, which make spherical harmonics the preferred representation formula for planetary gravity fields, even though, from a geophysical point of view, elliptical harmonics or other concepts based on the equilibrium figure of the planets would appear more useful.

Because the Legendre functions change sign $l - m$ times from pole to pole, and $2m$ times along $0° \le \lambda < 360°$, the spherical harmonic coefficients may be divided into three groups. *Zonal harmonics,* for which $m = 0$, are longitude-independent and divide the sphere into l zones. Even zonals are symmetric about the equator, while odd zonals are asymmetric about the same plane. An often used short-hand notation for the amplitude of spherical harmonic coefficients of a given degree and order is $\bar{J}_{lm} = \sqrt{\bar{C}_{lm}^2 + \bar{S}_{lm}^2}$, with the sign convention $\bar{J}_{l,0} = \bar{J}_l \doteq -\bar{C}_{l,0}$ for zonal terms. *Sectorial harmonics* are terms for which $l = m \ne 0$, which divide the sphere into sectors where the potential is latitude-invariant for a given longitude. *Tesseral harmonics* are the general case $l \ne m$.

By nature, the use of global basis functions on the sphere, like in the case of spherical harmonics, facilitates the modelling of smooth phenomena. Adequate modelling of localised features, like local gravity highs, using spherical harmonics requires a significant number of spherical harmonics. Moreover, as illustrated in the introduction, spherical harmonics may prove inefficient in the case of severely unevenly sampled data. It was realised early on in lunar gravimetric research that spherical harmonics might not be the ideal representation, given the significant gravity highs found at the ringed near-side maria [*Wong et al.*, 1971]. While globally supported basis functions are suited for global modelling, compactly supported functions are likely to perform better for small-scale phenomena. Furthermore, the use of extensive spherical harmonic series to exhaust the near-side data with no direct far-side measurements tends to lead to spurious far-side models, with no resemblance to reality. Therefore, *point-mass* and *disk* models were proposed, which intended to combine a global-basis and smooth long-wavelength potential with discrete model parameters, of compact basis, describing the high-frequency part of the potential.

The gravitational potential U due to arbitrarily shaped mass anomalies, *e.g.* point-masses, disks or other types of surface layer elements, for which the anomalous potential can be computed [*e.g.*, *Blakely*, 1995], is given in a general sense by

$$U = GM \left[\frac{1}{r} + \sum_{i=1}^{N} \delta m_i F_i \right] \tag{2.6}$$

where the first term simply describes the central part of the potential, δm_i the ratio of anomalous mass i to the lunar mass and F_i the potential function of the ith mass anomaly. In the case of a discrete point-mass model, the potential function is simply given by

$$F_i = \frac{1}{\|\mathbf{r} - \hat{\mathbf{r}}_i\|_2} \tag{2.7}$$

where

> $\mathbf{r}$ = the vector from the lunar centre of mass to the satellite; and
> $\hat{\mathbf{r}}_i$ = the vector from the lunar centre of mass to the ith point-mass.

Combination models, *i.e.* models that combine mass anomaly representations with low-degree spherical harmonics, have also found their applications, in particular in the early phases of lunar gravimetry.

Evidently, other representation formulae describing the gravitational attraction of a celestial body could be used, like *single-layer* and *double-layer* surface elements used in *boundary element* solution techniques [*Klees*, 1993, 1997; *Lehmann and Klees*, 1999], simple LOS accelerations [*Barriot and Balmino*, 1992; *Barriot*, 1994; *Barriot and Balmino*, 1994; *Barriot et al.*, 1998; *Moreaux et al.*, 1999] or gridded gravity anomalies [*Hajela*, 1979; *Vonbun et al.*, 1980; *Kahn et al.*, 1982], but these are not described here for the very reason that either i) they are not directly and conveniently related to the satellite motion, or, ii) they are not fully dynamic, like for the LOS accelerations, which is a problem when the orbit quality is poor. Moreover, methods like LOS accelerations derived from Doppler tracking data yield gravity field profiles in the direction of the tracking station on Earth, and not along the vertical on the Moon. This does not, however, imply that LOS methods were unimportant for lunar gravity modelling in the past, as will become clear from the following sections.

The relationship between the selenopotential and the lunar satellite flight dynamics is given by the inertial acceleration due to the gravitational attraction. For a body-fixed, non-rotating coordinate frame, the acceleration $\ddot{\mathbf{r}}$ due to the selenopotential is given by the gradient of U, which in spherical coordinates reads

$$\ddot{\mathbf{r}} = \nabla U = \frac{\partial U}{\partial r}\mathbf{e}_r + \frac{1}{r}\frac{\partial U}{\partial \phi}\mathbf{e}_\phi + \frac{1}{r\cos\phi}\frac{\partial U}{\partial \lambda}\mathbf{e}_\lambda \tag{2.8}$$

where $\mathbf{e}_r$, $\mathbf{e}_\phi$ and $\mathbf{e}_\lambda$ are the unit vectors in the $\{r, \phi, \lambda\}$ body-fixed coordinate basis. Denoting the matrix that describes the rotation from an inertial basis to a body-fixed basis by $\mathbf{E}$, cf. Appendix C, the transformation

$$\mathbf{r}_{BF} = \mathbf{E}\,\mathbf{r}_I \tag{2.9}$$

relates the body-fixed coordinates subscripted by "BF" to inertial coordinates subscripted by "I". Accordingly, after evaluation the gradient (2.8), the acceleration in the inertial frame, which is used for force computation and orbit integration, is given by

$$\mathbf{a}_I = \mathbf{E}^T \mathbf{a}_{BF} \qquad (2.10)$$

2.2.2 Lunar satellite tracking data and orbital characteristics of previous missions

Most western models have been developed on the basis of Doppler tracking by NASA/DSN [*e.g.*, *Kinman*, 1992] using stations at Goldstone, California; Madrid, Spain; and Canberra, Australia, and also, in the earlier phases, the Manned Space Flight Network (MSFN) [*Sjogren et al.*, 1974a]. All lunar gravity models developed to date are *satellite-only*, *i.e.* no data other than spacecraft tracking data enter the solutions, and the models are developed on the basis of least squares fits to the available data. The primary data sources of the first era of lunar gravity research were Lunar Orbiter I–V and the Apollo spacecraft, in particular the Apollo 15 and 16 *sub-satellites*, all of which are described in Table 2.1. While the data of the Command Service Module of an Apollo configuration were frequently utilised for line-of-sight profiling of gravity anomalies over maria, the Lunar Module data were generally too short (rapid descent, rapid ascent) to have any practical impact on global gravity field modelling.

The Lunar Orbiters (LO) were a series of missions that were in orbit around the Moon from August 1966 to January 1968 [*Lorell and Sjogren*, 1968; *Ridenoure*, 1991]. All Lunar Orbiters were placed into elliptical orbits with low periapsis altitudes, roughly 40–210 km above the lunar surface, in support of photo-reconnaissance for the Apollo landings. With the exception of LO-III and LO-IV, the orbital period was about 200–210 min. While the first three missions were in near-equatorial orbits (12° to 21°), LO-IV and LO-V were placed in near-polar orbits of about 85°, thereby providing the first high-inclination data. The data consist of mostly 60 s 2-way and 3-way measurements[1], with an encouraging overall noise level better than 1 mm/s and some data even at 0.35 mm/s. A major disappointment, however, was the LO-V data, which would have been the best of the Lunar Orbiter data for global modelling due to the high inclination, but suffered from an increased data noise level, by a factor of five to ten [*Konopliv et al.*, 1993; *Lemoine et al.*, 1997].

A significant issue involved with the use of Lunar Orbiter data is the fact that each of the five missions consisted of two phases: a primary mission phase devoted to photographic mapping, with extended tracking, and an extended phase, when tracking was sparse. During both phases, but most frequently in the primary phase, attitude and orbit control manoeuvres were executed in order to line up the

---

[1]In tracking terminology, 2-way Doppler measurements are measurements over two link paths: an up-link (ground station – spacecraft) and a down-link (spacecraft – ground station), where the transmitting and the receiving antenna are one and the same; 3-way measurements are similar, but utilise different stations to transmit and receive, cf. Fig. 5.3

Mission	Date	Period min	a km	e	i °	Periapsis km
LO-I	14/08/1966	218	2766	0.3031	12.2	189
	21/08/1996	209	2693	0.3335	12.0	57
	25/08/1966	206	2669	0.3336	12.3	41
LO-II	10/11/1966	218	2772	0.3021	12.0	196
	15/11/1966	208	2689	0.3353	11.9	50
	08/12/1966	210	2702	0.3408	17.6	43
	14/04/1967	209	2693	0.3289	17.2	69
	27/06/1967	212	2716	0.3184	16.5	113
	24/07/1967	212	2716	0.3240	16.1	98
LO-III	09/02/1967	215	2744	0.2899	21.0	210
	12/02/1967	209	2689	0.3331	20.9	55
	12/04/1967	207	2680	0.3294	21.2	59
	17/07/1967	212	2722	0.3088	21.0	144
	30/08/1967	131	1968	0.0435	20.9	145
LO-IV	08/05/1967	721	6150	0.2773	85.5	2706
	06/06/1967	501	4820	0.6238	84.9	75
	09/06/1967	344	3753	0.5163	84.4	77
LO-V	05/08/1967	505	4852	0.6017	85.0	195
	07/08/1967	501	4822	0.6185	84.6	100
	09/08/1967	191	2537	0.2760	84.7	100
	10/10/1967	225	2832	0.3155	85.2	198
A15SUB	29/08/1971	118	1858	0.0167	151.0	89
A16SUB	27/04/1972	119	1849	0.0205	169.3	73
Clementine	19/02/1994	474	4651	0.5400	89.5	402
	21/02/1994	299	3418	0.3755	89.3	398
	22/02/1994	299	3414	0.3789	89.4	383
	11/03/1994	298	3414	0.3738	89.8	400
	27/03/1994	298	3414	0.3598	90.1	447
	11/04/1994	298	3415	0.3657	90.0	427
Lunar Prospector	16/01/1998	118	1838	0.0005	90.0	99
	29/01/1999	111	1768	0.0085	90.0	15
	24/02/1999	111	1768	0.0076	90.0	16
	24/03/1999	111	1768	0.0083	90.0	15
	21/04/1999	111	1768	0.0086	90.0	15
	05/05/1999	111	1768	0.0090	90.0	14
	01/06/1999	111	1768	0.0087	90.0	15
	29/06/1999	111	1768	0.0070	90.0	18

Table 2.1   Summary of U.S. tracking data sources. In addition to the Lunar Orbiters (LO) and the Apollo 15 and 16 sub-satellites (A15SUB/A16SUB), tracking of the Apollo Command Service Modules (CSM) contributed to LOS acceleration profiling over the ringed maria. Post-Apollo data include the Clementine and Lunar Prospector missions

spacecraft instrumentation for detailed photographic mapping of designated areas. Because the attitude control system of the Lunar Orbiters were uncoupled in pitch and yaw, each manoeuvre imparted spurious orbital accelerations. Each of these manoeuvres were time-tagged, and, hence, it is possible to estimate the manoeuvres and thereby accommodate for their orbit effects. In general, however, this estimation is not very robust, and the multitude of such three-axis acceleration estimations is an obstacle for gravity field modelling. The total of up to 14 sets of attitude manoeuvres per day during the primary phase limited the useful arc length of the Lunar Orbiter data to a maximum of about one day [*Konopliv et al.*, 1993], which clearly does not benefit the estimation of low-frequency spherical harmonic constituents.

The tracking data of the Apollo missions cover short time spans and, moreover, suffer from high attitude and orbit control activity, which makes it unsuitable for incorporation in global gravity field modelling. Furthermore, the data was exclusively low-inclination. However, the Apollo data were used to profile line-of-sight accelerations over near-side basins, thereby confirming and expanding the knowledge of mascon existence, formation and support mechanism [*Sjogren et al.*, 1972a, 1972b; *Muller et al.*, 1974; *Sjogren et al.*, 1974b, 1974c].

The so-called Apollo 15 and 16 Particles and Fields Sub-satellites were small, passively spin-stabilised spacecraft that were deployed from the Command Service Modules (CSM) of an Apollo configuration into near-circular, retrograde orbits [*Sjogren et al.*, 1974a]. The spacecraft, designed to study the magnetic field, the plasma and the flux of energetic particles in the vicinity of the Moon, also carried S-band transponders to support high-precision gravity mapping [*Coleman et al.*, 1972a, 1972b]. Due to the pure spin-stabilisation, no attitude manoeuvres were performed, making the data cleaner, and hence more suited for gravity modelling purposes, than the Lunar Orbiter data. Both sub-satellites were tracked by the MSFN, but despite their name, there was no tracking between the sub-satellites and the Command and Service Modules of Apollo 15 and 16. In other words, there was no satellite-to-satellite tracking while passing over the lunar far-side. The Apollo 15 sub-satellite was released on August 4, 1971, in a 151.0° inclination 2-hour orbit (approx. 100 km altitude) and remained in orbit for about 2 years. However, not all the Apollo 15 sub-satellite data has been recovered, as the satellite was sparsely tracked, with roughly one orbital revolution being tracked daily, except at dedicated periods when about every third orbit was tracked [*Konopliv et al.*, 1993]. The Apollo 16 sub-satellite had a lifetime of only 35 days in its 169.3° inclination orbit [*Russell et al.*, 1973], and was thereby the first demonstration of the truly amazing variations in satellite lifetimes as a function of initial orbital elements.

Other U.S. data have been processed in the past as well, such as the Explorer 35 and 49 missions. These spacecraft were placed in high-altitude orbits, and could therefore be applied in the determination of low-frequency harmonics as well as the lunar moments of inertia [*Bryant and Williamson*, 1974; *Gapcynski et al.*, 1975; *Blackshear and Gapcynski*, 1977]. However, for higher degree and order analysis, these data are of no additional benefit to the Lunar Orbiter and Apollo sub-satellite data.

Evidently, tracking during the sixties and early seventies was also performed by the Soviet Union, resulting in, roughly speaking, two camps of developers: the U.S. modelers on one side, and the Soviet modelers on the other. Early Soviet efforts to determine the gravitational field of the Moon include *Akim* [1966] who solved for four zonal coefficients and second and third degree tesserals up to order two; *Akim and Vlasova* [1977] who developed a first Russian model from a set of Luna spacecraft; *Akim and Vlasova* [1983] who made the first effort to exhaust all Luna data, which frequently had an orbit geometry distribution superior to the Apollo spacecraft; and *Kascheev* [1988] who based his estimation on a line-of-sight acceleration technique, rather than the fully dynamic approach, much like the initial Lunar Orbiter investigations on the U.S. side. Unfortunately, most of the Soviet data at that time was processed as a single stream of data, in a single-shot estimation of both the initial state vector of the Luna spacecraft as well as a low order gravitational field [*Akim and Vlasova*, 1977, 1983; *Ridenoure*, 1991], which seriously hampered the possibility to combine US and Soviet data. It also would pose non-trivial technical difficulties to combine these Soviet punch card data with data recovered from the same period in the U.S. Moreover, the data was hampered by a higher noise level than the DSN S-band measurements. More recent Russian model developments include *Sagitov et al.* [1986], who developed a 16x16 model largely similar to the model of *Bills and Ferrari* [1980], and *Akim and Golikov* [1997] in an on-going effort to reprocess the historical data, much like was done in the U.S.. Russian modelling efforts are further detailed in *Sagitov et al.* [1986] and *Akim and Golikov* [1997].

The Clementine spacecraft, launched on 24 January 1994, became the first U.S. spacecraft to return to the Moon in nearly 20 years [*Nozette et al.*, 1994]. Carrying an S-band transponder and placed into an elliptical polar orbit with a period of approximately 300 min and a mean periapsis altitude of 415 km [*Lemoine et al.*, 1997], it remained in lunar orbit from 19 February 1994 to 4 May 1994. Unlike the previous U.S. missions, whose periapses were located close to the lunar equator, the periapsis of the Clementine orbit was located near 30°S during the first month of mapping and near 30°N the following month. Tracking of Clementine was performed by 26 m and 34 m DSN antennae, as well as by a 30 m antenna operated by the U.S. Naval Research Laboratory at Pomonkey, Maryland. The data were predominantly 10 s integrated 2-way Doppler measurements, with typical noise levels of 0.25–0.30 mm/s (DSN) and 2.5–3.0 mm/s (Pomonkey) [*Lemoine et al.*, 1997]. The high periapsis of Clementine, combined with a relatively high eccentricity reduces the sensitivity to gravity field variations, and hence makes the Clementine orbit less than ideal for gravity modelling purposes.

Lunar Prospector indicated NASA's return to the Moon, and was launched on 7 January 1998, to enter lunar mapping orbit on 16 January of the same year [*Binder*, 1998]. During the one-year primary mission phase it was on a 118 minutes, 100 km mean altitude, polar orbit. From January 1999 its orbit was gradually lowered toward a 25 km level, as a means to increase the spatial resolution of the mapping instruments and to increase the sensitivity to gravity-induced orbit perturbations, thereby facilitating further improvement in the gravity field modelling. The end of mission for Lunar Prospector occurred at the end of July 1999, when the spacecraft was commanded to perform a hard landing in a permanently dark crater near the south pole of the Moon in support of the search for lunar volatiles.

In addition to radio tracking, *lunar laser ranging* (LLR) has also given information about the low-degree lunar gravity field. In the early 1970s *Williams et al.* [1973] employed laser ranging to retro-reflectors emplaced by Apollo spacecraft and Lunakhod-2 in combination with gravity field solutions to determine the three principal lunar moments of inertia. Lunar laser ranging data are sensitive to second-degree and third-degree spherical harmonics through the lunar physical librations. This combination is over-determined in the sense that the two observations of the physical librations plus gravimetric determination of C_{20} and C_{22} (unnormalised counterparts of $\overline{C}_{20}$ and $\overline{C}_{22}$) yields a set of four equations for the determination of three moments of inertia. As such, the combination of laser ranging and gravimetry also allows internal consistency checks to be performed [*Bills*, 1995]. Furthermore, continued lunar laser ranging has since led to a three orders of magnitude improvement in accuracy of the lunar ephemeris, a several orders of magnitude level improvement in the measurement of the variations of the Moon's rotations, and finally additional information on the Moon's tidal acceleration and rotational dissipation [*Dickey et al.*, 1994].

The last type of tracking data that has been used to infer properties of the mass density distribution of the Moon are very-long-baseline interferometric (VLBI) observables [*King et al.*, 1976]. The tedious data processing, combined with only limited data sets available, have, however, prevented such tracking from making a significant impact. Nevertheless, the method remains popular, mainly because it may provide an augmentation to the Doppler data information, *i.e.* information perpendicular to the line-of-sight direction, if used in a differential mode. This is particularly true for short data arcs. Therefore, the use of VLBI has been studied for the investigation of the lunar core [*Hanada et al.*, 1993], and is also foreseen for future lunar missions [*Namiki et al.*, 1999; *Sasaki et al.*, 1999].

2.2.3   Selenopotential models

Milestone developments in lunar gravity modelling, summarised in Table 2.2, were initiated by *Akim* [1966] who, on the basis of Luna–10 data, was able to determine the asymmetry of the lunar gravity field, as well as the first and realistic values for the oblateness and the pear-shape of the gravitational potential. *Lorell and Sjogren* [1968] were the first to publish a spherical harmonics model on the basis of Lunar Orbiter data, and these results supported the view that the Moon was gravitationally "rougher" than anticipated, and that a relatively large number of harmonic coefficients would be needed to describe a highly accurate selenopotential model.

A real scoop was the discovery of the mascons by *Muller and Sjogren* [1968]

Reference	Description
Akim [1966]	First effort; proof of asymmetry
Lorell and Sjogren [1968]	4x4 SH + zonals to degree 8
Muller and Sjogren [1968]	Discovery of the lunar mascons
Liu and Laing [1971]	8x8 SH + zonals to degree 15
Wong et al. [1971]	Surface-layer + low-degree SH
Michael and Blackshear [1972]	13x13 SH solution
Sjogren et al. [1972a, 1972b, 1974b, 1974c]	Acceleration profiles of mascons
Williams et al. [1973]	Moment of inertia based on LLR
Muller et al. [1974]	Theory of mascon mechanism
Bryant and Williamson [1974]	3x3 long-arc SH solution
Blackshear and Gapcynski [1977]	Zonals $J_2 - J_6$ + moment of inertia
Ananda [1977]	Mascon model, 117 point-masses
Ferrari [1977a, 1977b]	16x16 SH solution
Bills and Ferrari [1980]	16x16 SH solution
Sagitov et al. [1986]	16x16 SH solution
Konopliv et al. [1993]	60x60 SH solution, re-processing old data
Dickey et al. [1994]	Improved estimate of moments of inertia
Akim and Golikov [1997]	8x8 SH solution, reprocessing Russian data
Lemoine et al. [1997]	70x70 SH solution, Clementine data
Konopliv et al. [1998]	75x75 SH solution, Lunar Prospector data
Konopliv and Yuan [1999]	100x100 SH solution, Lunar Prospector data
Konopliv et al. [2001]	165x165 SH solution, Lunar Prospector data

Table 2.2   Summary of developments in lunar gravity field modelling. In general, subsequent solutions make use of most available data sets at that time. The latest Lunar Prospector-based solutions, for example, also include both Lunar Orbiter, Apollo sub-satellite and Clementine data. "SH" denotes *spherical harmonic models* and l x m denotes the maximum degree and order of the same solution

later that year, and their work sparked off a significant amount of interest both in the direct modelling of the lunar gravity field and on the interpretation side. Questions like: "What could the mascons be?" and "How could the Moon support them over geological time scales of billions of years?" popped up and sparked off a significant interest in lunar science.

Lunar laser ranging solutions started to become available from the early 1970s on and mainly played the role of improving the estimation of second degree harmonics as well as providing an independent means of verifying the solutions derived from satellite dynamics [*Williams et al.*, 1973; *Dickey et al.*, 1994]. This dualism has proved to be useful as the estimates tend to diverge more than the error associated with any individual solution, and, *vice versa*, errors in the satellite-derived coefficients also hampered the estimation of the lunar moments of inertia [*Bills*, 1995].

While the short-arc data from the Apollo spacecraft were mainly used to produce acceleration profiles over the lunar near-side, thus confirming and improving the findings from the Lunar Orbiter missions, they were also implemented in point-mass and surface layer solutions, which attempted to embed the localised gravity highs, without creating spurious far-side effects that would follow from moderate degree and order spherical harmonic expansions [*Wong et al.*, 1971]. *Ananda* [1977] derived a mass points model consisting of 117 masses from Apollo 15 and 16 sub-satellite data. *Sjogren et al.* [1974a] applied both LOS acceleration estimation techniques as well as point-mass (mascon) estimation near the lunar limbs using Apollo 15 and 16 sub-satellite data. These were comparatively long-arc, unperturbed orbits (no manoeuvres), and were valuable additions to LO-IV and LO-V high-inclination orbits. These sub-satellites also confirmed the suspicion that unfilled maria are usually associated with negative gravity anomalies and that mascons had to be near-surface features and not mass anomalies buried deep in the Moon.

Ferrari [1977a, 1977b] were among the first to derive models showing plausible correlations between far-side gravity anomalies and surface features, like the craters Mendeelev, Korolev and Moscoviense. His method was based on mean elements, which in turn were derived from short-arc fits to the tracking data. By using mean elements, instead of a direct method, they were able to dampen spurious far-side effects. *Bills and Ferrari* [1980], in turn, combined most of the data available at the time, like Lunar Orbiter, Apollo, Apollo sub-satellite and LLR, to develop a 16x16 spherical harmonic model. By combination of the short-arc data used for acceleration profiling, and the mean elements used to develop a smooth field by Ferrari three years earlier, they hoped to achieve a near-optimal combination of all available data sets. This model remained the best lunar gravity field model for 13 years.

A significant recommendation of the Lunar Observer study [*Cook et al.*, 1990; *Ridenoure*, 1991] was that reprocessing of the historic data from all Lunar Orbiters plus the Apollo 15 and 16 sub-satellites would be a worthwhile effort. New developments in geodetic theory and computer technology would enable development

Selenoid Heights From GLGM-2

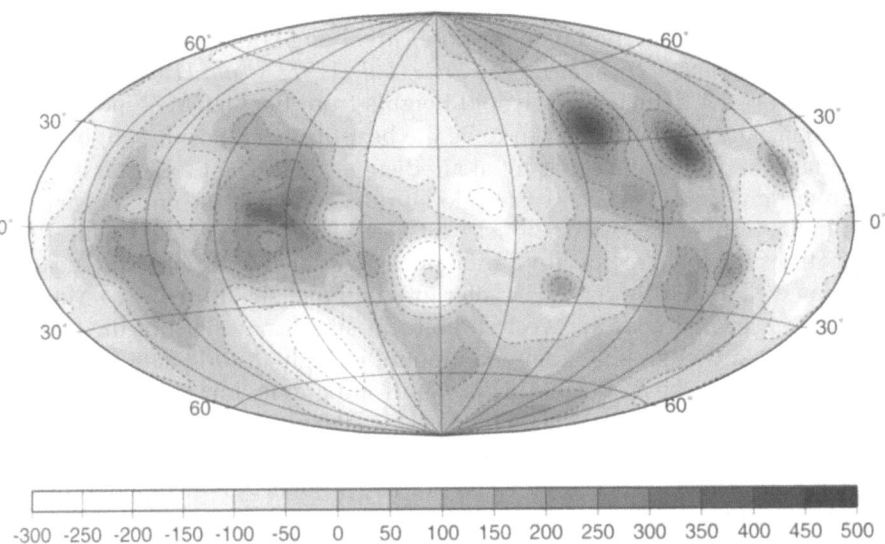

Figure 2.1   Selenoid heights in metres from GLGM–2. The reference field consists of the central term of the selenopotential plus the second-degree zonal, and the interval between the dashed contour lines is 100 m. The projection is a Hammer projection, centred around 270° eastern longitude

of a model that would make maximum use of the high-resolution near-side sampling that was available. Combined with an empirical constraint, like a Kaula rule, to smoothen the far-side, new analysis would undoubtedly yield a model much better for navigation purposes of future space missions. These efforts led to the development of *Lun60D*, a 60th degree and order model, by *Konopliv et al.* [1993]. This model is indeed tailored towards orbit determination and long-term orbit prediction, and therefore exhibits excellent performance in terms of its root-mean-square (RMS) fit to the tracking data.

A problem concerning geophysical interpretations using Lun60D was the excessive power at high degrees [*Lemoine et al.*, 1997]. When Clementine data became available, it was decided to emphasise long-wavelength selenophysical modelling rather than orbit fit. This implies tuning of the constraints and data weighting scheme until the power spectrum would be "selenophysically meaningful". As a result, *GLGM–2*, a 70x70 model by *Lemoine et al.* [1997] exhibits a smoother lunar far-side, and a power spectrum largely below the associated Kaula constraint. In other words, GLGM–2 focuses on large-scale selenophysical patterns, *e.g.* in terms of long- and medium-wavelength crustal thickness estimation, at the cost

of a dampening of the actual signal power in the data. An adjoint product from
the Clementine mission is the development of a global model of the lunar topography from Clementine laser altimetry measurements [*Smith et al.*, 1997], which
at present serves all interpretation purposes involving both topography and gravity data sets [*Neumann et al.*, 1996; *Von Frese et al.*, 1997; *Arkani-Hamed*, 1998, 1999*a*,
1999*b*]. Figure 2.1 depicts the selenoid heights from the GLGM–2 model, where
the reference gravity field is taken to be the best-fitting ellipsoid, and, hence, the
central term and second degree zonal are disregarded. The projection is a Hammer projection centred at 270° eastern longitude, such that the entire near-side is
found to the right of the centre and, *vice versa*, the far-side is seen to its left. Promi-
nent features are the large positive anomalies associated with the major near-side
basins, as well as the concentric circular gravitational signature of the Mare Orientale crater located near the western limb. A significant merit of the GLGM–2
model is that, although only a limited amount of high-altitude data are available
for the polar areas, the model predicts a significant gravity low associated with the
South Pole Aitken basin, located on the far-side, just north of the lunar south pole.
The gravity highs of the far-side equatorial regions are by no means supported by
observations, and care should be taken not to acknowledge them as undisputed
true features of the selenopotential.

The current state-of-the-art lunar gravity field models are dominated by Lunar
Prospector data. By virtue of the extensive tracking in its low-altitude polar orbit,
Lunar Prospector delivered more information than any other previous mission.
The development procedure is, much like for Lun60D, to a good extent based on
orbit determination fits, rather than the desire to limit the degree power. In other
words, the modelling efforts are much more data-driven than was the case for
GLGM–2. As discussed before, there are several ways to look at selenophysics,
by emphasising various physical scales (*e.g.* global and medium-scale features in
the case of GLGM–2). A basic principle behind the Lunar Prospector-based gravity field modelling efforts has been to make sure that the power in the data is
adequately represented, in other words not to dampen any high-frequency information. From a local or regional selenophysical point of view, this gives an
interesting perspective, because, for this type of analysis, it is exactly the high-
degree selenopotential terms which provide the key information [*Simons et al.*,
1997]. Series of 75x75 models, the *LP75n* models, where *n* denotes version number,
100x100 models, the *LP100n* solutions, as well as 165x165 models have been produced [*Konopliv et al.*, 1998; *Konopliv and Yuan*, 1999; *Hubbard et al.*, 1999; *Konopliv
et al.*, 2001]. The merit of two *published* models analysed in this book[2], LP75G and
LP100J, is found in the fact that the orbit error is reduced by at least one order of
magnitude compared to pre-Lunar Prospector solutions [*ibid.*] and that they have
finally confirmed that mare-filled far-side basins have associated mascons, albeit
less pronounced than for the major near-side basins.

For these reasons, GLGM–2 and the new series of models derived from Lunar

---

[2]The (very recent) mid-2001 publication of the LP165P model effectively prevents this model from
being discussed in any detail in this book.

Selenoid Heights From LP75G (m)

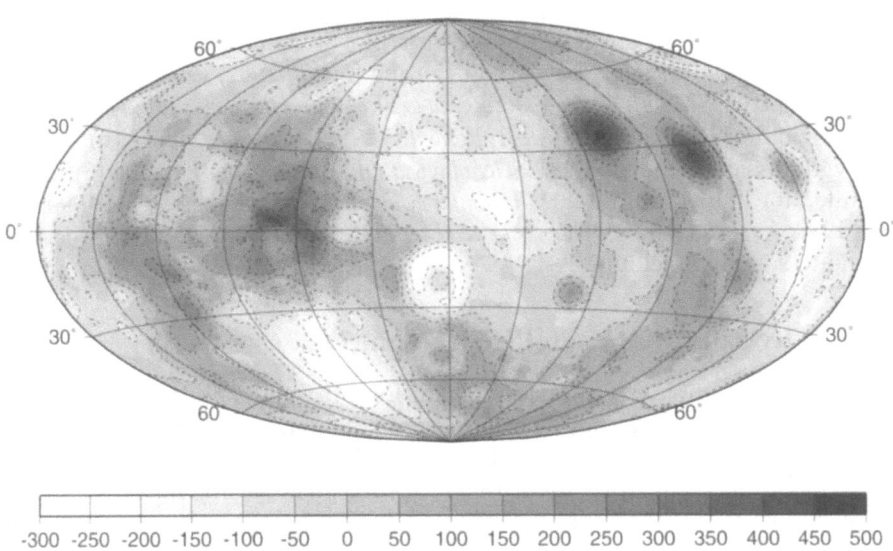

Figure 2.2   Selenoid heights in metres from LP75G. The reference field consists of the
central term of the selenopotential plus the second-degree zonal. The
contour interval and the projection is the same as for GLGM–2. Compared to
GLGM–2, the model displays larger excursions, as well as more pronounced
high-frequency variations. Both of these facts are attributed to the improved
quality of the Lunar Prospector data and the choice of different data weighting
schemes, as compared to GLGM–2 and other earlier non-JPL models

Prospector form the present basis for both geophysical interpretation and future
lunar satellite mission design. The equipotential surface of the lunar gravity field
model LP75G is depicted in Fig. 2.2. As is the case for GLGM–2, the model repro-
duces the expected negative correlations between topography and gravity for the
near-side basins. The Mare Orientale ring structure is more pronounced than was
the case for pre-Lunar Prospector models, and, moreover, the model indicates the
presence of larger peak variations of the selenoid as well as more high-frequency
signatures. All these are direct indicia of the better quality of the LP75G data set,
compared to GLGM–2.

2.3   Future solutions

Future gravimetric mapping experiments of the gravitational field of the Moon will include tracking between satellites for global data collection. Such missions have been studied extensively in Europe [*Flury*, 1981; *Coradini et al.*, 1996; *Floberghagen*, 1995; *Milani et al.*, 1996; *Racca et al.*, 1996; *Häusler*, 1998; *Häusler et al.*, 1998] and in the U.S. [*Ridenoure*, 1991; *Konopliv*, 1991], however, at present the only approved mission is the Japanese *Selenological and Engineering Explorer* (SELENE), scheduled for launch in mid-2004 [*Namiki et al.*, 1999; *Matsumoto et al.*, 1999; *Heki et al.*, 1999; *Kaneko et al.*, 1999; *Sasaki et al.*, 1999]. The obvious question is: why satellite-to-satellite tracking?

2.3.1   Why satellite-to-satellite tracking?

The first point is obviously the requirement for a truly global data set, with uniform and highly precise sampling of the selenopotential (through some functional related to the gravitational potential through differential or integral operators). Such a mapping would provide a much needed data set for lunar gravimetry and related selenophysical interpretation. It would largely remove the need for global power law constraints, *e.g.* in the formulation of a *Kaula's rule of thumb*, and it would allow direct access to local and medium-sized far-side features.

However, there is more. Any satellite method for gravity field determination has one fundamental disadvantage. Due to the satellite altitude, the effect of individual mass inhomogeneities of the Moon, such as mascons, transition zones between maria and highlands, smaller craters, ridges and rilles, as well as deeper mass anomalies, is damped, cf. (2.2). The amount of damping is, for the very low-altitude Lunar Prospector data, not very pronounced, but for satellites at Clementine altitude it poses a serious problem. Furthermore, assuming that the goal of the experiment is to determine the selenopotential at the highest level of accuracy over a vast range of scales (spherical harmonic wave numbers), the design of the global tracking experiment must provide a good sensitivity (signal-to-noise ratio higher than one) over the same range of harmonic frequencies.

The broader context of gravity sensors and gravity quantities and their respective information content is provided by the *Meissl scheme* [*Meissl*, 1971; *Rummel and Van Gelderen*, 1995; *Rummel*, 1997], depicted in Fig. 2.3, which outlines the principles of a *spectral theory* of gravity quantities on a sphere. The figure shows the disturbing potential along with its radial first and second derivatives (in a local satellite coordinate frame, the z-coordinate points in the radial direction) at two horizontal levels, one at satellite altitude and one at the planet surface. Eigenvalues connect the two levels, and the arrows indicate the directions in which they apply. The scheme is a spectral scheme and therefore applies per degree and order of the spherical harmonic expansion. The arrows moreover correspond to directions of smoothing, or attenuation of constituents of the series expansion, while the reciprocal eigenvalues correspond to unsmoothing, or amplification, in the other direc-

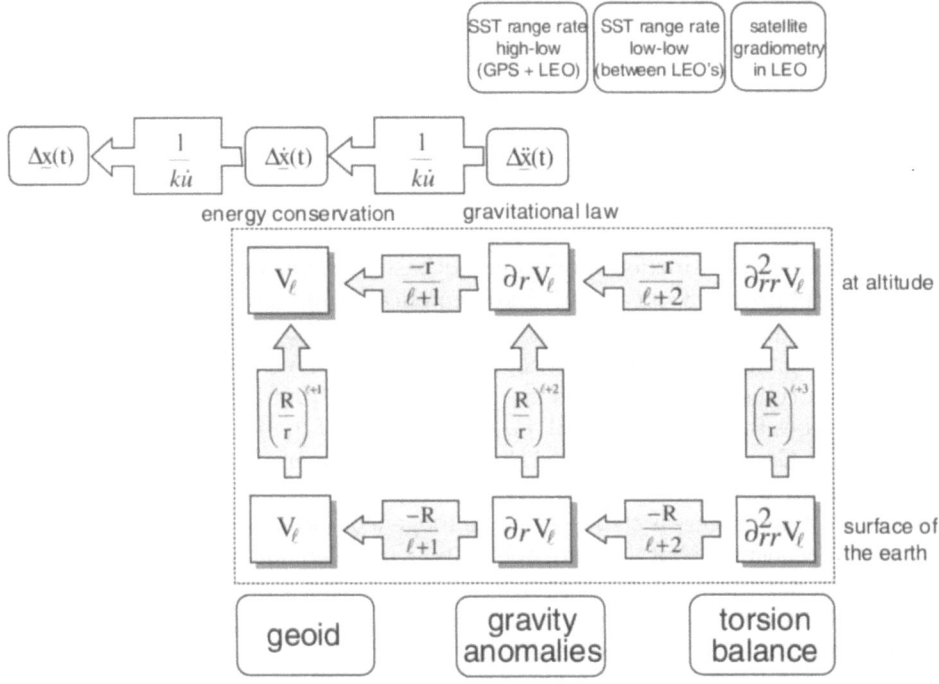

Figure 2.3   The Meissl scheme, or pocket guide of satellite geodesy, which relates the
anomalous geopotential, here denoted by V, and its first and second order
derivatives to tracking concepts proposed for future missions. The
selenodetic equivalent is trivial. Sources: *Meissl* [1971], *Rummel and Van
Gelderen* [1995] and *Rummel* [1997]

tion. In the space domain smoothing corresponds to integral operations, while unsmoothing corresponds to differentiation; hence it is readily seen that observation
of differentials of the gravitational potential effectively counteracts the damping
effect of the satellite altitude by amplification of the higher frequencies of the disturbing potential. Along a similar line, observation of selenopotential differentials
may be used to extract high-frequency gravity field information.

All quantities of the Meissl scheme can therefore be associated with tracking
principles, *e.g.* SST or SGG, or gravity quantities, *e.g.* selenoid heights, surface
layers (mascons, satellite velocities, etc.). In the case of research on the gravitational field of the Earth, this has led to two complementary directions. Medium-
wavelength gravity field modelling and modelling of temporary gravity field variations are envisaged with SST [*e.g., Watkins et al.*, 1995; *Balmino et al.*, 1998; *Visser*,
1999], while the high degree and order stationary gravity field is to be determined
with SGG techniques [*ESA*, 1996, 1999, 2000; *Balmino et al.*, 1998; *Woodworth et al.*,
1998; *Visser*, 1999; *SID*, 2000]. In the case of the Moon, the nature of the geophysical processes, and, hence, the scale of the problems of interest may be satisfactorily

addressed by satellite-to-satellite tracking, and hence, it is the observation method of choice. It also happens to be the cheaper solution, in terms of overall hardware costs.

The Meissl scheme not applies only to the gravity field quantities themselves, but also to the contained observation errors. Noise and systematic errors are therefore amplified if *downward continued* to the lunar surface. As a consequence, even in the presence of a fully global data set acquired at satellite altitude, regularisation may nonetheless be required. However, the extension of the Meissl scheme from deterministic function models to stochastic models describing the errors is not a straightforward operation. The theoretical foundation is given by *Rummel* [1997], who uses SGG experiments to exemplify the use of the Meissl scheme for error propagation. Principally, what is required is a large number of repetitions of the data-collecting experiment and the choice of a probability density function that describes the ensemble of all these repetitions. If the probability density distribution is Gaussian (normal), the statistics of the experiment are fully described by the first two statistical moments (mean value and variance). If furthermore the data collection process is *stationary*, *i.e.* homogeneous and isotropic over the lunar sphere, it is ensured that the error spectrum is invariant and therefore representative for all repetitions of the experiment. Finally, if the measurement collection is *ergodic*, the stochastics can be deduced from a single realisation of the experiment, and not from the whole statistical ensemble.

2.3.2   Prospects and future missions

The SST tracking principle was first demonstrated spectacularly by *Muller and Sjogren* [1968] in their mapping of the near-side lunar gravity field and consecutive discovery of the mascons. Considering the Earth as a satellite of the Moon, their line-of-sight Doppler residuals basically define what today would be called a high–low satellite-to-satellite configuration. Soon after, *Wolff* [1969] established the first mathematical framework for direct mapping of the Earth's gravitational potential using a satellite pair, with further improvements [*e.g.*, *Rummel et al.*, 1978; *Rummel*, 1979], including drag-free control to remove non-conservative drag effects, introduced in the early 1970s [*e.g.*, *Comfort*, 1973]. Satellite-to-satellite tracking comes in two flavours: i) the *low–low* configuration, in which both satellites are orbiting closely, separated only by a small in-plane angular distance, or, additionally, a small angular separation in the longitude of the ascending node or inclination to include cross-track information in the measurements; and ii) the *high–low* configuration, in which one satellite is orbiting at high altitude, or in a highly eccentric orbit, with the second remaining in a low orbit. Evidently, this second type resembles the more conventional Earth-based Doppler tracking of planetary spacecraft as the orbit of the *high* satellite is significantly less affected by variations in the gravity field, and is therefore much less of a dynamic platform than the *low* satellite. Furthermore, high–low SST frequently implies tracking over larger distances (long baselines) which reduces the sensitivity at shorter wavelengths of the gravity field.

Although a number of SST experiments involving satellites in Earth orbit have been performed to date, they have only had a limited impact on gravity field modelling. It is only recently that SST data sets were included in global geopotential models [*Lemoine et al.*, 1998], and earlier efforts were limited to orbit determination and to regional gravity mapping. The first experiments include the 1975 high–low range rate tracking of about 30 orbits of Apollo spacecraft in Earth orbit by the Applications Technology Satellite–6 (ATS–6) flying at geosynchronous altitude [*Vonbun et al.*, 1980; *Kahn et al.*, 1982], followed by ATS–6/GEOS 3 range rate tracking later that same year [*Hajela*, 1979; *Kahn et al.*, 1982]. Both these experiments aimed at the determination of local 5° mean gravity anomalies from the SST data and documented the value of SST techniques for mapping the structures of the Earth's gravity field as well as for locating problems in existing terrestrial data sets [*Douglas et al.*, 1980; *Kahn et al.*, 1982].

More recent developments include *Global Positioning System* (GPS) tracking of low Earth-orbiting satellites such as TOPEX/Poseidon for precise orbit determination [*Yunck et al.*, 1990, 1994; *Yunck and Melbourne*, 1996; *Bertiger et al.*, 1994; *Melbourne et al.*, 1994; *Lemoine et al.*, 1998] and gravity field modelling [*Lemoine et al.*, 1998]; the Explorer Platform/Extreme Ultraviolet Explorer for orbit determination [*Gold*, 1994; *Lemoine et al.*, 1998] and gravity field improvement [*Olson*, 1996; *Lemoine et al.*, 1998]; and Microlab–1 carrying the so-called GPS/MET experiment for both orbit and gravity field modelling [*Lemoine et al.*, 1998]. Similarly, tracking by the *Tracking and Data Relay Satellite System* (TDRSS) of TOPEX/Poseidon and Explorer Platform/Extreme Ultraviolet Explorer have been applied in both orbit determination [*Marshall et al.*, 1995, 1996; *Visser and Ambrosius*, 1996] and gravity field improvement [*Marshall et al.*, 1995; *Lemoine et al.*, 1998]. Both GPS and TDRSS tracking of the Space Shuttle have furthermore been performed in support of high-precision experiments of the manned space programme [*Rowlands et al.*, 1997]. Finally, throughout the 1980s and 1990s, several studies of SST experiments for dedicated global terrestrial gravity field experiments were conducted [*e.g.*, *Kaula*, 1983; *Wagner*, 1983, 1987b; *Colombo*, 1984; *Schrama*, 1986; *Reigber et al.*, 1987; *Wiejak et al.*, 1991; *Watkins et al.*, 1995]. On-going or approved SST-based terrestrial gravity field missions include *Challenging Minisatellite Payload* (CHAMP) [*Reigber et al.*, 1996; *Sehnal et al.*, 1997; *Schmitt and Bauer*, 2000], the *Gravity Recovery and Climate Experiment* (GRACE) [*Watkins et al.*, 1995; *Mazanek et al.*, 2000] and the *Gravity and Steady-State Ocean Explorer mission* (GOCE) [*SID*, 2000; *ESA*, 2000]. All of these missions employ high–low SST by means of a GPS receiver. GRACE also employs low–low SST radar ranging in combination with drag-free control, *i.e.* compensation of non-conservative forces through accelerometers that measure the induced acceleration and actuators that actually correct the orbit. The primary limitations of SST systems for gravity field recovery are the errors in the non-conservative surface force models, *e.g.* solar radiation pressure, albedo and drag (the latter not applicable to the lunar problem, since the Moon's atmosphere is insignificant).

In terms of lunar missions, the potential benefit of SST was first suggested by

Kaula [1969]. *Ananda et al.* [1976] did the first global study, emphasising exactly the obvious far-side improvement. In Europe *Flury* [1981] performed the first analysis of global as well as regional gravity field recovery on the basis of Doppler tracking involving a relay satellite orbiting at high altitude and a low-altitude near-polar orbiter as part of the Polar Lunar Orbiter (POLO) mission proposal [*ESA*, 1979]. The long-running Lunar Observer study [*Cook et al.*, 1990; *Ridenoure*, 1991; *Konopliv*, 1991] did never materialise in the United States, despite obvious scientific merit and a strong support by the scientific community. A major European study was the *Moon Orbiting Observatory* (MORO) mission [*Coradini et al.*, 1996; *Floberghagen et al.*, 1996; *Milani et al.*, 1996; *Racca et al.*, 1996], which unfortunately also did not survive the selection process. The intention of MORO was a global mapping of the gravity field by means of continuous low–low 2-way Doppler tracking. A similar fate denied the flight opportunity of *Lunarstar*, a proposed follow-on mission to Lunar Prospector. The Lunarstar proposal [*Häusler*, 1998; *Häusler et al.*, 1998] envisaged SST in its simplest and most cost-effective way possible: 1-way Doppler from a mother spacecraft, relayed to Earth stations by a small sub-satellite. A much more ambitious project is the approved Japanese SELENE mission, which will carry a range of selenodetic instruments [*SELENE*, 1996; *Araki et al.*, 1999; *Heki et al.*, 1999; *Matsumoto et al.*, 1999; *Nagae et al.*, 1999; *Namiki et al.*, 1999; *Sasaki et al.*, 1999]: first, there is conventional 2-way Earth-based Doppler tracking of the main orbiter as well as of two small relay satellites; second, there is 4-way Doppler tracking (station-orbiter-sub-satellite-orbiter-station) which effectively provides the inter-satellite radio links; and, third, differential very-long-baseline-interferometry (ΔVLBI) is foreseen using artificial radio sources on one of the small sub-satellites and on the main spacecraft, which provides tracking information perpendicular to the LOS direction from Earth-based stations, and therefore completes a near three-dimensional tracking system. SELENE is scheduled for launch in July 2004 and will, if successful, finally provide direct access to the thus far gravity-wise dark lunar far-side.

Chapter 3

Assessment of modern lunar gravity field models through orbit analysis

'*...the greatest usefulness of linear perturbation theories for orbital motion error studies is probably in understanding what information various satellite tracking data contribute to a particular gravity model solution. To this end, it can be seen that an accurate determination of the gravity coefficient error covariance is as important as the determination of the actual coefficient values.*'

G. W. Rosborough, 1987

3.1   Introduction

All present-day lunar gravity field models are – with the exception of a minor contribution from lunar laser ranging data to the second-degree harmonics – *satellite-only* models. Orbit analysis is therefore a fundamental tool for quality assessment, alongside arguments of compatibility imposed by selenophysics. A general problem in selenodesy and selenophysics is the lack of independent data to serve purposes of calibration and validation of basically any result. Model assessment techniques may therefore be quasi-circular and far from independent. Regarding the use of prior information from lunar physics, primarily the assumptions on coefficient amplitudes and the distribution of the selenopotential signal power over the degrees, the indicia of their correctness are for the larger part provided by the induced improvement in measurement fit of the satellite orbits. One problem of satellite orbit and orbit error measures, on the other hand, is that they are "global" measures of the selenopotential quality, as opposed to a direct function of location on the lunar sphere, and that RMS-of-fit values over several-day satellite arcs actually contain very little spatial information on the intrinsic selenopotential model quality. Nevertheless, gravity field models capable of predicting reasonable mass anomaly structures and estimates of crustal thickness tend to be considered reliable. It should therefore be clear that it is only through continued iteration over the available data sets, along with improved modelling in several branches of selenoscience, that solid quality assessment and statistically significant quality improvement of lunar gravity field models may be achieved.

This impasse is likely to persist until one is in the position to independently and rigorously determine the unique gravity field up to a significant degree and order, and therefore disregard *a priori* requirements from lunar physics. The present use of prior knowledge in tuning the model towards "meaningful" coefficient values is therefore to be considered a symbiotic interplay between satellite orbit determination and selenophysics from the derived gravity field, and applies to all models based on Earth-based measurements alone.

As a result, it is possible to tune the gravity field model towards the needs of global-scale selenophysical studies, largely by means of a significant dampening of the higher frequencies, or, *vice versa*, to emphasise the fit of the observational data, and hence the orbit determination process, by choosing constraints that produce optimum fits. The latter approach obviously retains the power in the observed tracking information, and is therefore also suitable for fine-scale selenophysical studies of well-sampled areas, like the lunar near-side. Hence, lunar gravity models are sensitive to regularisation (constraint) schemes in several ways: i) for pure stabilisation of the least squares analysis; ii) to achieve a certain global smoothness of the model, which enables far-side studies as well as global orbit modelling, and; iii) for tuning of the model towards a possible type of application. Models therefore inherit features directly related to the regularisation schemes used in their development. Whereas the GLGM–2 model is deliberately tuned towards long-wavelength selenophysical analysis needs [*Lemoine et al.*, 1997], the most recent Lunar Prospector models, the LP75n and LP100n series to a large extent focus on orbit determination performance and higher-degree selenophysical interpretation [*Konopliv and Yuan*, 1999; *Carranza et al.*, 1999].

Figure 3.1 depicts the degree-wise RMS amplitude spectra of GLGM–2 and LP75G, along with the associated variances as well as constraints applied in their development. The constraint formalism is given in the form of a scaled Kaula's rule. It is readily seen that the two models yield different amplitude spectra, with a nearly one order of magnitude difference at the higher end of the degree scale. The tuning of GLGM–2 towards the needs of global-scale selenophysics therefore apparently contradicts the higher power needed for improved satellite orbit fits, a conclusion which will be supported in Chap. 4, where it is shown that the difference in power cannot be attributed to the additional data provided by Lunar Prospector alone. In other words, there is a strong link with the data process-

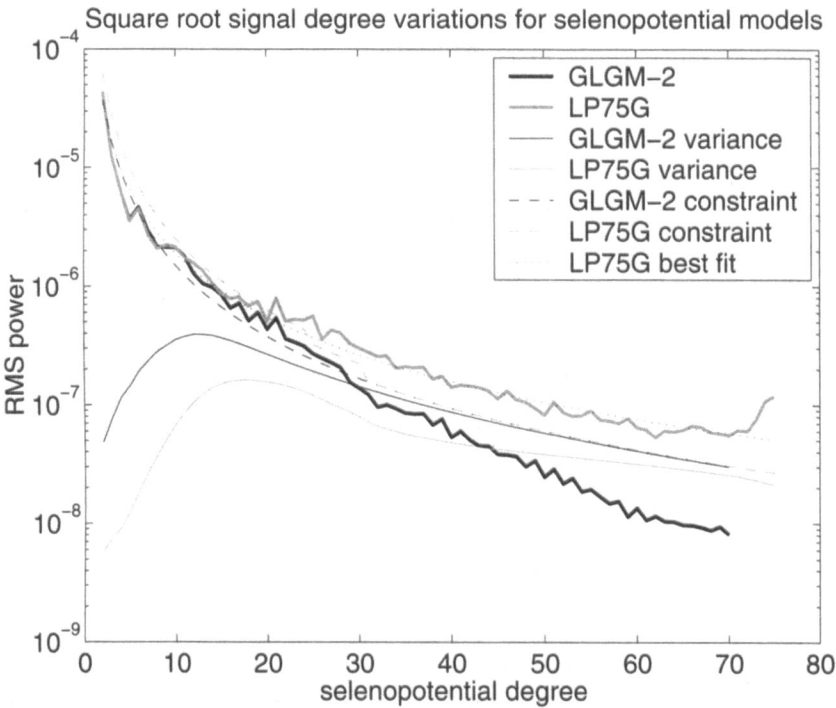

Figure 3.1   Degree-wise RMS amplitude spectrum of GLGM–2 and LP75G, the
corresponding error variances and the applied Kaula constraints. While the
constraints are identical above $l = 30$, one observes a significantly higher
power for LP75G than for GLGM–2, culminating at nearly one order of
magnitude difference. The corresponding variance curves show that,
whereas the GLGM–2 errors asymptotically approach the applied Kaula rule,
the LP75G variances remain below its constraint. The "best fit" curve for
LP75G corresponds to an *a posteriori* fit of an exponential function to its
degree-wise variations

ing strategy. Given the situation of adequate observability of only a limited number of linear combinations of harmonics and no far-side data, the impression that
the degree-wise signal-to-noise ratio of LP75G is higher than one for the complete
range of harmonics must be considered an artefact of the modelling strategy. That
is, it is only through extensive regularisation that the error remains bounded for
the higher degrees, and asymptotically approaches the constraint curve. Based on
the data alone, a signal-to-noise ratio (SNR) of one would be met around $l \simeq 10$
and $l \simeq 15$ for GLGM–2 and LP75G, respectively. Moreover, the optimistic signal-
to-noise ratio is a result of optimistic data weighting schemes. Nevertheless, an
obvious merit of LP75G, in terms of orbit determination performance, is the ability to observe certain higher frequency orbit perturbations. The peak amplitude

near the maximum degree 75 obviously indicates aliasing of higher frequencies in the model. In other words, given the improved sampling of the lunar near-side provided by the Lunar Prospector data, combined with the processing strategy of the JPL-derived models, the observability of a limited number of high frequency constituents is gradually improved accordingly. However, the overall SNR based on the tracking data remains poor, and it is therefore a tremendous challenge for the gravity field model developer to extract this high-frequency information with an acceptable statistical reliability.

A suitable tool for the assessment of the intrinsic selenopotential model accuracy is a qualitative analysis of the resulting orbit error. Furthermore, while the orbit accuracy achieved from selenopotential models is in first instance indicative of the *sec* quality of the models, orbit accuracy is also directly linked to the quality of selenophysical products derived from orbit data, *e.g.* laser altimetry (in direct or crossover mode), as already flown on Clementine [*Zuber et al.*, 1994; *Smith et al.*, 1997] and envisaged for SELENE [*Araki et al.*, 1999; *Nagae et al.*, 1999; *Namiki et al.*, 1999; *Sasaki et al.*, 1999], or imaging experiments requiring knowledge of spacecraft position and attitude. Orbit accuracy figures therefore form a product which directly enters the statistics of all science derived from orbit data. The present chapter therefore presents covariance analysis results based on the variance-covariance matrices of GLGM–2 and LP75G. A linear perturbation theory (LPT) [*Kaula*, 1966; *Rosborough*, 1986; *Schrama*, 1989; *Visser*, 1992] is applied to address the accuracy of low lunar orbits computed from the two models. Second, the LPT method provides a direct means to characterise the orbit errors in both a time-wise and a space-wise sense. The results of the covariance analysis are compared with results derived from coefficient differences (between the two models). For the covariance analysis, the calibrated error variance-covariance matrices of GLGM–2 and LP75G are chosen to represent the true error level of the models. Further considerations on the usefulness of the covariance matrix as the error measure for lunar gravity field solutions are given in Chap. 4.

Furthermore, both models are applied in an analysis of long-term behaviour and lifetime predictions, which are of due importance for satellite mission analysis. The existence of frozen orbit geometries as well as periodic orbits is investigated. It will be the conclusion of basically all types of analysis that GLGM–2 and LP75G exhibit widely different orbit behaviour, both deterministically and in terms of mapping of the associated error variance-covariance. In other words, a clear convergence of lunar gravity field model behaviour is yet to be established.

3.2   The role of model calibration

A problem of concern for the assessment of selenopotential models is that of calibration. It has, *e.g.*, been shown in the previous section that LP75G displays signs of optimistic data weighting, which shows up in the results as overly optimistic signal-to-noise ratios. Moreover, with little or no external data available for the verification of the constraint schemes, other than orbit determination fits, the mere fact that the formal errors asymptotically approach the constraint curve would only be a proof of the internal consistency of the solution. Such a feature is therefore nothing but a mathematical consequence of the regularisation scheme and the appropriate use of data weighting factors. In other words, this result alone does not prove the validity of the selenopotential degree-wise amplitude curve, and care must be taken not to view this as a decisive argument. With limited data sets available (only a few lunar gravity mapping missions have been flown), covering merely a limited range of inclinations and other orbital characteristics, lunar gravity field models are not expected to be general purpose models. Rather, orbit errors and selenophysical products derived from orbit data are expected to exhibit systematic error patterns, where the lowest error levels are associated with inclinations present in the data sets used to infer the respective potential model, and with the regions of the Moon best covered with measurements. Second, as selenopotential models are largely sensitive to regularisation schemes, and there are few independent data, *e.g.* in situ ground data, available for verification of their validity, it is likely that the constraints may influence the error measures (*e.g.* the *a posteriori* error variance-covariance matrix). Third, although the error measures used in lunar gravity modelling thus far are usually merely formal, in the sense that systematic effects are disregarded in the error propagation model, empirical calibration of the per-arc or per-satellite least squares equation systems (normal equations or square root information filters) is frequently performed. The purpose of this – usually significant – effort is to derive a single error measure which contains both the propagated measurement precision (noise) as well as systematic shortcomings in the modelling of the measurements and the satellite dynamics. Another purpose of the calibration is to improve the numerical properties of the combined least squares equation system, which is fundamentally the weighted sum of arc-wise normal equations, after taking due account of the correlations between the arc-dependent parameters and the so-called global parameters common to all arcs, like the gravity field coefficients. In particular for situations involving vastly different data sources, and, hence, possible datum problems, it may be that a straightforward summation of equations leads to inferior properties, for which calibration acts as a countermeasure.

A frequently used method of calibration involves subset solutions, where a single set of data, *e.g.* one arc or all data from one satellite, is removed from the master solution, which of course includes all data available [*Lerch*, 1991]. Under the assumption of zero correlation between measurements, the *a priori* covariance Q_y of the measurements is diagonal and simply contains the inverse data weighting factors w^{-1} of the individual data sets. The objective of the calibration is then to adjust the initial data weight, for the data set not included in the subset solution, in such a way that the aggregate differences between the coefficients of the master and subset solutions match the aggregate differences between the coefficient formal standard deviations (sigmas) of the same two solutions. Since the observation errors in any set of data not included in the subset solution contain all errors (random and systematic, measurement-related and dynamical) this type of calibration aims to derive an error variance-covariance matrix that reflects the intrinsic overall potential coefficient error, as opposed to the straightforward propagation of random measurement errors into the potential model (or any linear function thereof). In fact, this is the merit of the method, and by its application it has been possible, without explicit stochastic and systematic error models, to arrive at meaningful error measures for terrestrial gravity fields [*Marsh et al.*, 1988; *Lerch et al.*, 1991, 1994; *Nerem et al.*, 1994; *Tapley et al.*, 1996; *Lemoine et al.*, 1998]. However, the method fails to distinguish between the effects of the various types of errors, and, moreover, does not take due account of correlations between coefficient estimates, nor the fact that the estimation of the lunar gravity field is severely biased by the regularisation. This realisation is important, for the obvious reason that the modelling would benefit from a better understanding of the error associated with a lunar gravity field solution. This is further elaborated in Chap. 4, which deals with aspects of information content and regularisation of lunar gravity field solutions, and where also an approach for quantification of the various types of error parameters is discussed.

The calibration of GLGM–2 follows the Lerch procedure [*Lemoine et al.*, 1997]. In this case, the assumption that the expected difference between the subset solution and the master solution must be predicted by the two solution covariance matrices requires that, for any calibrated subset solution, it holds

$$E\left\{(\bar{x} - x)(\bar{x} - x)^T\right\} = Q(\bar{x}) - Q(x) = Q(\bar{x} - x) \tag{3.1}$$

where $E\{\bullet\}$ is the expectation operator and $Q(\bar{x})$ and $Q(x)$ denote the covariance matrices of the subset solution $\bar{x}$ and the master solution x, respectively. Practically, this is achieved by restricting the error measure to the standard deviations and then require the calibration factor K^2 for any subset solution $\bar{x}$ to fulfill

$$\|\bar{x} - x\|^2 = K^2\left(\text{trace}\left[Q(\bar{x})\right] - \text{trace}\left[Q(x)\right]\right) = K^2\left(\|\sigma(\bar{x})\|^2 - \|\sigma(x)\|^2\right) \tag{3.2}$$

Returning from the abstract vector notation x to spherical harmonic coefficients, this implies that the calibration factor K_l for each degree l reads

$$K_l = \left[\frac{\sum_{m=0}^{l}\left(G_{lm} - \overline{G}_{lm}\right)^2}{\sum_{m=0}^{l}\left(\overline{\sigma}_{lm}^2 - \sigma_{lm}^2\right)}\right]^{1/2} \tag{3.3}$$

where $G_{lm} \in \{\overline{C}_{lm}, \overline{S}_{lm}\}$, σ_{lm} obviously represents the formal standard deviation of G_{lm} and the bar again denotes the subset solution. Notice that this approximation procedure for the definition of calibration factors fails to take into account the correlations between the gravity field coefficient estimates. The average calibration factor $\overline{K}$ for a given set of data is straightforwardly computed as the average over a span of degrees $l \in \{2, L\}$, *i.e.*

$$\overline{K} = \sum_{l=2}^{L} \frac{K_l}{L - 1} \tag{3.4}$$

where the summation is usually truncated at $L \leqslant l_{max}$, for the simple and obvious reason that the higher degree coefficients are usually dominated by the regularisation, rather than tracking data, which may produce less than meaningful calibration factors [*e.g.*, *Marsh et al.*, 1988; *Lemoine et al.*, 1997, 1998]. $\overline{K}$ defines a new scaling factor for the least squares equation to be calibrated. Given an *a priori* weight w_a, the new scaled weighting factor w_n reads

$$w_n = \frac{w_a}{\overline{K}^2} \tag{3.5}$$

Again, under the assumption that the inverse weights describe the *a priori* data errors, weighting factors and calibrated *a priori* data sigmas are related through

$$\sigma_{a_n} = \frac{1}{\sqrt{w_n}} \tag{3.6}$$

Calibration factors greater than one indicate that the data excluded from the master solution is given too much a weighting, and that this set of data affects the master solutions coefficients by more than what would be expected by the formal variance of the same estimates. In other words, the covariance is optimistic for the data set being calibrated. This of course suggests an iterative process, the aim of which being to arrive at unit values of $\overline{K}$ for all data sets. Obviously this is a major effort, and a primary reason to simplify the subdivision into larger sets of data, rather than maintain a per-arc calibration strategy. Calibration factors found in this way usually lead to *a priori* measurement sigmas significantly larger than the intrinsic measurement precision [*Marsh et al.*, 1990; *Lerch*, 1991; *Lemoine et al.*, 1997, 1998]. Such a down-weighting reflects the empirical reduction of the estimation accuracy needed to account for systematic error sources remaining in the processing.

The calibration of LP75G, on the other hand, is based on the quality of the measurements [*Konopliv*, private communication], in other words the desire to exploit the valuable power in the data for orbit analysis and higher degree selenophysical interpretation, together with a certain amount of scepticism towards the deliberate down-weighting of the information in the tracking data, for the sake of a more consistent error measure. Rather, the philosophy is that it is unnatural to attribute force model deficits and reference frame deficits to the quality of the measurements, which actually in terms of noise levels nowadays is pretty encouraging. Instead, the models derived in this way are frequently published with a footnote saying that the real selenopotential error probably is larger than what is predicted by the associated covariance matrix, due to the negligence of systematic effects. This approach goes back to the Lun60D model [*Konopliv et al.*, 1993], and is largely the same for all JPL-derived models to date.

Given the entangled interplay of constraints, calibration and limited observability of individual spherical harmonic coefficients, along with apparently conflicting requirements from orbit determination and lunar physics, it is obvious that assessment of the true value of the error measures associated with present-day selenopotential models is a delicate matter. Consequently, the same argument applies for any linear projection of the associated covariances, *e.g.* in terms of linearised errors in the satellite orbits, selenoid heights or gravity anomalies.

3.3 Covariance analysis of GLGM–2 and LP75G

3.3.1 Linear orbit perturbation theory

In terms of osculating Keplerian elements $\{a, e, i, \Omega, \omega, M\}$ the gravitational potential reads [*Kaula*, 1966]

$$U = \frac{GM}{a} \sum_{l=0}^{\infty} \sum_{m=0}^{l} \left(\frac{a_e}{a}\right)^l \sum_{p=0}^{l} \overline{F}_{lmp}(i) \sum_{q=-\infty}^{\infty} G_{lpq}(e) S_{lmpq}(\omega, M, \Omega, \theta) \qquad (3.7)$$

where θ is the hour angle of the prime meridian (the equivalent of the Greenwich Sidereal Time or Greenwich hour angle). In practise, the summation over spherical harmonic degrees l is limited to the central term $l = m = 0$ and $l \in \{2, \ldots, l_{max}\}$, given the discrete modelling of the gravity field parameters from a finite number of measurements, and the choice of the origin of the coordinate system coinciding with the Moon's centre of mass which causes all degree one terms to be zero. The normalised inclination functions $\overline{F}_{lmp}(i)$, which account for the fact that the orbit is inclined to the equator, can either be constructed from their unnormalised equivalents, given by *Kaula* [1966], like the normalised associated Legendre functions, or directly from the normalised Legendre functions using Fourier techniques [*Wagner*, 1983; *Sneeuw*, 1991]. These works furthermore make use of an index $k = l - 2p$, which opposite to p has an illustrative interpretation along the inclined orbit. The terms $G_{lpq}(e)$ are known as *eccentricity functions* or *Hansen functions* and relate to the non-circularity of the orbit. In general, the functions $G_{lpq}(e)$ are of the order $\mathcal{O}(e^{|q|})$ [*Kaula*, 1966], and, consequently, for nearly circular orbits ($e \ll 1$), it is sufficient to consider $q = 0$ and $q = \pm 1$ only [*Allen*, 1967]:

$$G_{l,p,0}(e) = 1 + \mathcal{O}(e^2)$$
$$G_{l,p,+1}(e) = \frac{e}{2}(3l - 4p + 1) + \mathcal{O}(e^3) \qquad (3.8)$$
$$G_{l,p,-1}(e) = \frac{e}{2}(-l + 4p + 1) + \mathcal{O}(e^3)$$

Furthermore, the phase function S_{lmpq} is defined as

$$S_{lmpq} = \tilde{C}_{lm} \cos \psi_{lmpq} + \tilde{S}_{lm} \sin \psi_{lmpq} \qquad (3.9)$$

where

$$\tilde{C}_{lm} = \begin{bmatrix} \overline{C}_{lm} \\ -\overline{S}_{lm} \end{bmatrix}_{\substack{l-m \text{ even} \\ l-m \text{ odd}}} \quad \text{and} \quad \tilde{S}_{lm} = \begin{bmatrix} \overline{S}_{lm} \\ \overline{C}_{lm} \end{bmatrix}_{\substack{l-m \text{ even} \\ l-m \text{ odd}}} \qquad (3.10)$$

and the phase argument ψ_{lmpq} is given by

$$
\begin{aligned}
\psi_{lmpq} &= (l - 2p)\omega + (l - 2p + q)M + m(\Omega - \theta) \\
&= f_{lmp}(\omega + M) + qM + m\lambda_a
\end{aligned}
\tag{3.11}
$$

where λ_a is the body-fixed longitude of an ascending node passage and basically serves as an initial condition for the lunar rotation. Hence,

$$
f_{lmp} = l - 2p - m\frac{\lambda_a - \Omega + \theta}{\omega + M}
\tag{3.12}
$$

For a given selenopotential model, S_{lmpq} is therefore merely a function of the argument of latitude, through M and ω, and the position of the ascending node relative to the hour angle of the prime meridian $\Omega - \theta$.

The linear perturbations theory used in the present treatise [*Rosborough*, 1986; *Rosborough and Tapley*, 1987; *Schrama*, 1986, 1989; *Visser*, 1992; *Balmino*, 1993] is based on the *Hill equations* [e.g., *Kaplan*, 1976; *Colombo*, 1984; *Schrama*, 1989] and the existence of true *repeat orbits*. Although a periodic orbit signal obviously is not essential for the spectral analysis of the orbit behaviour, it aids in decoupling the terms of different orders, see Sect. 3.3.3. The use of a true repeat orbit should therefore be seen in the same context as the assumption in Fourier analysis of non-periodic signals that the non-periodic signal equals a periodic signal over the interval in question. A main characteristic of repeat orbits is that the satellite ground-track (or equivalently, the orbit of the satellite with respect to the lunar surface) repeats itself after an integer number of orbital revolutions, by virtue of which the orbit perturbations become periodic signals and, hence, true Fourier series. The Hill equations are approximate equations, based on a circular reference orbit, which may be solved exactly, as opposed to the standard LPT which yields linearised solutions to the Lagrange planetary equations [e.g., *Kaula*, 1966]. It is implicitly assumed that the coordinate frame of integration is inertial and fixed, *i.e.* that effects of precession and nutation during the period of integration are neglected, and that the perturbations may be adequately described by a linear theory, *i.e.* the perturbations remain small with respect to the mean elements and, in addition, remain within the linear domain. The validity of the LPT approach for qualitative orbit analysis has furthermore been extensively tested through comparisons with numerical orbit integration [*Visser*, 1992; *Visser et al.*, 1995a].

Expressions for orbit perturbations relative to a reference orbit were developed by *Rosborough* [1986] based on earlier work on perturbations in the orbital elements by *Kaula* [1966]. In the framework of the Hill equations, only order zero terms of the eccentricity are retained. These dominant radial $\Delta r^{(0)}$, along-track $\Delta\tau^{(0)}$ and cross-track $\Delta c^{(0)}$ components of the perturbation yield [*Rosborough*, 1986; *Rosbor-*

ough and Tapley, 1987; Visser, 1992]:

$$\Delta r^{(0)} = \sum_{l=2}^{l_{max}} \sum_{m=0}^{l} \sum_{p=0}^{l} C_{lmp}^{r} S_{lmp0}$$

$$\Delta \tau^{(0)} = \sum_{l=2}^{l_{max}} \sum_{m=0}^{l} \sum_{p=0}^{l} C_{lmp}^{\tau} S_{lmp0}^{*} \qquad (3.13)$$

$$\Delta c^{(0)} = \sum_{l=2}^{l_{max}} \sum_{m=0}^{l} \sum_{p=0}^{l} C_{lmp}^{c+} S_{(l+1)mp0}^{*} - C_{lmp}^{c-} S_{(l-1)mp0}^{*}$$

where the *amplitude factors* are given by

$$C_{lmp}^{r} = a \left(\frac{a_e}{a}\right)^{l} \overline{F}_{lmp} \left[\frac{2(l-2p)}{f_{lmp}} + \frac{4p-3l-1}{2\left(f_{lmp}+1\right)} + \frac{4p-l+1}{2\left(f_{lmp}-1\right)}\right]$$

$$C_{lmp}^{\tau} = a \left(\frac{a_e}{a}\right)^{l} \overline{F}_{lmp} \left[\frac{2(l+1)-3(l-2p)\frac{1}{f_{lmp}}}{f_{lmp}} + \frac{4p-3l-1}{\left(f_{lmp}+1\right)} + \frac{l-4p-1}{\left(f_{lmp}-1\right)}\right]$$

$$C_{lmp}^{c+} = \frac{1}{2}a \left(\frac{a_e}{a}\right)^{l} \frac{1}{f_{lmp}} \left[\frac{\overline{F}_{lmp}}{\sin i}\left\{(l-2p)\cos i - m\right\} - \overline{F}_{lmp}'\right] \qquad (3.14)$$

$$C_{lmp}^{c-} = \frac{1}{2}a \left(\frac{a_e}{a}\right)^{l} \frac{1}{f_{lmp}} \left[\frac{\overline{F}_{lmp}}{\sin i}\left\{(l-2p)\cos i - m\right\} + \overline{F}_{lmp}'\right]$$

in which

$$\overline{F}_{lmp}' = \frac{d\overline{F}_{lmp}}{di}, \qquad (3.15)$$

and

$$S_{lmpq}^{*} = \int S_{lmpq}(\psi)d\psi = \tilde{C}_{lm} \sin \psi_{lmpq} - \tilde{S}_{lm} \cos \psi_{lmpq} \qquad (3.16)$$

This last integral merely amounts to a 90° phase shift as $S_{lmpq}^{*}(\psi_{lmpq}) = S_{lmpq}(\psi_{lmpq} - 90°)$. The above notation is slightly different than the one maintained by *Rosborough* [1986], due to the fact that for circular reference orbits the mean motion n equals the time-derivative of $\omega + M$. Notice that the cross-track solution has a different functional form than its counterpart derived from the Lagrange planetary equations [*Rosborough, 1986; Rosborough and Tapley, 1987*]. The equivalence of the two solutions is proven in *Balmino et al.* [1996].

Introducing the repeat condition of the lunar satellite orbit, prescribed by an integer number N_r orbital revolutions in a time period of N_m integer nodal months, or

$$N_m \frac{2\pi}{\dot{\Omega} - \dot{\theta}} = N_r \frac{2\pi}{\dot{\omega} + \dot{M}}, \qquad (3.17)$$

where N_m and N_r are relative primes, one has

$$f_{lmp} = l - 2p + m\frac{N_m}{N_r} \qquad (3.18)$$

The denominator terms f_{lmp} and $f_{lmp} \pm 1$ determine the frequency of the perturbations in units of cycles per orbital revolution (cpr), of which there are two major components:

- a once-per-revolution component $\dot{\omega} + \dot{M}$ responsible for short-periodic perturbations at $l - 2p$ cpr

- a 'monthly' component $\dot{\Omega} - \dot{\theta}$ yielding medium and long-periodic perturbations, the so-called m-monthlies

The algorithm for the determination of all $\{lmp\}$ combinations that contribute to a particular frequency is also given by *Rosborough* [1986]. One fundamental result is that while orders are uncoupled, degrees are not. Specifically, coefficients of different orders have distinct frequency constituents, whereas coefficients of different degrees may contribute to orbit perturbations of identical frequency, provided $l_{max} < N_r/2$.

The radial, along-track and cross-track perturbations relative to the reference orbit given by (3.13) correspond to the particular solution of the Hill equations for the non-resonant part of the gravitational potential. Resonant conditions, for which $f_{lmp} \in \{-1, 0, 1\}$, cf. (3.14), are evidently not modelled by the LPT. The homogeneous part of the solution is caused by errors in the epoch state vector (initial conditions) and may therefore, provided they are within the linear domain, be absorbed by epoch state-vector adjustment (orbit determination). A similar situation applies for the resonant part of the gravity field caused by the zonal constituents [*Schrama*, 1989; *Visser*, 1992].

Equation 3.17 illustrates one fundamental difference between lunar satellite orbits and Earth mapping orbits: due to the slow rotation of the Moon, the shortest possible repeat orbit has a cycle time of approximately one sidereal month, or 27.32166 days [*Seidelmann*, 1992]. Repeat cycles are therefore necessarily of long duration, with due consequences for orbit analysis over the full repeat period. In the present situation of large dynamical model errors, satellite orbit determination is performed using short-arc approaches, with arcs usually no longer than a few days, in order to avoid the contribution of large low-frequency terms. Orbit error analysis based on linear perturbation theory may therefore, while being correct, fail to resemble actual error levels encountered during missions, simply because of frequent orbit estimates in an actual mission situation. For orbit determination purposes it is also common practise to estimate empirical acceleration parameters[1] to improve fit of the observations and, hence, the orbit recovery. In this respect, the orbit errors following from the LPT approach simply equals the

---

[1]The technique of estimating empirical forge parameters is frequently referred to as *reduced-dynamic* orbit determination

orbit error expected from an arc of length equal to the repeat cycle, with the estimation of a single state vector. In the present context, however, the goal is to understand the qualitative behaviour of the gravity-induced errors, which so profoundly dominate present-day orbit error budgets, rather than to determine an absolute accuracy level of present-day low lunar satellite orbits. In a second instant, the derived orbit behaviour is used to understand and assess the quality of the respective selenopotential model.

Within the time interval of the shortest repeat cycle, a low lunar satellite will complete more than 300 revolutions, *e.g.* 334 revolutions for a 100 km orbit. Resonance problems related to central body rotation are therefore absent for present day models (these would only occur for $m \simeq 334$ or multiples thereof). Another consequence of the slow lunar rotation is that the frequency separation is small. With a minimum repeat close to one sidereal month, the basis frequency equals $1/N_r$ cycles per revolution, or $1/334$ cpr for the 100 km orbit test case.

3.3.2   Projection of the covariance matrix

The linear relationship between the orbit error and the spherical harmonic coefficients as given by (3.13) allows for a straightforward projection of the calibrated error variance-covariance matrices of GLGM–2 and LP75G onto the position coordinate of interest. Without loss of generality, the error in any given position coordinate $\alpha \in \{r, \tau, c\}$ may be represented by a linear function of the form

$$\Delta\alpha = \sum_{l=2}^{l_{max}} \sum_{m=0}^{l} \left(\Delta\overline{C}_{lm}c_{lm} + \Delta\overline{S}_{lm}s_{lm}\right) \tag{3.19}$$

where c_{lm} and s_{lm} are time-dependent series expansions in the (mean) elements of the reference orbit. If the linear functions c and s and the coefficient errors are denoted by the two vectors

$$
\mathbf{f} = \begin{bmatrix} \vdots \\ c_{lm} \\ \vdots \\ s_{lm} \\ \vdots \end{bmatrix} \qquad \text{and} \qquad \mathbf{g} = \begin{bmatrix} \vdots \\ \Delta\overline{C}_{lm} \\ \vdots \\ \Delta\overline{S}_{lm} \\ \vdots \end{bmatrix} \tag{3.20}
$$

then

$$\Delta\alpha = \mathbf{f}^T\mathbf{g} \tag{3.21}$$

The variance of α may be obtained in one of the following ways. Error propagation [*e.g., Moritz*, 1989; *Leick*, 1995] yields the covariance Q_α of orbit position coordinate α

$$\mathbf{Q}_\alpha = \mathbf{f}^T\mathbf{Q}_g\mathbf{f} \tag{3.22}$$

where $\mathbf{Q}_g$ is the error variance-covariance matrix of the gravity field model, hence, $\mathbf{Q}_g = E\{\mathbf{g}\mathbf{g}^T\}$. Alternatively, the expectation of $\Delta\alpha^2$ gives the variance of $\Delta\alpha$ as a function of the gravity field covariance,

$$\sigma_{\Delta\alpha}^2 = E\{\Delta\alpha^2\} = \mathbf{f}^T E\{\mathbf{g}\mathbf{g}^T\}\mathbf{f} = \mathbf{f}^T\mathbf{Q}_g\mathbf{f} \tag{3.23}$$

The square root of the diagonal part of (3.22) or the variances given by (3.23) yields the standard deviation of the orbit position coordinate resulting from the uncertainty in the modelling of the selenopotential coefficients.

Since the orbit error is generally a function of time, through ψ_{lmpq}, and, hence, f, the error is time-averaged, usually over ψ_{lmpq}. The square root of the thus derived time-averaged variance is naturally the RMS value of the standard deviation of the orbit error, and hence provides a single number characterising the intrinsic orbit error. Hence,

$$\mathrm{RMS}_{\sigma_{\Delta\alpha}} = \sqrt{\frac{1}{2\pi}\int_0^{2\pi}\sigma_{\Delta\alpha}^2(\psi)d\psi} = \sqrt{< E\{\sigma_{\Delta\alpha}^2\} >} \tag{3.24}$$

where $< \bullet >$ is a short-hand notation for the averaging operator.

3.3.3   Time-wise characteristics of low lunar orbit accuracy

The linear formulation of the orbit error due to the selenopotential provides a means for quantifying such effects in the time domain. Time-wise orbit error characteristics from the calibrated error variance-covariance matrices of present-day lunar gravity models (in short: covariance orbit analysis) may evidently be represented as a function of orbital frequency and any meaningful (sub)set of selenopotential coefficients, *e.g.* order m, degree l or maximum degree l for cumulative effects. The total error, being the sum over all orders, provides the global RMS characterisation of the error. Coefficient-wise orbit errors are also possible, but as will be shown, given the fact that spectral leakage and extensive use of constraints cause significant correlations between coefficient estimates, "lumped" or grouped coefficient measures are preferred. The coefficient-wise representation may nevertheless prove to be useful for lunar gravity field tailoring experiments in the future, where for example laser altimetry crossover measurements may be used to improve the models. The set of tailored coefficients is then frequently taken to be those for which the orbit error exceeds a certain threshold level plus all resonant orders (if present).

In the following, the radial orbit error is the prime component of interest, since it is directly related to altimetric measurements, and hence a possible future selenophysical application. However, the method is principally identical for all components, and the along-track and cross-track parts are occasionally presented to enhance the information provided by the radial component alone. Evidently, the orbit error characteristics depend on the orbital elements. In the sequel, the primary test case is a typical 100 km mapping orbit, described in Table 3.1, comple-

Reference orbit

Altitude	100.0 km
Mean semi-major axis	1838.0 km
Mean eccentricity	0.0
Mean inclination	90.0°
Mean motion	8.8859×10^{-4} rad/s
Period	117.849 min
Rate of ascending node	0.0 °/day
Rate of argument of perilune	-4.094°/day

Table 3.1   Orbital characteristics of the primary test orbit applied in the time-wise and spatial analysis of gravity-induced low lunar orbit errors

mented with information from other altitudes and inclinations when considered illustrative.

Radial accuracy spectrum

For a given $\{lmp\}$ combination corresponding to the lowest coefficient degree yielding a frequency f_{lmp} or $f_{lmp} \pm 1$, the total expected *mean square* radial amplitude, or mean variance, at that same frequency reads, for the zonals and non-zonals respectively [*Rosborough*, 1986; *Rosborough and Tapley*, 1987]:

$$E\{\Delta r^2\}_{m=0} = \frac{1}{2} \sum_{k=0}^{\frac{l_{max}-l}{2}} \sum_{j=0}^{\frac{l_{max}-l}{2}} \left(C^r_{(l+2k)0(p+k)} + (-1)^l C^r_{(l+2k)0(l-p+k)} \right) \qquad (3.25)$$

$$\left(C^r_{(l+2j)0(p+j)} + (-1)^l C^r_{(l+2j)0(l-p+j)} \right) Q^c_{(l+2k)0(l+2j)0}$$

$$E\{\Delta r^2\}_{m \neq 0} = \frac{1}{2} \sum_{k=0}^{\frac{l_{max}-l}{2}} \sum_{j=0}^{\frac{l_{max}-l}{2}} C^r_{(l+2k)m(p+k)} C^r_{(l+2j)m(p+j)} \qquad (3.26)$$

$$\left(Q^c_{(l+2k)m(l+2j)m} + Q^s_{(l+2k)m(l+2j)m} \right)$$

where $Q^c_{lmkn} = E\{\Delta \overline{C}_{lm} \Delta \overline{C}_{kn}\}$ and, hence, elements of the variance-covariance matrices $\mathbf{Q}^c$ and $\mathbf{Q}^s$ of the $\overline{C}_{lm}$ and $\overline{S}_{lm}$ coefficients, respectively. The RMS value of the error (the root mean square value of the standard deviation of the radial orbit error) is straightforwardly found by taking the square root of the variance, and the summation indices $\{k, j\}$ are obviously limited to parameters with common frequency constituents. Evidently, computation speed-up may also be achieved by making use of the symmetry of the covariance matrix in the algorithms for the summations.

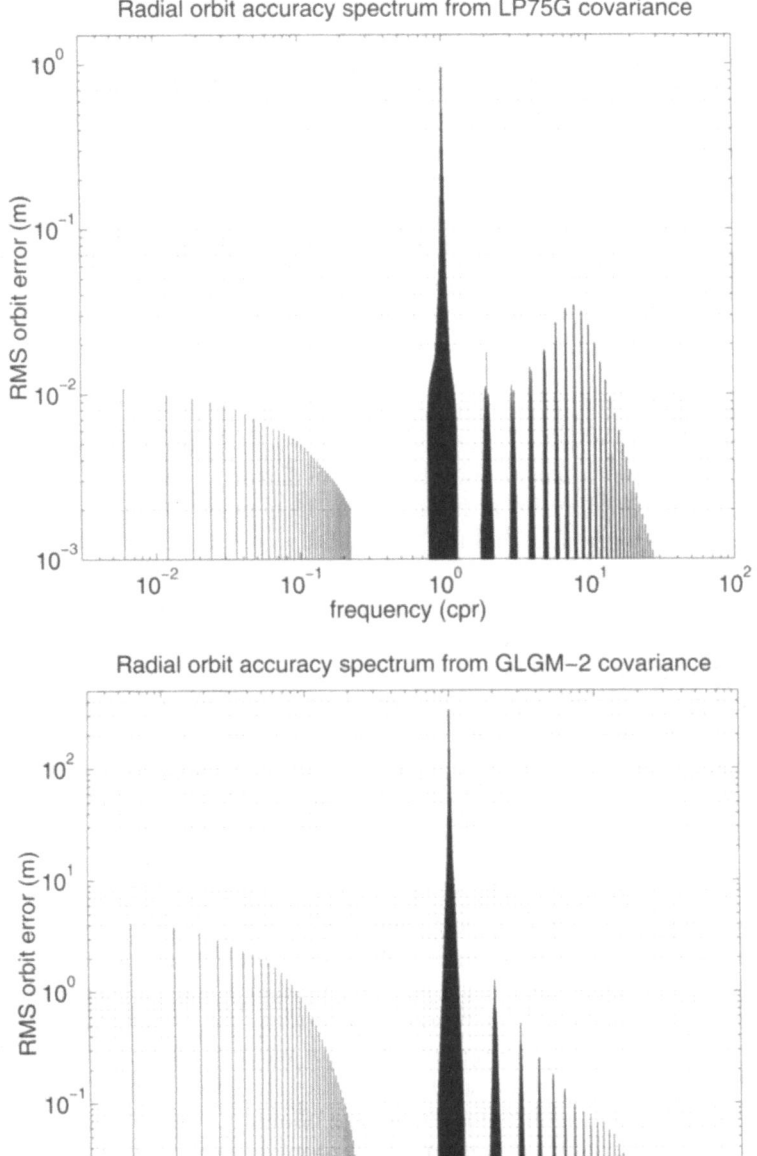

Figure 3.2   Radial orbit error frequency spectrum for a 100 km altitude polar orbit from the calibrated GLGM–2 (top) and LP75G (bottom) error variance-covariance matrices

The accuracy spectra for GLGM–2 and LP75G are depicted in Fig. 3.2. The GLGM–2 spectrum depicts all errors larger than 1 cm in amplitude, while that of LP75G depicts all errors larger than 1 mm. Immediately apparent in the plots is the order of magnitude difference between the two models. Given the low polar orbit of Lunar Prospector, the question is immediately triggered whether this tendency persists for non-polar inclinations. This question is further elaborated in a dedicated subsection. The high peaks near the resonant 1 cpr frequency, 335.3 m for GLGM–2 and 0.964 m for LP75G, indicate the problem of state-vector effects in orbit and gravity field modelling, as do the relatively significant amplitudes close to 0 cpr. For precise orbit computations, 0 and 1 cpr terms may be removed empirically, but, in gravity modelling, the introduction of orbit determination absorption parameters is usually not justified, given the natural goal of relating unmodelled orbit errors of all frequencies to gravity field coefficients and not to empirical parameters of unknown physical nature. Hence, a significant part of the overall error remains near the resonant frequencies. Other peak amplitudes are found at and near integer cpr values, unlike the case for Earth gravity models which also exhibit significant amplitudes at intermediate frequency values [$e.g.$, Smith et al., 1994]. This is due to the fact that the slow rotation of the Moon causes clustering of frequencies around integer cpr values. It can easily be shown that for all intermediate frequencies to be filled, the maximum degree and order of the model l_{max} must exceed $N_r/2$, which in the case of the 100 km test orbit amounts to $l_{max} = 167$, a condition evidently not met for present-day potential models. An additional striking feature of LP75G is that the errors clustered around integer cycles per revolution are not monotonically decreasing, in other words orbit mismodelling effects are not straightforwardly dampened with increasing orbital frequency. In an intermediate frequency interval, from about 4 to 13 cpr, the radial amplitudes exhibit larger errors than for example at 3 cpr, with peak amplitudes around 7 and 8 cpr. At the present point in time one may only speculate about the cause, but it is likely that this feature is related to mascon estimates in LP75G. $Konopliv$ et $al.$ [1998] report the discovery of several new far-side mass anomalies, features that also appear in the later LP100J model, albeit less pronounced [$Konopliv$ and $Yuan$, 1999; $Konopliv$ et $al.$, 2001]. In principle, this would be the first direct indication of mascons associated with far-side mare-fill, if confirmed by later analysis. The size of a surface feature corresponding with an 8 cpr wave is roughly 1500 km, which matches the size of several lunar mass anomalies. It is therefore not unlikely that the intermediate higher estimation errors are related to uncertainties in the estimation of far-side mare fills. A perhaps even more daring theory, which may only be confirmed by a combination of dedicated high-resolution gravimetry and improved seismic profiling, is that these orbit error frequencies are due to large-scale variations in the density distribution in the deeper parts of the Moon.

Orders resonant with the body rotation, a feature which dominates terrestrial orbit errors, are, as previously outlined, not present. Beyond 20 cpr the radial orbit error decreases rapidly, which indicates that only the low-frequency part of the higher order coefficients contribute to the orbit errors. For GLGM–2 all frequencies

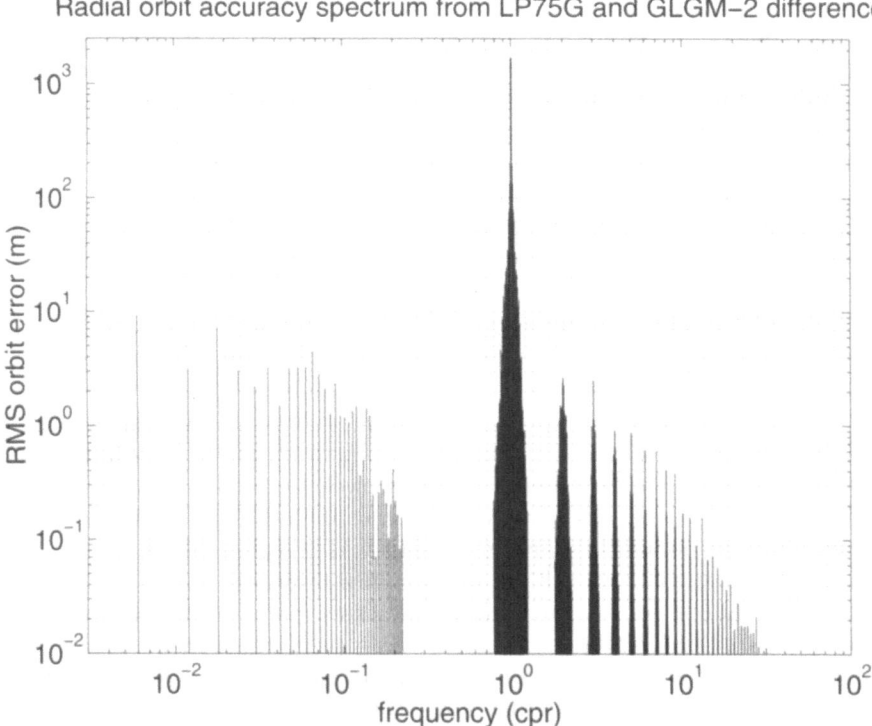

Figure 3.3   Radial orbit error frequency spectrum for a 100 km altitude polar orbit from LP75G and GLGM–2 coefficient differences

greater than 2 cpr have sub-metre amplitudes, while for LP75G all frequencies greater than 2 cpr even subside below the 2.5 cm level.

The orbit error spectra due to the selenopotential coefficient mismodelling as represented by the calibrated covariance matrices may be directly compared with the spectrum derived from coefficient differences LP75G vs. GLGM–2, shown in Fig. 3.3. When comparing coefficient values, it is crucial to take due account of systematic differences in the choice of body-fixed reference frames used in the model development. Whereas LP75G makes use of a principle axis system (as do LP100J and LP100K) based on numerically integrated lunar libration parameters from the JPL Development ephemeris files [Standish et al., 1995; Standish, 1998], GLGM–2 is derived in the IAU system, which is an international definition for planetary and natural satellite reference frames, updated on a tri-annual basis by a dedicated working group [Davies et al., 1992, 1996, 1997]. The IAU system basically uses the intersection of the Earth equator and the lunar equator, plus a choice of the prime meridian corresponding to a surface feature in the centre of the mean lunar disk as seen from Earth. In other words, this system is a "mean Earth-pointing system". Moreover, the analytical nature of the IAU system represents a truncated series of parameters. Consequently, while the zero meridian used in the JPL-derived models is the same as for the IAU system, the numerically integrated librations differ by approximately 60″ and 80″ seconds of arc in right ascension and declination, respectively. This systematic difference complicates the direct comparison of lunar gravity field models. However, it can be shown that the rotations on the lunar sphere involved in transformations between the two systems invoke coefficient corrections of the order 10^{-6} to 10^{-5} of the differences between the LP75G and GLGM–2 models. In other words, conclusions drawn based on coordinate system differences, although not strictly rigourous due to negligence of small time-dependent rotations, are valid.

It is readily seen that the coefficient differences predict larger errors than any of the two covariance matrices. The most important conclusion to be drawn from this is that neither of the covariances are able to accommodate the model differences, to within a 1σ level. In other words, the impression from the average degree variances, that the two models are far from converging towards a definite lunar gravity model, is confirmed. On the other hand, the peak factor six difference between the coefficient differences and the covariance matrix of GLGM–2, actually establishes faith in the calibration effort going into the GLGM–2 model. The fact that the covariance matrix predicts the orbit difference with respect to a more recent model within less than one order of magnitude indicates that the GLGM–2 covariance is a reasonable representation of the intrinsic selenopotential model error, including all error components. Also, the envelope of the spectrum resembles that of the GLGM–2 covariance. It may therefore not be concluded that GLGM–2 is optimistically weighted. From the derived spectra alone, with no external calibration data available, little can be said about the LP75G covariance. As indicated before, the degree-wise amplitude of the model and corresponding variance show signs of optimistic data weights, but this may not be confirmed by the difference spectrum alone. Moreover, as the characteristics of lunar gravity models are not only influenced by the available satellite data, but also by the way of applying Kaula constraints, it is at least questionable that the constraints that give semi-optimal satellite orbit quality also matches the degree-wise amplitude of the true gravity field. These considerations undoubtedly add to the feeling that the true quality of a gravity field solution only becomes apparent after the publication of a subsequent model, largely based on independent data.

Radial accuracy by degree and by order

Degree-wise and order-wise radial errors are easily derived from (3.13) as [*Rosborough*, 1986]

$$\Delta r_l = \sum_{m=0}^{l} \Delta r_{lm} = \sum_{m=0}^{l} \sum_{p=0}^{l} C_{lmp}^r S_{lmp0} \tag{3.27}$$

which leads to

$$E\{\Delta r^2\}_l = \frac{1}{2}Q_{l0l0}^c \sum_{p=0}^{l} \left(C_{l0p}^r C_{l0p}^c + (-1)^l C_{l0p}^r C_{l0(l-p)}^c\right) + \tag{3.28}$$

$$\frac{1}{2} \sum_{m=1}^{l} (Q_{lmlm}^c + Q_{lmlm}^s) \sum_{p=0}^{l} C_{lmp}^r C_{lmp}^c$$

where again, in the framework of the Hill equations, only order zero terms of the eccentricity, $q = 0$, are retained.

Similarly, the radial perturbation produced by all coefficients of a given order m follows from [*Rosborough*, 1986]

$$\Delta r_m = \sum_{l=m}^{l_{max}} \Delta r_{lm} = \sum_{l=m}^{l_{max}} \sum_{p=0}^{l} C_{lmp}^r S_{lmp0} \tag{3.29}$$

which yields, for zonals and non-zonal orders respectively

$$E\{\Delta r^2\}_{m=0} = \frac{1}{2} \sum_{l=2}^{l_{max}} \sum_{k=0}^{l} \sum_{j=0}^{l} \left[\left(C_{(l+2k)0k}^r + (-1)^l C_{(l+2k)0(l+k)}^r\right)\right. \tag{3.30}$$

$$\left.\left(C_{(l+2j)0j}^r + (-1)^l C_{(l+2j)0(l+j)}^r\right) Q_{(l+2k)0(l+2j)0}^c\right]$$

$$E\{\Delta r^2\}_{m\neq 0} = \frac{1}{2} \sum_{l=m}^{m+1} \sum_{p=0}^{l} \sum_{k=0}^{l} \sum_{j=0}^{l} \left[C_{(l+2k)m(p+k)}^r C_{(l+2j)m(p+j)}^r\right. \tag{3.31}$$

$$\left.\left(Q_{(l+2k)m(l+2j)m}^c + Q_{(l+2k)m(l+2j)m}^s\right)\right] +$$

$$\frac{1}{2} \sum_{l=m+2}^{l_{max}} \sum_{k=0}^{l} \sum_{j=0}^{l} \left[\left(C_{(l+2k)mk}^r C_{(l+2j)mj}^r +\right.\right.$$

$$\left.\left. C_{(l+2k)m(l+k)}^r C_{(l+2j)m(l+j)}^r\right) \left(Q_{(l+2k)m(l+2j)m}^c + Q_{(l+2k)m(l+2j)m}^s\right)\right]$$

The degree-wise and order-wise radial components of the typical 100 km polar lunar mapping orbit error due to the gravity field uncertainties are depicted in Fig. 3.4. Zonal coefficients are included in the analysis, their resonant parts exempted. The dominant characteristic for the degree-wise description is the sawtooth-pattern which actually persists for all degrees present in the models. Even degrees are better determined than odd degrees, with a steady amplitude difference of about one-and-a-half order of magnitude. The sawtooth-pattern in itself is an artificial effect caused by the truncation of the spherical harmonic series at $l = l_{max}$. By truncating the fundamentally indefinite series, and subsequently performing the time-wise analysis in a per-order fashion, since the orders have distinct frequency contributions, a distinction between even and odd degrees is automatically introduced. Indeed, for each order m there is one even

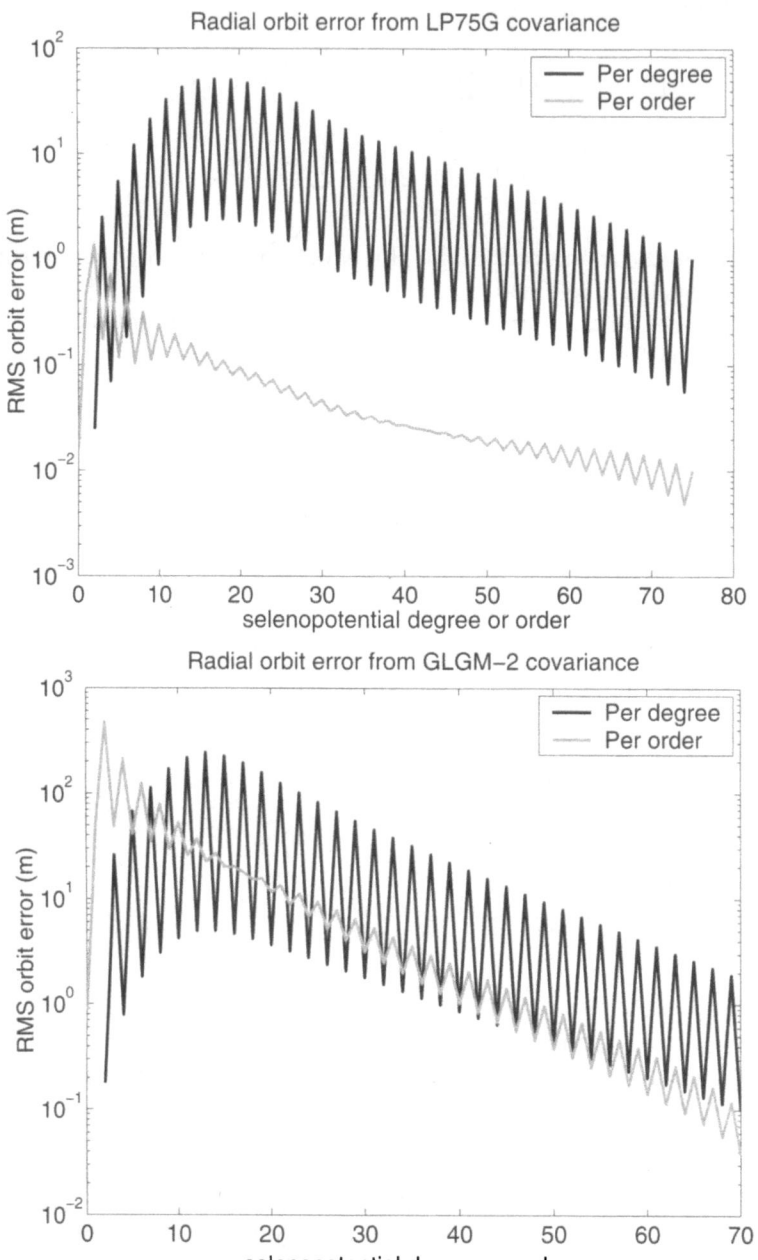

Figure 3.4   Radial orbit error for a 100 km altitude polar orbit as a function of
              selenopotential degree and order from the calibrated GLGM–2 (top) and
              LP75G (bottom) error variance-covariance matrices

degree more or one even degree less present than there are odd degrees (depending on whether $lmax$ is even or odd [*Koop*, 1993]. The peak-to-peak magnitude of the differences between even and odd degrees is rooted in the better observability of the m-monthly and resonant along-track effects which are caused by the even degree coefficients. Such m-monthly and resonant along-track perturbations are dominating the residual in a situation where the satellite tracking data do not adequately sample the short-periodic satellite perturbations of ω and M [*Kaula*, 1966; *Rosborough*, 1986]. Otherwise put, through the $l - 2p$ factor, even and odd degrees get separate treatments in the LPT. Empirical reduction of these effects may be achieved through precise orbit determination. For the order-wise description a reciprocal situation occurs, where, for the lower orders, the largest errors are found at even m-values. Both models furthermore exhibit a turning point for the order-wise error, beyond which the even values of m have smaller associated errors than the odd ones. For GLGM–2 this turning point is at $m = 17$, while for LP75G it is found at $m = 43$. Although no solid proof has been established for this behaviour, one hypothesis is that it is related to the domination of the constraints at the higher orders. Given the decoupling of selenopotential orders, the per-order radial orbit error never exceeds the overall error. For GLGM–2 this means that the by far largest part of the orbit error is found for the $m = 2$ coefficients, with an amplitude of 475 m. For LP75G the tendency of a two orders of magnitude improvement due to the inclusion of Lunar Prospector data is continued, with a peak amplitude at $m = 2$ of 1.36 m. Other large contributions to the overall error are found at even orders $m \in \{4, 6, 8\}$ for both models, beyond which the order-wise error is asymptotically reduced to zero.

Again, the errors as derived from the calibrated *a posteriori* error variance-covariance matrices are compared with those derived from coefficient differences, shown in Fig. 3.5. Again, there is a dominating sawtooth pattern, along with error levels a factor 4–5 higher than those predicted by the GLGM–2 covariance. In its totality, all of this supports the conclusion, drawn from the orbit error spectra alone, that current gravity field model behaviour still needs verification.

A very interesting aspect of LP75G is that while the total radial error of the complete 75×75 model can be considered small, with an amplitude of 1.80 m, the amplitude at odd degrees 15, 17 and 19 exceed 50 m. This is indicative of significant negative correlations between coefficients of different degrees and parity $l - m$, causing the overall orbit error to be significantly less than that found at individual degrees. Such behaviour has previously also been signalled for Earth models [*Smith et al.*, 1994]. Furthermore, it should be noted that around the same degrees the role of the prior information in the derivation of a lunar gravity field solution becomes important, and it is a result of extensive use of such *a priori* information that the total orbit error due to a gravity field model covariance may be reduced.

The presence of negative correlations obviously is related to the observability of the individual coefficients, and hence the value of the individual harmonic constituent estimates. Two further ways to enlighten the problem are given here.

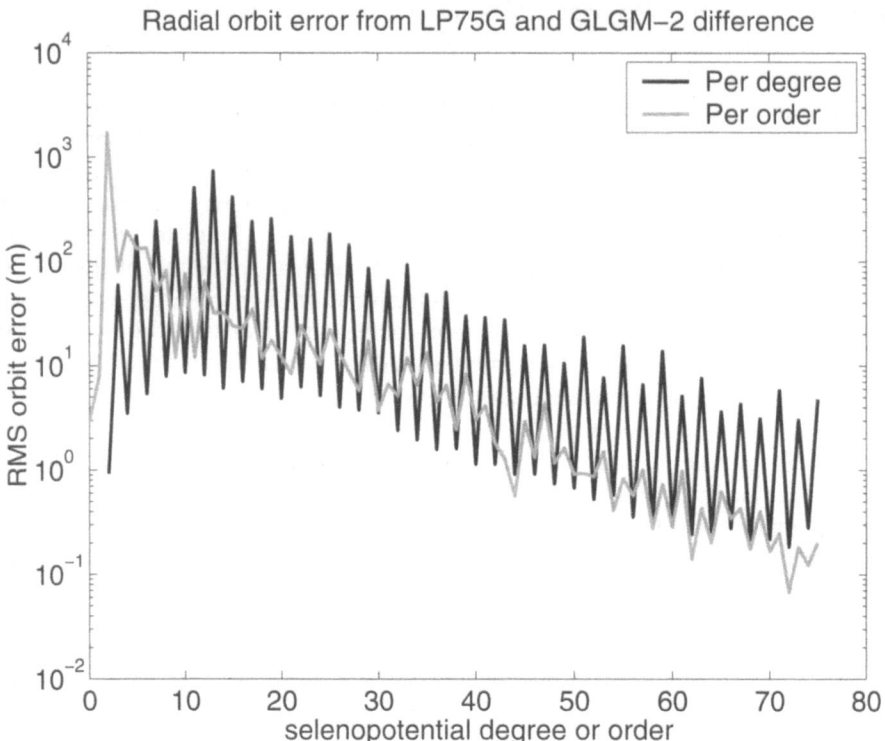

Figure 3.5   Radial orbit error for a 100 km altitude polar orbit as a function of
selenopotential degree and order from LP75G and GLGM–2 coefficient
differences

First, cumulative radial errors, *i.e.* errors that arise from the use of a selenopotential
model complete to degree *l*, are presented in Fig. 3.6. Three issues are of immediate
importance when interpreting the plots: first, the type and level of regularisation
applied to the least squares equations derived from the satellite tracking data information; second, the quality of that same information in terms of geometry, data
distribution and measurement precision; third, the calibration of the final covariance product. All factors directly affect the error amplitude as well as its time-wise
and spatial behaviour.

In this respect GLGM–2 clearly depicts a cumulative radial orbit error pattern
consistent with an over-constrained solution. First, the higher order terms (approx.
l > 30) hardly contribute to the orbit error, which is indicative of a significant bias
towards zero. In other words, smoothing of the high-degree terms has dampened
the error spectrum to such a level that there is hardly any more orbit information
present. This leaves the impression that the Clementine data were not exhausted
in terms of orbit fit, and that with a different goal in mind, a model based on
Clementine and historical data sets could produce slightly better orbits than those

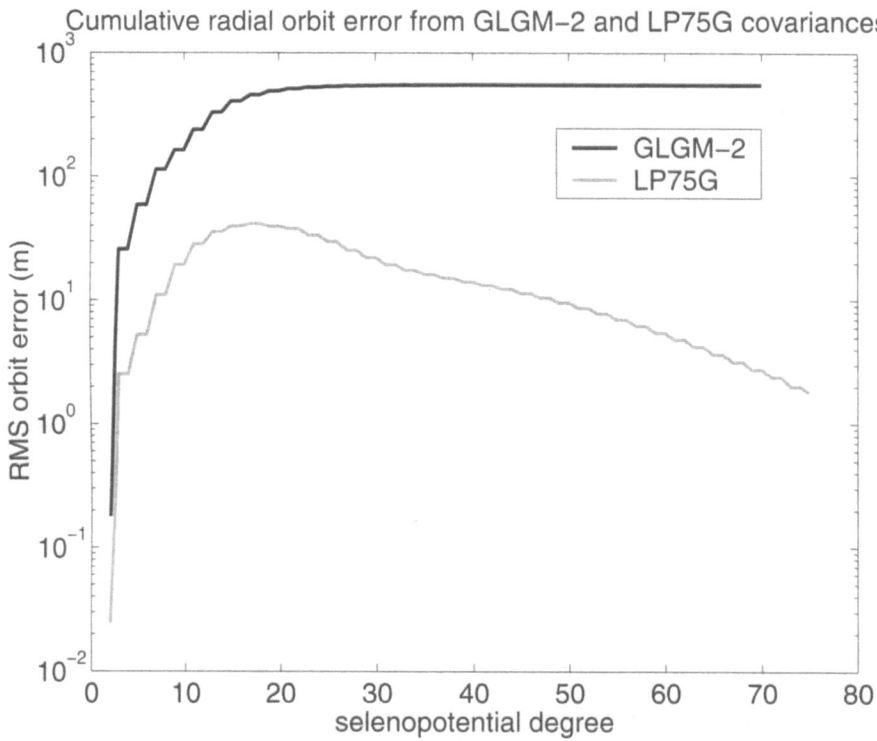

Figure 3.6   Cumulative radial orbit error for a 100 km altitude polar orbit as a function of
               selenopotential degree from the GLGM–2 and LP75G error
               variance-covariance matrices. Notice the difference in scale

emerging from the application of the GLGM–2 potential model. Second, this figure
illustrates clearly the role of constraints. Non-orbit related analysis on the basis of
GLGM–2 has proven the usefulness of the model for selenophysical studies [*e.g.,*
Zuber et al., 1994; *Neumann et al.,* 1996; *Von Frese et al.,* 1997; *Arkani-Hamed,* 1998,
1999*a*, 1999*b*], which therefore appears to be a modelling goal slightly conflicting
with that of producing satellite orbits with optimum tracking data fits.

     LP75G on the other hand obviously suffers from large (negative) correlations.
The cumulative radial orbit error clearly shows that all coefficients of the model
are needed to produce optimal satellite orbit results. GLGM–2 on the other hand
predicts that a limited coefficient set could be used for orbit determination. This
behaviour can be explained by a lumped coefficient representation of both the potential and the satellite orbit perturbations. In the frequency domain, the Doppler
data contains information on the lumped effect at each frequency, rather than direct information on each harmonic constituent. Without good sensitivity over the
whole frequency range (0 – 75 cpr), the potential coefficient estimation process is
unable to decorrelate the higher degree terms from the lower degree terms, all of

which containing low-frequency information. In other words, the problem of significant correlation between spherical harmonic coefficients will remain until a low lunar satellite tracking experiment is realised that also features a high sensitivity to short-wavelength (high frequency) orbit perturbations over the entire Moon. This is, as such, an additional argument for a dedicated lunar gravity field determi-. nation experiment, as the present situation leaves little doubt that the estimation of medium-range harmonics is deteriorated by the wish to exhaust the data over the well-sampled near-side, or, in other words, the wish to expand the spherical harmonic series to a relatively high order.

The second illustration of the significant correlations between harmonic coefficient estimates is that of the correlations themselves, depicted in Fig. 3.7. Out of necessity the correlations are made absolute, since a linear scale fails to illustrate the level of correlations present in the normal matrix or square root information filter, respectively. For illustration purposes the matrix is ordered by selenopotential order, beginning in the lower left corner with $m = 0$ coefficients, and progressing with $\{\overline{C}_{lm}, \overline{S}_{lm}\}$ coefficient pairs towards the upper right corner of the array. Evidently, the LP75G solution should not be applied for interpretation purposes based on individual coefficient measures, but only be used as a whole. The wide band of significant correlations spans several orders, and decorrelation of harmonic orders, which is a natural result of adequate sampling and measurement sensitivity does not occur. In other words, through the natural lumping of frequencies in the Earth-based Doppler data individual coefficient measures are not very enlightening. This problem is only artificially circumvented by the application of tight constraints, and it is the somewhat looser constraints of LP75G (with respect to GLGM–2), combined with the estimation of an additional ~ 750 gravity field parameters, that cause the increased correlation. It deserves mentioning that the LP100J and LP100K models, as well as possible subsequent Lunar Prospector gravity products, are anticipated to show a slightly better degree of decorrelation between harmonic coefficient estimates. However, given the relative freedom of the estimation process needed for optimal orbit computation, given by the somewhat relaxed constraint relations with respect to previous solutions, plus the significant data weighting factors, no significant change can be expected on the basis of near-side data alone. As long as the far-side gravity-induced satellite orbit perturbations are only indirectly measured through the integrated effect on the global orbit, and, furthermore, there is no tracking instrument amplifying the high-frequency information relative to long-wavelength information, spectral leakage and correlations will remain significant problems.

Error variations with orbital height and inclination

So far, the error analysis of low lunar satellite orbits has focused on the radial component of a single case of polar orbits at 100 km altitude. One interesting facet of this type of orbit was the nearly two orders of magnitude improvement of LP75G with respect to GLGM–2. A natural question is therefore whether this improve-

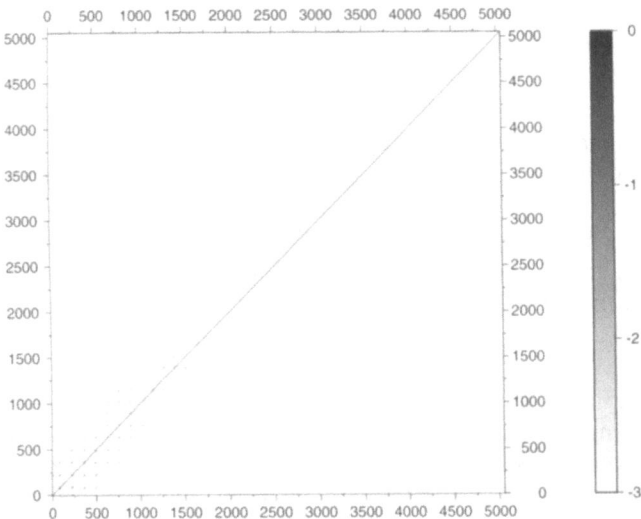

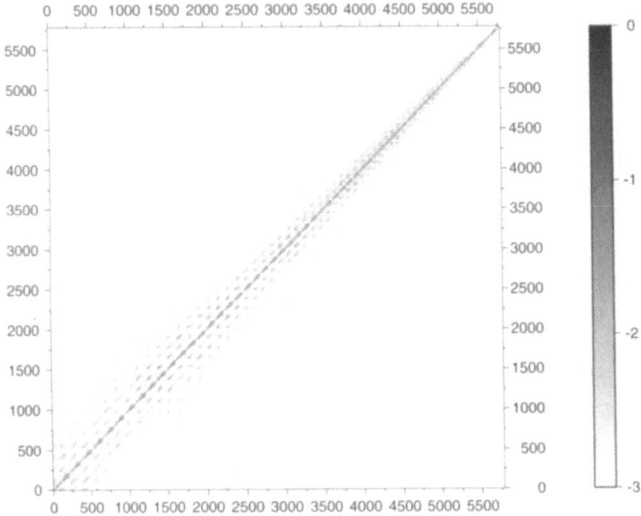

Figure 3.7   Log$_{10}$ absolute correlation coefficients for GLGM–2 (top) and LP75G (bottom)

ment is limited to inclinations close to 90°, and, hence, directly related to the orbit of Lunar Prospector. Proper calibration of the LP75G a posteriori error variance-covariance matrix will ensure that realistic errors are being predicted for the Lunar Prospector orbit. However, there is a severe scarcity of tracking data of low lunar orbiters, and it is therefore expected that the orbit errors show patterns clearly correlating with the amount and quality of tracking data available. In other words, it is illustrative to assess the performance of LP75G at non-polar inclinations. Also of interest is the intrinsic improvement provided by the Clementine data included in GLGM–2. While the Clementine spacecraft orbited at polar inclination, its altitude was much higher, with perilune at approximately 400 km and apolune at 2500 km.

Figures 3.8, 3.9 and 3.10 depict the predicted orbit errors for all three components (radial, along-track and cross-track) for all inclinations, covering satellite orbit altitudes from 50 to 500 km. Consistent error lows (with respect to surrounding inclinations) are found for inclinations close to those of actually flown orbits, compare Table 2.1. For intermediate inclinations, *e.g.* the complete range from 30° to 80° (and complementary inclinations, symmetrically distributed about $i = 90°$), smoother patterns are indications that the errors are largely determined by regularisation. Perhaps equally striking is the relatively poor performance of both models at equatorial or near-equatorial inclinations. Obviously, short-arc tracking of Apollo era spacecraft in the 12 to 21 degrees inclination range does not induce significant improvement of equatorial orbits.

The GLGM–2 gravity field covariance model predicts a radial orbit error at 100 km polar altitude of 554 m, which is about a factor 2.3 less than the error found at 50° inclination. The improvement relative to the 85° inclination of LO-IV and LO-V is, however, limited to about 350 m, which in turn illustrates that the benefit of the Clementine tracking data is, although significant, no quantum leap in lunar satellite orbit prediction. Consistent results are found for the along-track and cross-track components. GLGM–2 predicts 100 km orbit errors in the order of 1965 m at $i = 90°$ to 5008 m at $i = 50°$ along-track and 700 m at $i = 85°$ to 1493 m at $i = 50°$ cross-track, respectively, at 100 km altitude. Intermediate inclination variations are furthermore determined by the available tracking data, in combination with the smoothing effect of the Kaula constraint.

The LP75G results demonstrate the fact that there – at present – is no such thing as a general purpose lunar gravity field model. Comparison with non-Lunar Prospector orbit inclinations tells that the orbit errors are strongly dominated by the availability of high-precision tracking data at polar inclination. Strong gradients near $i = 90°$ illustrate the deteriorated performance at neighbouring inclinations, with peak errors again found for non-tracked orbit inclinations. This is obviously an intrinsic weakness of lunar gravity field modelling, next to the single-hemisphere data coverage. As a result, LP75G radial errors, for the 100 km orbit case, vary between 1.80 m at $i = 90°$ and 391 m at $i = 50°$. Similarly, the along-track and cross-track variations are 5.58 m at the polar inclination to 1479 m at $i = 47°$, and 2.18 at $i = 90°$ to 446 m at $i = 50°$, respectively. In other words, achieving a global measure of orbit accuracy (and a greater conformity in lunar

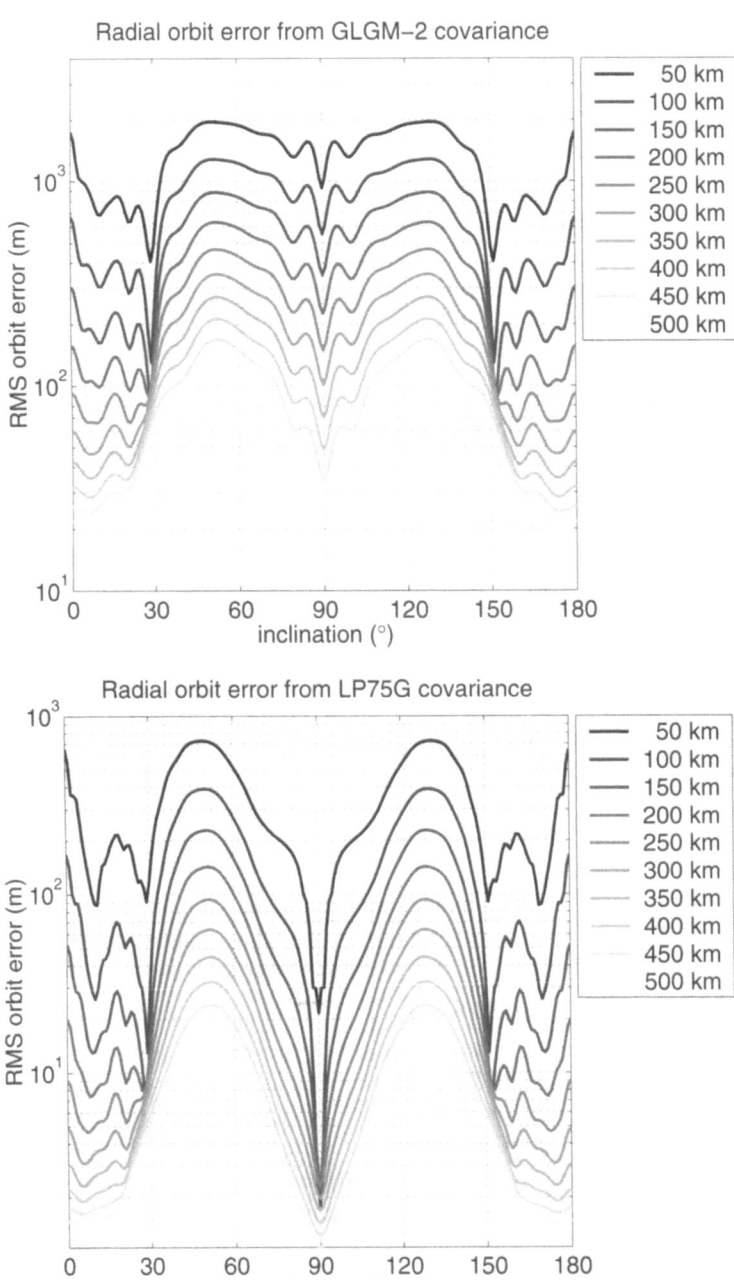

Figure 3.8   Radial orbit error as a function of orbit inclination and altitude from the GLGM–2 (top) and LP75G (bottom) error variance-covariance matrices

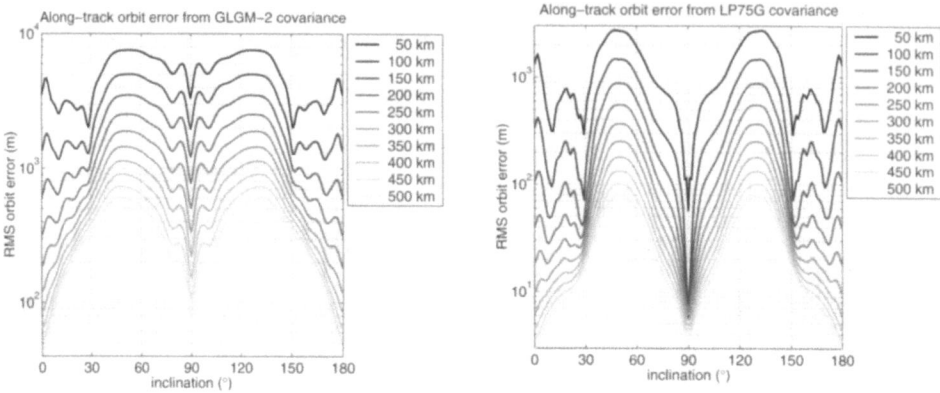

Figure 3.9　Along-track orbit error as a function of orbit inclination and altitude from the GLGM–2 (left) and LP75G (right) error variance-covariance matrices

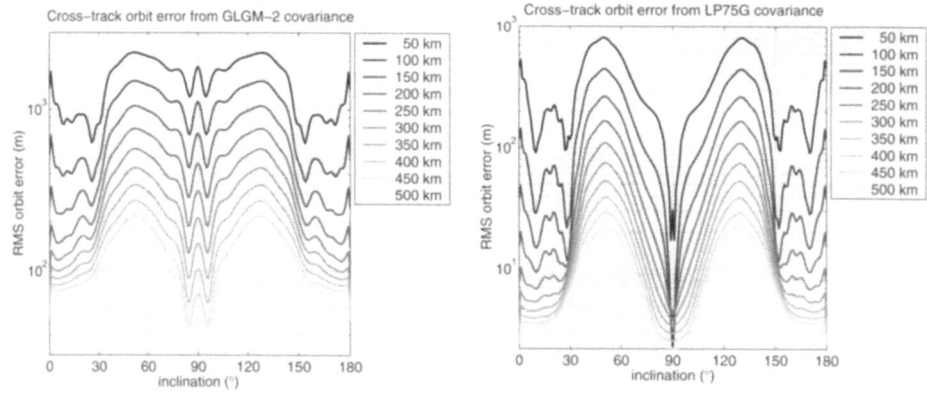

Figure 3.10　Cross-track orbit error as a function of orbit inclination and altitude from the GLGM–2 (left) and LP75G (right) error variance-covariance matrices

gravity field model assessment) will benefit from tracking data at inclinations not covered thus far.

These results furthermore trigger the question of the optimal orbit choice for future lunar gravity field mapping missions. Straightforward thinking might motivate an interest in an approximately 50° inclination for, *e.g.* SELENE, in order to enhance the orbit information in the selenopotential models. This is, however, a premature conclusion, as the optimal orbit choice for a global gravity field experiment depends on a multitude of factors. First, the spherical harmonic functions form an orthogonal basis on the full sphere. Aiming at a global solution using spherical harmonics, anything but a polar orbit is therefore adding to the incompleteness of the sampling of the basis functions, and therefore to the ill-conditioning of the inverse problem, cf. Chap 4. Such issues are very real issues for any gravity field mission, since there might be appealing technical reasons to opt for non-polar orbit inclinations, *e.g.* a sun-synchronous orbit, which is frequently preferred for Earth observation satellites [*e.g.*, *ESA*, 2000]. Second, and related to the first remark, the mapping of the potential from a polar orbit has not been exhausted, and the orbit error picture of present-day models are biased by the fact that they are developed from single-hemisphere data coverage, using Earth-based observations. The present comparison of performance at a range of inclinations is therefore far from fair to the polar orbit case. Therefore, while high-precision tracking data at non-polar inclinations surely will improve the localised improvement at respective inclinations, it will not solve the problem of lack of globally distributed data. In other words, the optimum orbit choice may not be straightforwardly determined, but requires careful study.

Finally, it deserves mentioning that the LPT analysis of 50 and 100 km polar orbits is near-resonant for a model of size 75x75. In other words, frequencies close to $f_{lmp} \in \{-1, 0, 1\}$ are the cause of the apparent crossing of altitude curves for LP75G. Such a result does not affect the validity of the conclusions, but merely illustrates the fact that the numbers must be regarded as qualitative, rather than absolute figures for present-day lunar orbit accuracy.

3.3.4 Spatial characteristics of low lunar orbit errors

The space-wise analysis of the orbit errors in a Moon-fixed reference frame complements the frequency domain analysis tools available for the error analysis of gravity field models. Such orbit error predictions are instrumental in assessing the gravity field model quality and behaviour in a spatial sense, as they directly relate relatively large orbit errors to the areas of the Moon where the selenopotential is weakly determined. Gravity-induced orbit errors are therefore obviously selenographically correlated. Particularly relevant for altimetric missions is the reduction of radial ephemeris errors down to a level at least comparable to the (altimeter-specific) error in the altimetric measurement, in order to maximise the quality of altimetric data products. While altimetric mapping of the Moon has a history [*Kaula et al.*, 1972, 1973, 1974], the first real attempt to perform a global topographic mapping, *i.e.* altimetry in a direct mode, was done by a laser altimeter on-board the Clementine spacecraft in 1994 [*Smith et al.*, 1997]. Crossover analysis, one of several key technologies for high-precision orbit determination and gravity field model tailoring for Earth observation satellites [*e.g.*, *Scharroo and Visser*, 1998] using laser altimeter measurements, however, has yet to be performed. Although a thorough analysis of the use of crossover techniques for selenodetic purposes is outside the scope of this book, in the following results are presented which illustrate their principal benefit, based on the observable part of the orbit error.

The spatial representation of the radial orbit error due to the uncertainty in the selenopotential is derived from the time-wise formulation. Only the order zero terms (in eccentricity) allow an explicit formulation in terms of selenographical longitude and latitude. To this order, the total error produced by the errors in all coefficients of the selenopotential is given by *Rosborough* [1986] as:

$$\Delta r^{(0)} = \sum_{l=2}^{l_{max}} \sum_{m=0}^{l} \Phi_{lm}^c \left(\Delta \overline{C}_{lm} \cos m\lambda + \Delta \overline{S}_{lm} \sin m\lambda \right) \pm$$
$$\sum_{l=2}^{l_{max}} \sum_{m=0}^{l} \Phi_{lm}^s \left(\Delta \overline{C}_{lm} \sin m\lambda - \Delta \overline{S}_{lm} \cos m\lambda \right) \tag{3.32}$$

or, equivalently,

$$\Delta r^{(0)} = \sum_{l=2}^{l_{max}} \sum_{m=0}^{l} \Delta \overline{C}_{lm} \left(\Phi_{lm}^c \cos m\lambda \pm \Phi_{lm}^s \sin m\lambda \right) +$$
$$\sum_{l=2}^{l_{max}} \sum_{m=0}^{l} \Delta \overline{S}_{lm} \left(\Phi_{lm}^c \sin m\lambda \mp \Phi_{lm}^s \cos m\lambda \right) \tag{3.33}$$

where Φ_{lm}^c and Φ_{lm}^s are explicit functions of latitude and the plus sign is used if the satellite is on an ascending (northbound) track and the minus sign for descending (southbound) tracks. The $\left\{ \Delta \overline{C}_{lm}, \Delta \overline{S}_{lm} \right\}$ notation is used to denote errors in the potential field coefficients.

With the introduction of a crossover location, defined as a crossing point on the lunar sphere of an ascending and a descending satellite ground track, the spatial form of the radial orbit error (3.32) may be conveniently decomposed into a fully *selenographically correlated* component

$$\Delta \gamma^{(0)} = \sum_{l=2}^{l_{max}} \sum_{m=1}^{l} \Phi_{lm}^c \left(\Delta \overline{C}_{lm} \cos m\lambda + \Delta \overline{S}_{lm} \sin m\lambda \right) \tag{3.34}$$

and a *selenographically anti-correlated* component

$$\Delta v^{(0)} = \pm \sum_{l=2}^{l_{max}} \sum_{m=1}^{l} \Phi_{lm}^s \left(\Delta \overline{C}_{lm} \sin m\lambda - \Delta \overline{S}_{lm} \cos m\lambda \right) \tag{3.35}$$

The former is common to the ascending and descending passes, and is therefore also known as the *mean regional error*. Since the potential field is sensed in the same way during each repeat cycle, this part of the error may not be observed from residual crossover analysis of altimetric measurements, nor from other statistical orbit analysis, like collinear track analysis, altimeter height residuals or orbit differencing. The anti-correlated part, equal in magnitude, but of opposite sign, for the ascending and descending passes, is observable from altimetry residuals, and it may be attempted to reduce this component of the error in the future by including residual crossover observations in both orbit determination and gravity field recovery. In the literature, the anti-correlated part is often termed the "variable" radial orbit error [*e.g., Tapley and Rosborough, 1985; Rosborough, 1986*]. Following

Scharroo and Visser [1998], this convention shall not be used, since there is no variation in the anti-correlated component other than with selenographical location.

The radial variance given the full variance-covariance matrix of the gravity model is obtained by squaring and taking the expected value of (3.33) [*Rosborough*, 1986]:

$$
E\left\{\Delta r^{(0)^2}\right\} =
$$

$$
\sum_{l=2}^{l_{max}} \sum_{j=2}^{l_{max}} \sum_{m=0}^{l} \sum_{k=0}^{l} Q^c_{lmjk} \left(\Phi^c_{lm}\cos m\lambda \pm \Phi^s_{lm}\sin m\lambda\right)\left(\Phi^c_{jk}\cos k\lambda \pm \Phi^s_{jk}\sin k\lambda\right)
$$

$$
+ Q^s_{lmjk}\left(\Phi^c_{lm}\sin m\lambda \mp \Phi^s_{lm}\cos m\lambda\right)\left(\Phi^c_{jk}\sin k\lambda \mp \Phi^s_{jk}\cos k\lambda\right)
$$

$$
+ 2Q^{cs}_{lmjk}\left(\Phi^c_{lm}\cos m\lambda \pm \Phi^s_{lm}\sin m\lambda\right)\left(\Phi^c_{lm}\sin k\lambda \mp \Phi^s_{lm}\cos k\lambda\right)
$$

(3.36)

As before, the square root yields the radial standard deviation. Similarly, the variance of the selenographically correlated and anti-correlated orbit errors are given by

$$
E\left\{\Delta\gamma^{(0)^2}\right\} = \sum_{l=2}^{l_{max}} \sum_{j=2}^{l_{max}} \sum_{m=0}^{l} \sum_{k=0}^{l} \Phi^c_{lm}\Phi^c_{jk}\left(Q^c_{lmjk}\cos m\lambda \cos k\lambda\right.
$$

(3.37)

$$
\left. + Q^s_{lmjk}\sin m\lambda \sin k\lambda + 2Q^{cs}_{lmjk}\cos m\lambda \sin k\lambda\right)
$$

$$
E\left\{\Delta\nu^{(0)^2}\right\} = \sum_{l=2}^{l_{max}} \sum_{j=2}^{l_{max}} \sum_{m=0}^{l} \sum_{k=0}^{l} \Phi^s_{lm}\Phi^s_{jk}\left(Q^c_{lmjk}\sin m\lambda \sin k\lambda\right.
$$

(3.38)

$$
\left. + Q^s_{lmjk}\cos m\lambda \cos k\lambda - 2Q^{cs}_{lmjk}\sin m\lambda \cos k\lambda\right)
$$

As a consequence, by computing the selenographically correlated and anti-correlated part of the radial orbit error, one does not only obtain a selenographical distribution of the radial error component, but one can also address the question of the possible benefit of future altimeter crossover analysis, provided the covariance matrices are truly representative of the actual error level in present-day models.

Figures 3.11 and 3.12 depict these errors for GLGM–2 and LP75G, for the same 100 km altitude, polar test orbit used for the time-wise analysis, cf. Table 3.1. GLGM–2 yields RMS values for the correlated and anti-correlated errors of 10.98 m and 552.83 m, respectively, which clearly shows that the larger portion of the error is reducible by application of crossover techniques on the Clementine altimeter data. This situation is unlike that of the Earth, where the mean and variable part are found to be roughly equal RMS-wise [*Smith et al.*, 1994]. Although the major signal peaks are found over the unsampled lunar far-side, it is found, somewhat surprisingly, that the near-side exhibits relatively high selenographically correlated error levels. The anti-correlated error, on the other hand, appear to be symmetrically distributed over the near-side and far-side hemispheres, with the two peaks situated at the equator for longitudes of 0° and 180°, and gradually decreasing towards the poles.

Selenographically Correlated Radial Orbit Error from GLGM-2 Covariance (m)

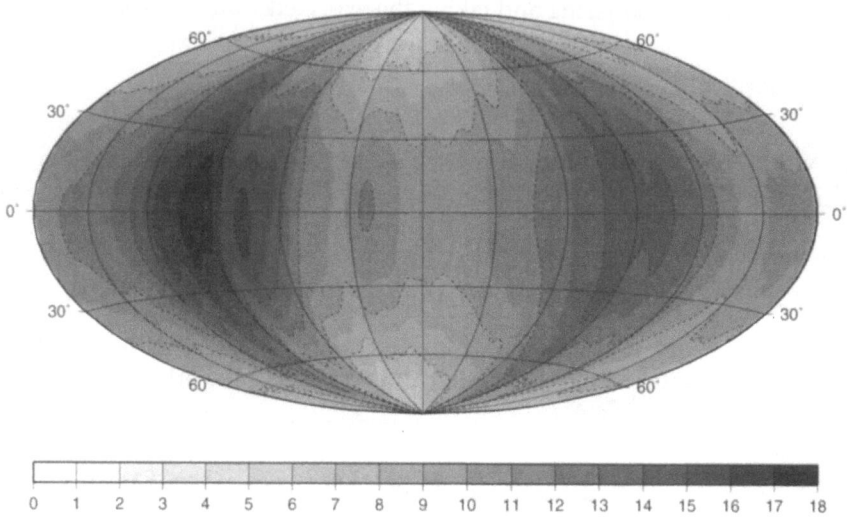

Selenographically Correlated Radial Orbit Error from LP75G Covariance (m)

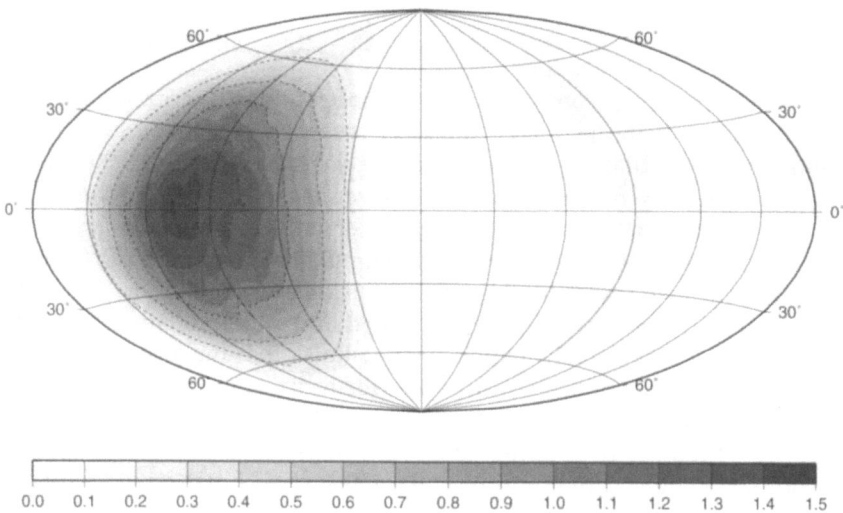

Figure 3.11   Selenographically correlated orbit error in metres for a 100 km altitude polar orbit from the GLGM–2 (top) and LP75G (bottom) error variance-covariance

Selenographically Anti-Correlated Radial Orbit Error from GLGM-2 Covariance (m)

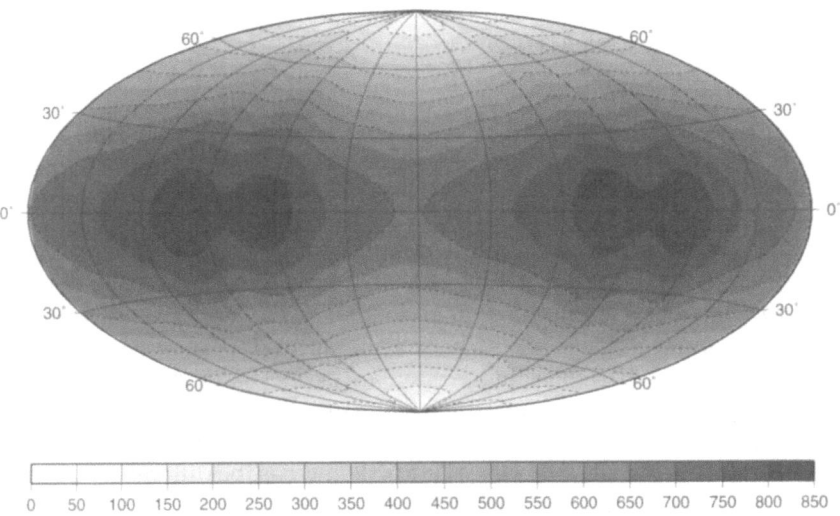

Selenographically Anti-Correlated Radial Orbit Error from LP75G Covariance (m)

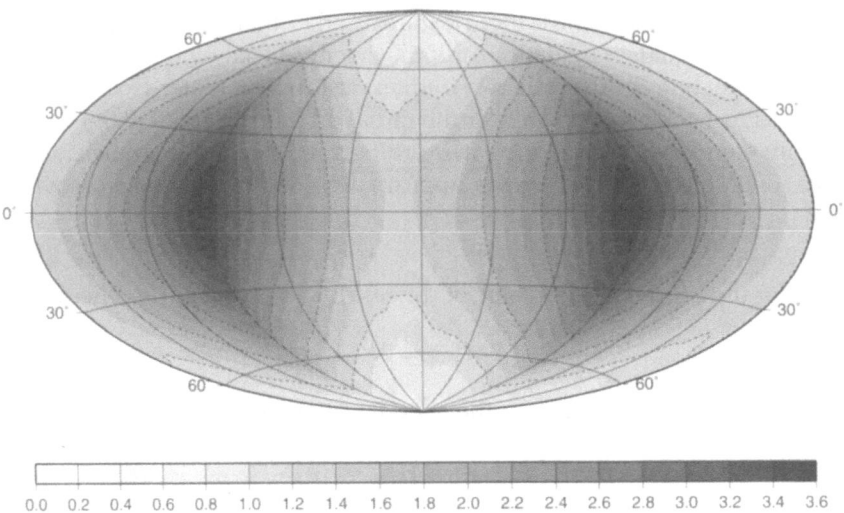

Figure 3.12 Selenographically anti-correlated orbit error in metres for a 100 km altitude polar orbit from the GLGM–2 (top) and LP75G (bottom) error variance-covariance

The large error difference between the correlated and anti-correlated component can be related to the correlation between the orbit errors at ascending and descending passes, as well as to the relationship between altimeter residual crossover differences and the radial orbit error. Given the definition of the selenographically correlated and anti-correlated orbit errors,

$$\Delta\gamma(\lambda, \phi) = \frac{\Delta r_a(\lambda, \phi) + \Delta r_d(\lambda, \phi)}{2} \tag{3.39}$$

$$\Delta v(\lambda, \phi) = \frac{\Delta r_a(\lambda, \phi) - \Delta r_d(\lambda, \phi)}{2}$$

where Δr_a and Δr_d indicate the radial orbit errors at ascending and descending passes, respectively, the relationship with crossover residuals is easily derived. An altimeter crossover difference is defined as the difference between two altimeter height observations made at the same selenographical location (λ, ϕ), one observation being made when the satellite is on an ascending track, the other on a descending track. Clearly, all orbit errors which are common to both passes are eliminated from the difference measurement, like the selenographically correlated orbit error. A residual crossover difference Δxo is defined as the difference between two altimeter residuals ϵ at the same location,

$$\Delta xo = \epsilon_a(\lambda, \phi) - \epsilon_d(\lambda, \phi) \tag{3.40}$$

Residual crossovers therefore provide information on the orbit error, tidal errors and uncertainties in altimeter corrections, but not errors common to passes in the two directions. Consequently, if error sources other than gravity-induced orbit errors may be adequately modelled, or if the error induced by the selenopotential is clearly dominating the overall error budget, residual altimeter crossovers are ideally suited to establish estimates for the radial orbit error. From (3.39) and (3.40) it is readily seen that the anti-correlated orbit error, disregarding non-gravitational sources of altimeter residual errors, appears twice in the residual crossover difference:

$$\Delta xo = 2\Delta v(\lambda, \phi) \tag{3.41}$$

Hence, for the expected RMS value of the residual crossover differences, one has

$$RMS_{xo} = 2RMS_v \tag{3.42}$$

while the expected RMS value of the overall radial orbit error reads

$$RMS_r^2 = RMS_\gamma^2 + RMS_v^2 \tag{3.43}$$

Introducing the ratio of the RMS correlated and anti-correlated orbit errors as

$$\kappa = \frac{RMS_\gamma}{RMS_v} \tag{3.44}$$

one also finds

$$RMS_r = \frac{1}{2}\sqrt{1 + \kappa^2} RMS_{xo} \qquad (3.45)$$

From (3.39) one also finds for the expected RMS of the correlated and anti-correlated orbit errors, respectively,

$$E\left\{RMS_\gamma^2\right\} = \frac{1}{4}\left(RMS_a^2 + RMS_d^2 + 2RMS_a RMS_d\right) \qquad (3.46)$$

$$E\left\{RMS_\nu^2\right\} = \frac{1}{4}\left(RMS_a^2 + RMS_d^2 - 2RMS_a RMS_d\right)$$

Hence, if the ascending and descending radial orbit errors are uncorrelated, these errors will have an equal expectation, and $\kappa = 1$, and, moreover, the orbit error due to the selenopotential will propagate into the crossover residual RMS by a factor $\sqrt{2}$. On the other hand, for a positive correlation, the expectation of the correlated error will be greater than the expectation of the anti-correlated component, while, for a negative correlation, the expectation of the correlated error is less than the expectation of the anti-correlated part. Given the large difference between the correlated and anti-correlated parts of the radial orbit error due to the full GLGM–2 covariance, one may therefore conclude that ascending and descending orbit errors at crossover locations are heavily negatively correlated, and that a significant improvement of the GLGM–2 gravity field model may be achieved by application of residual laser altimetry crossover analysis.

For LP75G a similar reasoning applies, albeit on a smaller scale. The overall RMS of the selenographically correlated component amounts to 0.38 m, while the anti-correlated part yields 1.43 m. The factor 3-4 RMS difference indicates that improved orbit accuracy may be achieved through laser altimetry crossover analysis. Given the distribution of the anti-correlated component, with the highest error levels associated with the low-latitude areas centred around longitudes of 0° and 180°, both near-side and far-side improvement appears possible. Furthermore, the selenographically correlated part displays a pattern strongly correlated with the tracking data distribution, or, equivalently, the error in the selenoid undulation, which will be further elaborated in Chap. 4.

Although a rigourous analysis of the benefit of the inclusion of crossover measurements in lunar orbit and gravity field, *e.g.* in terms of accuracy of the available Clementine laser altimetry observations, is not within the scope of the present analysis, it appears appealing. First, Clementine altimetry was performed between 81° northern latitude and 79° southern latitudes [*Smith et al.*, 1997], and, hence, includes far-side information. Second, altimeter crossovers contain localised information, which may help decorrelate the estimation of coefficients containing identical frequencies. Finally, it has been observed for Earth observation satellites that the inclusion of altimeter crossovers for one satellite may improve the orbit accuracy for other satellites orbiting at different inclinations and altitudes [*Nerem et al.*, 1994; *Tapley et al.*, 1996; *Lemoine et al.*, 1997; *Scharroo and Visser*, 1998]. In other words, depending on proper calibration of the crossover data set, it is likely that crossover techniques may lead to better estimates of, in particular, the medium degree coefficients. Practical problems related to such applications are likely to be rooted in a possible laser altimeter bias, the error in the altimetry-derived estimate of the lunar topography, as well as in the treatment of steep slopes characterising the lunar terrain [*e.g., Smith et al.*, 1997, 1999*a*].

3.4   Long-term orbit behaviour

Applications of gravity fields other than studies of the underlying density distri- bution and spatial variations in crustal thickness are mainly found in the field of satellite flight dynamics. Prediction of the long-term behaviour of low altitude near-circular orbits is important for mission planning, where the optimal orbit choice for a given mission goal obviously depends on the dynamic behaviour. Fur- thermore, manoeuvre planning aiming at minimisation of propellant cost depends on the same dynamics. In the case of low lunar orbits, predictions based on im- proved gravity field models are important because previous models have shown to predict widely different behaviour. Ability to predict trends in the long-term behaviour is therefore not only a direct product of the gravity field modelling pro- cess, but also a tool that may be used for assessment of the models. In other words, significant dispersions found with respect to previous models must be explicable and plausible by the latest tracking data. Additionally, failure of consecutive mod- els to predict similar orbit behaviour must be regarded as further signs of the sen- sitivity of the lunar gravity field solution to the modelling approach, as outlined in the previous sections.

The most significant orbital variation in a conservative force field are the long- term variations in eccentricity and argument of perilune, caused by the zonal terms of the selenopotential, which map directly into variations in periapsis altitude. These variations only average out over periods of hundreds of days. As low alti- tude lunar mapping orbits require tight control on these variations, careful atten- tion needs to be paid to the effect of the zonal terms. Along with Venus and on the contrary to the Earth, the Moon is one of the few terrestrial bodies which does not exhibit any sizeable difference between J_2 and other zonal harmonics. Logi- cally, no single zonal coefficient therefore dominates the orbit behaviour. Short- and medium-periodic perturbations due to tesseral and sectorial terms are gen- erally less significant, and also average out over much shorter periods. Hence, apart from short-periodic variations with respect to the mean elements, the long- term analysis is perfectly able to demonstrate, without time-consuming numerical integration, the major orbit effects over significant time spans.

The approach used to simulate the long-term behaviour of near-circular or- bits under influence of the selenopotential originates from *Cook* [1991]. This solu- tion is obtained by linearising the singly averaged Lagrange planetary equations and eliminating one degree of freedom with an integral of motion. The resulting long-periodic solution is expressed in semi-equinoctial elements $k = e \cos \omega$ and

$h = e \sin \omega$. Orbits of particular interest are *frozen orbits*, which are equilibrium solutions of the linearised differential equations for $\{h, k\}$, defined as orbits with no long-term changes in eccentricity, argument of periapsis and inclination, and *periodic orbits* which are repetitive in $\{h, k, i\}$. Periodic orbits are therefore not necessarily identical to repeat orbits investigated in the previous sections. Previous work in this field includes, in chronological order: *Cook and Sweetser* [1992] who investigated aspects of orbit maintenance for the long-running lunar polar orbiter initiative based on historical data; *Konopliv et al.* [1993] who addressed the orbit behaviour of the high degree and order gravity field model Lun60D; *Meyer et al.* [1994] who focused mainly on satellite lifetime predictions; *Milani and Knežević* [1995] who developed long-term orbital element analysis tools for the (now cancelled) MORO mission; *Eckstein and Montenbruck* [1995] who further detailed the long-term orbit prediction problem for MORO; *Park and Junkins* [1995] who mainly focused on aspects of lunar mission analysis; *Vasile* [1996] and *Finzi and Vasile* [1997] who performed a thorough analysis of the orbital behaviour, including aspects of optimisation, from all models up to and including GLGM–2; and *Goossens* [1999] and *Goossens et al.* [1999] who investigated the long-term behaviour of all recent gravity models, including the initial Lunar Prospector model LP75G, and also combined the gravity-induced effects with solar radiation pressure. The results presented here follow the line of *Goossens* [1999], but are restricted to the gravity-induced effects only. The validity of the linearised approach for long-term orbit analysis has been demonstrated by several authors in the past [*Milani and Knežević*, 1995; *Vasile*, 1996; *Finzi and Vasile*, 1997; *Goossens*, 1999]. In the following, frozen orbit characteristics are presented along with more general periodic orbits.

3.4.1 Frozen orbits

Starting from the Lagrange planetary equations, a relatively simple first-order equation system for the long-term behaviour of the eccentricity and argument of perilune may be derived. Since the disturbing potential is averaged over the mean anomaly M, the semi-major axis does not undergo long-term perturbations. After averaging, only the zonal terms of the potential are retained. Moreover, because variations in the right ascension of the ascending node Ω do not affect any other coordinates, and the rate of change of Ω is negligible for low lunar orbits, this coordinate may be ignored. In a zonal field the spacecraft angular momentum about the rotational axis of the Moon is conserved, hence leaving only two unknowns of interest for orbit evolution over long time spans [*Cook*, 1991]:

$$\frac{dh}{dt} = (\eta + \epsilon)k \qquad (3.47)$$

$$\frac{dk}{dt} = \rho + (\eta - \epsilon)h$$

where the coefficients η, ϵ and ρ are given by [*Goossens*, 1999]

$$\eta = -n \sum_{\substack{l=4 \\ \text{even}}}^{\infty} \left(\frac{a_e}{a}\right)^l \bar{J}_l \frac{1}{4}(l-1)(l-2)\left(\bar{F}_{10\frac{l-2}{2}}(i) + \bar{F}_{10\frac{l+2}{2}}(i)\right) \tag{3.48}$$

$$\epsilon = -n \sum_{\substack{l=2 \\ \text{even}}}^{\infty} \left(\frac{a_e}{a}\right)^l \bar{J}_l \left[\frac{1}{2}l(l+1)\bar{F}_{10\frac{l}{2}}(i) - \cot i \frac{d\bar{F}_{10\frac{l}{2}}(i)}{di}\right] \tag{3.49}$$

$$\rho = n \sum_{\substack{l=3 \\ \text{odd}}}^{\infty} \left(\frac{a_e}{a}\right)^l \bar{J}_l(l-1)\bar{F}_{10\frac{l-1}{2}}(i) \tag{3.50}$$

This notation deviates from *Cook* [1991] in that it utilises stable recurrence relations for the inclination functions, which in turn allow computations up to high harmonic degrees. After transformation of the semi-equinoctial variable h through $h^* = h + \rho/(\eta - \epsilon)$, (3.47) is easily solved with standard techniques for first-order linear differential equations with constant coefficients. The equilibrium solution of the dynamical system is found by setting the right-hand-side equal to zero and solving for the orbital elements. After conversion into classical Keplerian elements, the frozen eccentricity e_f and argument of perilune ω_f are found to be

$$e_f = \left|\frac{\rho}{\eta - \epsilon}\right|, \text{ and} \tag{3.51}$$

$$\omega_f = \begin{cases} 90° & \text{if } \frac{-\rho}{\eta-\epsilon} > 0 \\ 270° & \text{if } \frac{-\rho}{\eta-\epsilon} < 0 \end{cases}$$

Figure 3.13 depicts the frozen orbit characteristics for GLGM–2 and LP75G for a mean semi-major axis of 1838.0 km, whereas Fig. 3.14 shows the corresponding sensitivity to the maximum degree applied in the computation, *i.e.* the cumulative effect of the zonal terms of the gravity field on the determination of the frozen eccentricity. Obviously, frozen near-polar orbit predictions are vastly different for the two models, with the frozen perilune being predicted over the north pole for GLGM–2 and over the south pole for LP75G. The low frozen eccentricity for GLGM–2 at $i = 90°$ also indicates a real possibility of long-duration free-flying satellite operation, which would be beneficial for several types of lunar mapping, and in particular gravity field mapping, for which frequent satellite manoeuvres are parasitic effects. LP75G, on the other side, in line with previous JPL lunar gravity field model developments, *e.g.* Lun60D [*Konopliv et al.*, 1993], predicts widely different frozen eccentricities. In fact, low polar frozen orbits seem unlikely, since e_f in this case even exceeds the crash eccentricity, *i.e.* the eccentricity that results in hard impact on the lunar surface. In terms of sensitivity of the frozen eccentricity results, it appears clear that the first approximately 16 zonals have largely the same influence for the two models. Above this degree, and, hence, where the difference in constraint schemes start playing a significant role, GLGM–2 quickly converges towards its low frozen eccentricity, whereas the higher degree zonals of

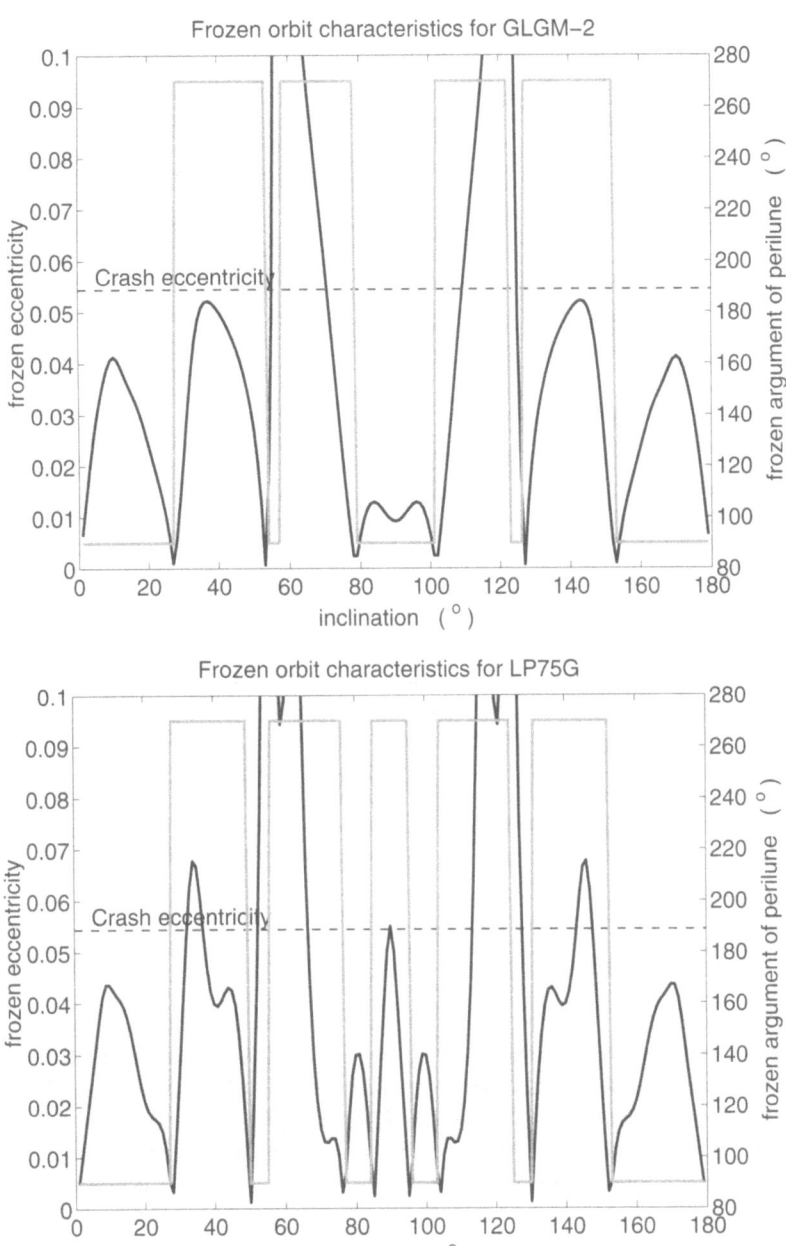

Figure 3.13   Frozen eccentricity (black curve) and argument of perilune (gray curve) as a
function of inclination for GLGM–2 (top) and LP75G (bottom). The mean
semi-major axis is 1838.0 km

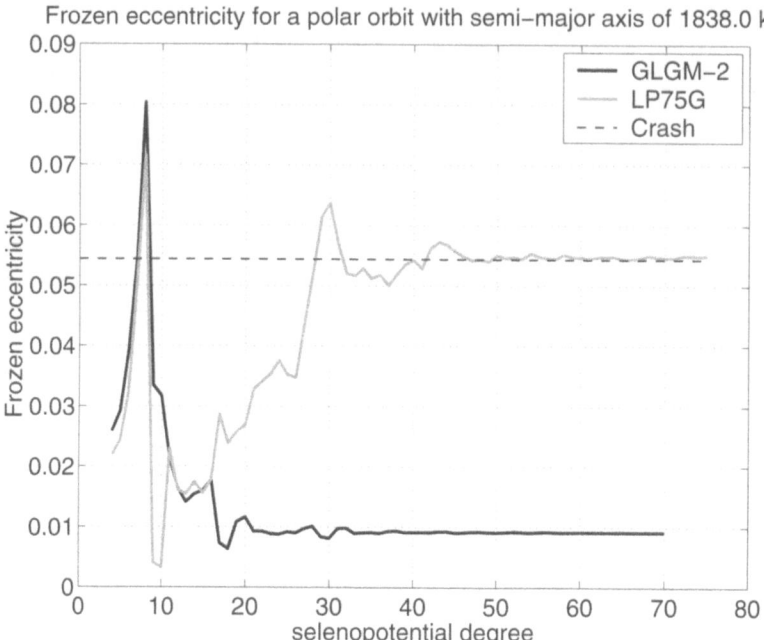

Figure 3.14   Frozen eccentricity as a function of maximum degree for GLGM–2 and LP75G, for a polar orbit with a semi-major axis of 1838.0 km

LP75G enforce a rapid increase in e_f, all the way up to approximately degree 45. In other words, a significantly larger number of zonals is in this case necessary for a definite overall long-term orbit prediction. The fact that LP75G converges to the crash eccentricity is a pure coincidence.

3.4.2   Periodic orbits

A generalisation of the frozen orbit, which appears as a point in the $\{h, k\}$ plane is the case of periodic orbits, described by the periodic solutions of (3.47). Such periodic orbits, manifested as closed loops in the phase plane, are of similar interest as frozen orbits due to their repetitiveness. As long as the perturbed orbit remains within the flight regime described by the mission requirements, also in this case one has the advantage of little or no active satellite control. Moreover, in the case that the characteristics of a periodic orbit as predicted by a given gravity model would exceed the approved flight envelope, algorithms for orbit maintenance and propellant consumption may be derived.

Table 3.2 presents the periods of initially circular polar orbits using GLGM–2 and LP75G. The existence of frozen orbits naturally indicates the possibility of neighbouring periodic orbits, as confirmed by the indefinite lifetimes predicted

Model	a = 1838.0 km		a = 1878.0 km	
	Period	*Lifetime*	*Period*	*Lifetime*
GLGM–2	872.3	∞	908.8	∞
LP75G	897.4	175.6	872.6	260.6

Table 3.2   Period and lifetime prediction in days for GLGM–2 and LP75G for initially circular polar orbits at mean altitudes of 100 km and 140 km

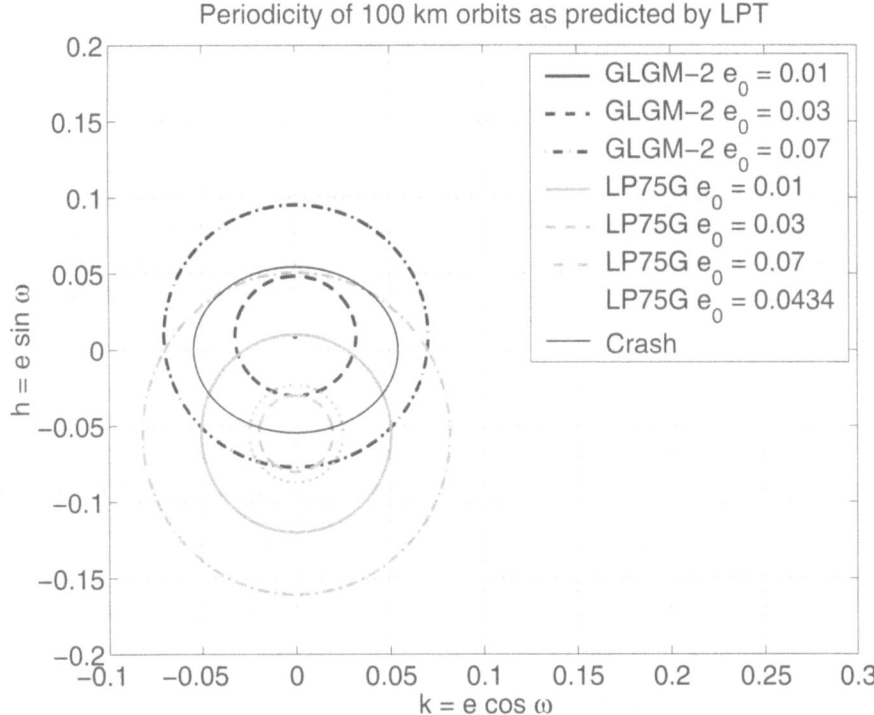

Figure 3.15   Periodic orbits in semi-equinoctial elements as a function of initial eccentricity e_0 for GLGM–2 and LP75G. All eccentricity vectors evolve clock-wise. The longest lifetime expectancy found for LP75G is 338 days, for $e = 0.0434$ and $\omega = 241°$

by GLGM–2. For LP75G, the tendency is that the satellite impacts long before completing a single period, indicating sizeable growth in eccentricity. A general overview of possible ranges of initial conditions yielding periodic orbits is given in Fig. 3.15. The motion of the eccentricity vector along the periodic orbits presented in this figure is in all cases clock-wise. While GLGM–2 provides attractive frozen orbit opportunities, and, hence, a safe range of periodic orbits, LP75G provides no possibility for indefinite satellite lifetimes. The longest lifetime accommodated by LP75G, found for $e_0 = 0.0434$ and $\omega_0 = 241°$, is 338 days.

Further illustration of practical mission analysis consequences of the different gravity models may be achieved by addressing the question on the ΔV manoeuvring associated with re-circularisation of the satellite orbit when impacts threatens [*Cook and Sweetser*, 1992; *Goossens*, 1999; *Goossens et al.*, 1999]. It is also possible to project the full gravity model covariance (as done for orbits in general) into errors in frozen and periodic orbit characteristics instead of limiting the error discussion of frozen and periodic orbits to the variance.

3.5   Discussion and outlook based on later Lunar Prospector products

In this chapter it has been shown that gravity models up to and including the Clementine mission predict widely different orbit behaviour, both deterministically and in terms of the associated error variance-covariance matrices, than models including the later Lunar Prospector data. This goes for time-wise and spatial characteristics of the orbit as well as long-term orbit predictions. The difference is, however, not solely attributed to the data sets applied in the inference of the models, but also to the "philosophy" behind the data reduction. As such, there is a clear distinction between "west coast" models developed by JPL and the "east coast" model GLGM–2 developed by NASA/GSFC. Regardless of the fact that all gravity models including Lunar Prospector data are developed by one and the same group using one and the same software, the major finding is that the GLGM–2 and LP75G gravity models are consistent with two different types of modelling strategy. While GLGM–2 focuses on non-orbit related applications, and strives to deliver meaningful data for global-scale selenophysical interpretation through tight constraints of the high-degree coefficient amplitudes, the merit of LP75G is found in the increased lumped high-frequency information intrinsically present in the low-altitude Lunar Prospector Doppler data. A related characteristic of the "west coast" approach to the gravimetric problem is that the retained power in the data allows for selenophysical studies that require higher degree harmonics, namely regional-scale and local analysis.

In terms of calibration of the error variance-covariance matrices associated with the models, it is found that solid calibration efforts have gone into the GLGM–2 development. The mismatch between the GLGM–2 covariance and the difference between GLGM–2 and the later LP75G solution is not of such a magnitude that the

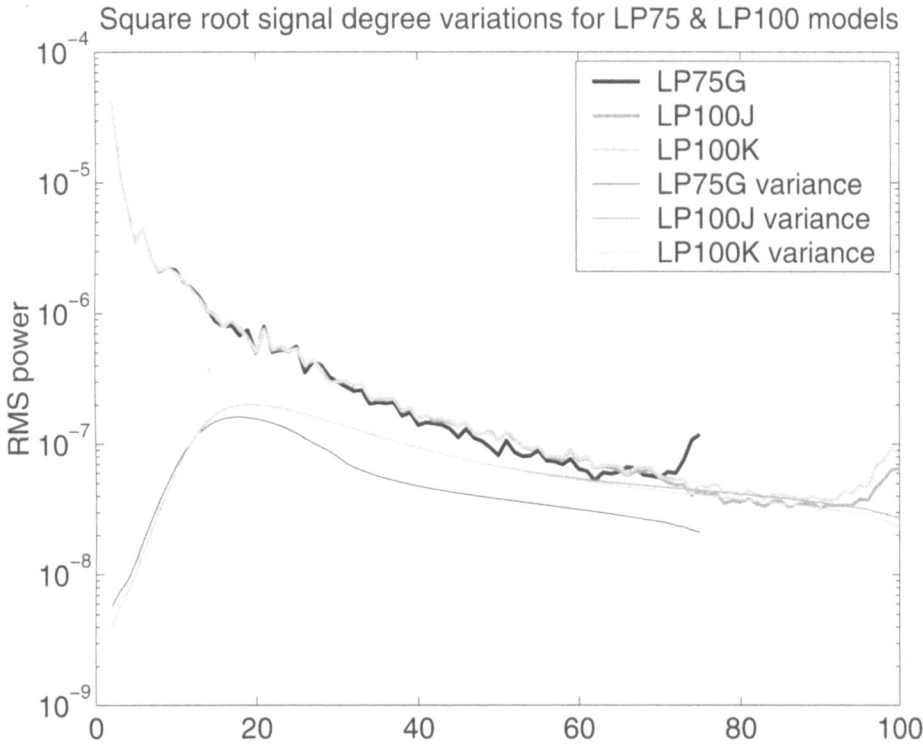

Figure 3.16   Degree-wise RMS spectral amplitudes of the Lunar Prospector-derived gravity field models LP75G, LP100J and LP100K

GLGM–2 covariance may be regarded overly optimistic. In this sense, the striking fact that Clementine and Lunar Prospector derived models predict such widely different orbit (error) behaviour may be partially attributed to the contribution of the Clementine data, or, equivalently, the different characteristics of the relatively high-altitude (and hence, low-sensitivity) Clementine data and the better Lunar Prospector data. However, data handling and processing strategies are found to be of monumental importance. LP75G, on the other hand, shows signs of optimistic (co)variances, and, moreover, increased correlations due to the relaxation of the constraints relative to GLGM–2. In other words, care should be taken to apply this model for selenophysical studies based on individual coefficients, as in fact the single proof delivered for the correctness of LP75G is its ability to deliver the most consistent lunar satellite orbits, when applied as a whole. However strong this might seem, verification through independent data is highly desired. Finally, it deserves mentioning that the orbit modelling improvement at non-polar inclinations arising from Lunar Prospector is limited. In other words, the two orders of magnitude in orbit prediction capability found at $i = 90°$, compared to previous models, is not to be used as a rule of thumb for other inclinations. In particular, for the range of inclinations not flown by lunar satellite missions, predominantly the $30° - 80°$ range (and obviously $100° - 150°$), the orbit behaviour is still largely dominated by the constraint schemes.

One promising aspect of LP75G is the aliasing of high-degree information in the degree range 70–75. Since LP75G is based on merely three months of nominal mission Lunar Prospector tracking, and, hence, is devoid of the extremely low-altitude data acquired during the extended mission as well as the final nine months of nominal 100 km data, it is interesting to conclude this chapter with an outlook towards the final products delivered by the Lunar Prospector mission. The aliasing of high-degree information in LP75G is a sign that valuable high-degree information is available in the extended mission data, which might be useful for further orbit determination improvement. Two additional models based on the extended mission of Lunar Prospector are available for this analysis, being LP100J and LP100K, both 100x100 spherical harmonic expansion derived using similar tools and methods as LP75G, and by the same analysts. The former is based on the first two months of the extended mission [*Carranza et al.*, 1999], whereas the latter has all Lunar Prospector data up to 30 July 1999, and is yet to be published [*Konopliv*, private communication]

Figure 3.16 depicts the degree-wise RMS amplitude spectra of all gravity models including Lunar Prospector tracking data as a function of spherical harmonic degree l. Compared to Fig. 3.1, it may be concluded that, amplitude-wise, subsequent Lunar Prospector products are similar to LP75G. The LP100J and LP100K models even indicate that the well-sampled near-side data include information beyond degree 100. Moreover, the variances associated with the latest solutions are consistent with the constraint schemes applied. In other words, the internal consistency of the Lunar Prospector models is high, and with the inclusion of more data as they have become available, the preliminary signs of optimistic calibration are largely removed. Another intriguing feature of the latest Lunar Prospector-based models is a generally higher power at the 35 to 70 degree range with respect to LP75G. Such a difference is attributed to the distribution of the better high-frequency tracking data information. However, the overall unambiguous determination of the amplitude spectrum of the selenopotential still requires globally sampled data, and few conclusions may be drawn concerning the correctness of the spectra at this point.

Chapter 4

Ill-conditioning of the lunar gravimetric inverse problem

> '*I am not talking about how I solved my problems, but how I posed them*'
>
> Umberto Eco, Postscript to The Name of the Rose

Throughout the previous chapters, the role of prior information in the determination of the selenopotential has been stressed. It has been shown that such information, frequently in the form of *a priori* upper bounds on the coefficient amplitudes, may be used to tune the gravity field solution, and, hence, arrive at models for different types of applications. Although there obviously is a unique gravity field, a major conclusion of the previous chapters is that, at present, no such thing as a general-purpose lunar gravity field model exists, satisfying the needs of all users with sufficient accuracy.

Obviously, this situation is directly related to the ill-conditioning of the inverse problem at hand, combined with the very definition of the lunar gravimetric problem. A formulation of the lunar gravimetric problem in terms of global spherical harmonic basis functions, while knowing that one half of the Moon basically is unaccessible by standard observation techniques, at first glance appears to be a typical example of a naively posed problem, given the inherent under-sampling of the potential field. On the other hand, the excellent near-side sampling actually invites the use of extended series expansions to exhaust the available orbit perturbation information. In such situations, the high-frequency near-side information will induce high-frequency counterparts in their far-side complements, and, hence, induce fictitious larger oscillations in the far-side gravity field solution.

Nevertheless, there are at least two reasons to stick with the spherical harmonic framework for lunar gravity field solutions. First, tradition in space geodesy and planetodesy is deeply rooted in the spherical harmonic framework. Major analysis tools are therefore strongly related to similar application software for terrestrial gravity field modelling. Moreover, BVP formulations of planetary gravity fields have yet to emerge, since the available observables (LOS Doppler measurements of satellites) are complicated functions of the spherical harmonic coefficients at satellite altitude, and therefore do not fit in a traditional gravimetric BVP formulation. For example, if the final gravity field solution arises from a mix of satellite orbit characteristics, the definition of the boundary surface of the measurements is by no means straightforward. This situation is obviously different for *in situ* gravimetric measurements on the boundary surface of the Earth.

Furthermore, also for the Earth, it is only in the past decade or so that compactly supported basis functions start to win terrain. Prime examples of such basis functions are splines [*Moreaux et al.*, 1999] or spherical wavelets [*Freeden et al.*, 1997; *Schneider*, 1997; *Freeden and Schreiner*, 1998; *Pereverzev and Schock*, 1999; *Schröder and Sweldens*, 2000] and also so-called band-limited Slepian functions [*Albertella et al.*, 1999], which are effectively orthogonal linear combinations of spherical harmonics on the area of interest. Such formulations are, however, far from commonplace, and, being in their early phase of development for gravimetry, they are not considered in this work. Second, when deriving gravity fields from orbit perturbation data, one is naturally interested in high quality orbits, since these are intrinsically related to the quality of the selenopotential estimate. Invoking prior constraints on the spherical harmonic functions ensures that the amplitude of the coefficients is bounded and may therefore, by appropriate choice, lead to better global orbit determination results. In other words, individual coefficient accuracy is traded for improvement in a lumped sense. The practical importance of these rather straightforward considerations should not be underestimated, because it is only by means of such empirical constraints that lunar gravity field models today are useful in orbit analysis as well as, in a more limited sense, for selenophysical applications. However, such approaches lead to gravity field solutions whose quality assessment becomes involved and nearly impossible to carry out in rigour. This is particularly true as long as little or no data are available for purposes of verification, and solutions may only be proven to be internally consistent. Consequently, it is not surprising that the application in mind may invite or require dedicated types and levels of regularisation.

It is therefore interesting to devote research efforts to the very nature of the ill-conditioning of the lunar gravimetric inverse problem. Analysis tools from modern numerical and functional analysis may be applied to the ill-posed problem arising from the processing of the available lunar satellite tracking data, and therefore shed light on the severity of the numerical issues involved, along with their physical implications. In other words, next to the orbit error analysis presented in the previous chapter, one is in the position to investigate the information content and aspects of quality assessment of gravity field models directly from the equation systems from which the models are derived. In this chapter, such analysis is further complemented with a review of present-day practice in lunar gravity field modelling. It is indeed shown that the indisputable knowledge level of the selenopotential is limited and that the characteristics of the gravity field models are dependent on the invoked type and amount of regularisation.

Another important issue is the understanding of the least squares estimator for gravity field applications so severely depending upon prior information. A key issue is that if the prior information is wrong, the solutions will be biased. The role of the bias and the bias computation is therefore of paramount interest. In this chapter it is argued that the frequently used collocation framework for error propagation and general quality analysis may not be suitable, as in many cases the bias is non-zero, and that the estimator should rather be viewed as biased. Such considerations are non-trivial, since they affect the error analysis and, possibly, also other quality measures for the selenopotential solution.

Finally, this chapter deals with pragmatic parameter choice methods for the selection of the regularisation parameter for discrete ill-posed problems. Such methods present an alternative to regularisation (or constraints) using empirical or assumed information. Frequently, such empirical rules of thumb are derived using a very limited data set, and they are not known or verified until the solution itself is known. Obviously, there is a danger of circular argumentation in this. The application and merit of pragmatic parameter choice methods is therefore to by-pass prior information, and rather extract the necessary level of regularisation from the linear equation system itself. Although such deterministic methods for the selection of the optimal regularisation of a linear equation system are based on the minimisation of the overall error, it is not expected that they will compensate for the severe under-sampling of the lunar gravimetric problem as it is posed at present. Therefore, for unsampled regions of the Moon, such methods applied to present-day data sets are not expected to yield solutions comparable to those for which lunar physics was used in the constraints selection. However, the methods will produce numerically optimal solutions, describing the knowledge of the selenopotential as provided by the measurements alone. Furthermore, a fundamental advantage of such methods is that they provide a framework for regularisation of future lunar gravity field solutions, based on *e.g.* SST, independent of the lunar physics one actually would like to deduce from the gravity field.

4.1   Setting the stage

The concept of *well-posed* and *ill-posed* problems dates back to the beginning of the previous century. *Hadamard* [1923] essentially defined two conditions for a problem to be ill-posed: *i)* if the solution is non-unique, or *ii)* if it is not a continuous function of the data, *i.e.* if an arbitrarily small perturbation of the data could cause arbitrarily large perturbations of the solution. At the time it was believed that ill-posed problems were unable to describe physical systems. This idea was wrong, however, and today ill-posed problems arise quite naturally in the form of inverse problems in many areas of science and engineering. Famous works by *Phillips* [1962] and *Tikhonov* [1963a, 1963b] initiated an ever-continuing analysis into mathematical problems that are ill-posed of nature, along with methods for their transformation into a neighbouring well-posed problem, in other words, their regularisation. Typically, inverse problems are concerned with the determination of the structure of a physical system from the system's measured behaviour (system identification), or with the determination of the unknown input that gives rise to the measured output (parameter estimation or causation). In contrast, in so-called direct problems, or forward modelling, the interest is in the system's behaviour given a certain input and internal structure.

While any computation involving series of measurements of the system output naturally leads to discrete inverse problems, thorough understanding and treatment of the problem generally requires an understanding of the underlying mathematical model. In other words, the framework for the sound treatment of discrete inverse problems is provided by continuum theory, particularly functional analysis and the theory of integral equations. Furthermore, although the relationship between the orbit perturbations and orbit and gravity field parameters is nonlinear, the underlying theory is confined to linear problems. Since the practical parameter estimation process is generally characterised by successive iterations over linearised equations, and, hence, the fundamental unknowns at each step are linear corrections to a given *a priori* gravity field model, this is, however, of no direct concern. Nonetheless, linearisation errors will of course remain one of the error components of the solution of the differential equation underlying the physical problem at hand.

In the literature, the classical example of a linear ill-posed problem is a Fredholm integral equation of the first kind with a square integrable kernel [*Groetsch,* 1984, 1993; *Kress,* 1999]. This equation can always be written in the generic form

$$\int_a^b K(x,y)f(y)\,dy = g(x), \quad x \in [a,b] \tag{4.1}$$

where the right-hand-side g and the kernel K are known functions, while f is the unknown, sought solution. A kernel K is said to be square integrable if

$$||K||^2 \equiv \int_a^b \int_a^b K(x,y)^2\,dx\,dy$$

is bounded. In operator theory notation, the same problem may be defined as

$$Af = g \tag{4.2}$$

where $A : F \to G$ is a linear operator from a normed space F onto a normed space G, with $f \in F$ and $g \in G$. For the inverse problem to be *well-posed* three conditions have to be met [*Kress,* 1999]: for a solution to *exist*, A must be surjective; for the solution to be *unique*, A must be injective (hence, A must be bijective), and, finally; for the solution f to be *stable* the inverse operator $A^\dagger : G \to F$ must be a continuous function of the data g. If any of these three conditions is violated, the problem is said to be *ill-posed*. Unsurprisingly, the severe under-sampling in present-day lunar gravimetry implies that the issues of uniqueness and stability are primary concerns.

In parameter estimation applications, like orbit determination and gravity field estimation, K is known exactly by the underlying mathematical model, while g typically consists of measured quantities. Consequently, g is only known with a certain accuracy, and only in a finite set of points $x_1 \ldots x_m$. In lunar gravimetry, the residual orbit perturbations fit such a Fredholm model, but this picture is unfortunately severely complicated by the fact that a gravity field model is usually a result of the combination of several types of observations made at different locations and satellite altitudes. It should also be clear that although the present representation of the linear integral equation is one-dimensional, all applications on the lunar sphere are of course two-dimensional, and would require a 2D formulation should one wish to explicitly use the integral equation approach for satellite data processing. However, all relevant properties of the ill-posed problem, including their regularisation may be adequately described by the 1D model, and, moreover, the extension to 2D is straightforward.

Mathematically speaking, the difficulties with (4.1) are inseparably connected with the *compactness* of the operator A associated with the kernel K. If the operator A is compact, its inverse $A^\dagger$ is unbounded [*Groetsch*, 1993; *Kress*, 1999]. In the following, the practical ways of understanding and dealing with this fact are discussed.

Undoubtedly, the superior tool for analysis of first-kind integral equations with square integrable kernels is the *singular value expansion* (SVE) of the kernel :

$$K(x, y) = \sum_{i=1}^{\infty} \mu_i u_i(x) v_i(y) \qquad (4.3)$$

where the functions u_i and v_i are termed the *singular functions* of K and are orthonormal with respect to the usual inner product

$$\langle \phi, \psi \rangle \equiv \int_a^b \phi(t) \psi(t) \, dt$$

The numbers μ_i are the non-negative *singular values* of K, usually ordered in nonincreasing order. A fundamental relationship between the singular values and the singular functions is given by

$$\int_a^b K(x, y) v_i(y) \, dy = \mu_i u_i(x), \quad i = 1, 2, \ldots, \infty \qquad (4.4)$$

which shows that any singular function v_i is mapped onto the corresponding u_i, with the singular value μ_i acting as amplification factor. In other words, this relation governs the mapping of the basis functions of f onto the functionals spanning the space to which g belongs. Combined with the linear integral equation, this yields

$$\sum_{i=1}^{\infty} \mu_i \langle v_i, f \rangle u_i(x) = \sum_{i=1}^{\infty} \langle u_i, g \rangle u_i(x) \qquad (4.5)$$

and, hence, the SVE form of the solution f reads

$$f(y) = \sum_{i=1}^{\infty} \frac{\langle u_i, g \rangle}{\mu_i} v_i(y) \tag{4.6}$$

The overall behaviour of the singular values and singular functions is of course by no means arbitrary, but interlinked with the properties of the kernel K. Generally, it holds that [*Hansen*, 1998a]

- the smoother the kernel K, the faster the decay of the singular values μ_i. Smoothness is in this context measured by the number of continuous partial derivatives

- the smaller the singular values μ_i, the more oscillations (zero crossings) there will be of the singular functions u_i and v_i

Given the fact that the SVE triplets $\{\mu_i, u_i, v_i\}$ are characteristic and unique for a given kernel K, the solution to the integral equation of the first kind is completely characterised by the coefficients $\mu_i^{-1} \langle u_i, g \rangle$ and the singular functions v_i. Evidently, smoothness criteria exist for (4.6), as no solution may be expected if the right-hand side does not converge. This is further elaborated in the next section.

The practical implication of these observations is that (4.6) may be regarded as a spectral representation of the solution f, and that the spectral properties of this expansion are given by the coefficients $\mu_i^{-1} \langle u_i, g \rangle$. Furthermore, the smoothing effect of the integration with the kernel K are illustrated: the higher the spectral components in f, the more they are damped in g through the multiplication with μ_i. *Vice versa*, the inverse problem has the opposite effect on the oscillations in g, *i.e.* an amplification of the spectral components $\langle u_i, g \rangle$ with a factor μ_i^{-1}. The result of course is an amplification of the high-frequency components, including their errors. The decay rate of the singular values is for these reasons fundamental for the behaviour of ill-posed problems, as it directly provides insight into the severity of the ill-conditioning. A fundamental aspect of mathematical regularisation schemes is therefore to somehow counteract small spectral coefficients by means of filtering.

Generally speaking, there are two types of ill-conditioned problems. *Rank-deficient* problems, *i.e.* problems with ill-determined rank, are characterised by distinct clusters of singular values, with a distinct gap between the larger and smaller values. For such problems, regularisation generally filters out the small singular value components by means of truncation of the spectral coefficients. Ill-posed problems, on the other hand, are generally characterised by slowly decaying singular values, with no gap in the spectrum. Gravity field determination falls into this latter category, as it is commonly found that the observability of linear combinations of spherical harmonics, lumped in terms of orbital frequency, is gradually reduced with increasing frequency.

4.1.1 The Picard condition

As outlined, not every right-hand side g will lead to a smooth solution f, and certain smoothness criteria naturally arise for the right-hand side of the integral equation. The so-called Picard condition prescribes that the right-hand side g must be somewhat smoother than the desired solution f in order for (4.6) to actually converge. In other words, in order that there exists a square integrable solution f to (4.1), the right-hand side g must satisfy [*e.g.*, *Groetsch*, 1984, 1993; *Hansen*, 1990, 1998a; *Kress*, 1999]:

$$\sum_{i=1}^{\infty} \left(\frac{\langle u_i, g \rangle}{\mu_i} \right)^2 < \infty \tag{4.7}$$

The Picard criterion states that from some point on in the summation in (4.6) the absolute value of the coefficients $\langle u_i, g \rangle$ must decay faster than the corresponding singular values for a square integrable solution f to exist. Mathematically, it is identical to the requirement that the right-hand side g belong to $\mathcal{R}(\mathcal{K})$, the range of K. In other words, if g has any, arbitrarily small component outside $\mathcal{R}(\mathcal{K})$, then there is no square integrable solution. It is exactly this lack of stability that makes the first-kind integral equation ill-posed. Compliance with the Picard condition should therefore possibly be tested, in order to have a first hand graphical view of the problem at hand. A problem here, however, are measurement errors contaminating the observations g. The theoretical and practical consequences of this for lunar gravimetry are elaborated in Sect. 4.1.2, with emphasis on the GLGM–2 and LP75G gravity field models.

In practical situations, the right-hand side g consists of measured quantities and is effectively only an approximation of the exact data. Any naive attempt to compute the solution directly through (4.6) is therefore likely to diverge, or return a useless result with a large norm. This is where regularisation comes in, as it replaces the original ill-posed problem with a nearby regularised (well-posed) problem belonging to the class of second-kind Fredholm integral equations, which is known to have a stable solution. If the problem is not "too ill-posed", the aim is that the regularised solution will have a suitably small residual, will not be too far off from the desired, unknown solution to the unperturbed (error-free) problem, and will also satisfy the constraint relations. The choice of the regularisation parameter(s) is therefore a paramount issue in solving ill-posed problems. Statistically speaking, the introduction of the regularisation decreases the statistical error (*e.g.* the covariance matrix) at the cost of adding a bias to the solution. The consequences of the bias are elaborated in throughout this chapter, since it is one of the key points in understanding the quality assessment and solution strategy of present-day selenopotential models.

4.1.2 Discretisation and the singular value decomposition

In practical applications one is naturally confined with a finite set of measurements, which requires the solution of a finite set of parameters. In other words, the problem at hand needs to be discretised. Discretisation of the lunar gravimetric problem, formulated in terms of a finite set of spherical harmonic coefficients arises quite naturally by the linearised relationship between the measurement residuals and the correction to the finite number of *a priori* model parameters. If one, on the other hand, would like to explicitly use the integral equation framework, for measurements taken at some given boundary surface, methods like that of Galerkin or quadrature methods are commonplace [*e.g., Martensen and Ritter*, 1997; *Hansen*, 1998a]. In gravitational potential recovery applications, the natural discretisation leads to a square system, *i.e.* a normal equation system, some QR-factorised triangular system, like a square root information filter, or a bi-diagonalised system,

$$\mathbf{A}\mathbf{x} = \mathbf{y}, \qquad \mathbf{A} \in \mathbb{R}^{n \times n} \tag{4.8}$$

where $\mathbf{x} \in \mathbb{R}^n$ and $\mathbf{y} \in \mathbb{R}^n$ and n is the number of estimation parameters. Since the system is derived in a least-squares sense from a set of linearised observational equations, the final, but still not regularised, square system (4.8) is the end product of the minimisation problem

$$\min \|\widetilde{\mathbf{A}}\mathbf{x} - \widetilde{\mathbf{y}}\|_2^2 \tag{4.9}$$

where $\widetilde{\mathbf{A}} \in \mathbb{R}^{m \times n}$, $m \geqslant n$, is the design matrix containing the partial derivatives of the m measurements with respect to the n unknown estimation parameters and $\widetilde{\mathbf{y}}$ are the linearised measurements. The subscript "2" denotes the L_2 norm.

The superior tool for analysis of discrete ill-posed problems is the discrete variant of the singular value expansion, the *singular value decomposition*. The ordinary singular value decomposition (SVD) applies to a single matrix $\mathbf{A}$, while the *generalised singular value decomposition* (GSVD) operates on a matrix pair $(\mathbf{A}, \mathbf{L})$. In the following, the SVD will be discussed, while the GSVD is given in Appendix B. The SVD is used to analyse the non-constrained matrix $\mathbf{A}$; the GSVD plays an important role when regularisation is introduced. The resemblance of the (G)SVD with the SVE is obvious, as the singular values of the matrix $\mathbf{A}$ are strongly related to, and in many cases an approximation of, the singular values of the underlying kernel K [*Hansen*, 1998a].

The ordinary singular value decomposition of a square or rectangular matrix $\mathbf{A} \in \mathbb{R}^{m \times n}$, assuming $m \geqslant n$, is given by [*e.g., Press et al.*, 1992; *Björk*, 1996; *Golub and Van Loan*, 1996; *Hansen*, 1998a]:

$$\mathbf{A} = \mathbf{U}\mathbf{\Sigma}\mathbf{V}^T = \sum_{i=1}^{n} \mathbf{u}_i \sigma_i \mathbf{v}_i^T \tag{4.10}$$

where $\mathbf{U} = (\mathbf{u}_1, \ldots, \mathbf{u}_n) \in \mathbb{R}^{m \times n}$ and $\mathbf{V} = (\mathbf{v}_1, \ldots, \mathbf{v}_n) \in \mathbb{R}^{n \times n}$ are matrices with orthonormal columns, *i.e.* $\mathbf{U}^T\mathbf{U} = \mathbf{V}\mathbf{V}^T = \mathbf{I}_n$, and where the diagonal matrix $\mathbf{\Sigma} = \text{diag}(\sigma_1, \ldots, \sigma_n)$ has non-negative elements appearing in non-increasing order,

$$\sigma_1 \geqslant \sigma_2 \geqslant \cdots \geqslant \sigma_n \geqslant 0 \tag{4.11}$$

As for the SVE, the singular values σ_i of the SVD gradually decay to zero, an increase in the dimensions of $\mathbf{A}$ will increase the number of small singular values, and the elements of the left and right singular vectors $\mathbf{u}_i$ and $\mathbf{v}_i$ tend to show more sign changes as the index i increases, or, equivalently, σ_i decreases.

Ill-conditioning of the matrix $\mathbf{A}$ may be addressed by the SVD, by considering, for all $i = 1, \ldots, n$,

$$\mathbf{Av}_i = \sigma_i \mathbf{u}_i, \qquad\qquad \|\mathbf{Av}_i\|_2 = \sigma_i \qquad (4.12)$$
$$\mathbf{A}^T \mathbf{u}_i = \sigma_i \mathbf{v}_i, \qquad\qquad \|\mathbf{A}^T \mathbf{u}_i\|_2 = \sigma_i$$

A small singular value σ_i, as compared to $\sigma_1 = \|\mathbf{A}\|_2$, indicates the existence of a certain linear combination of the columns of $\mathbf{A}$, spanned by the elements of the right singular vector $\mathbf{v}_i$, such that $\|\mathbf{Av}_i\|_2 = \sigma_i$ is small. A similar argument holds for the left singular vectors $\mathbf{u}_i$ and the rows of $\mathbf{A}$. Small singular values σ_i in other words imply that $\mathbf{A}$ is nearly rank-deficient, and that the associated left and right singular vectors are effectively the numerical null vectors of $\mathbf{A}^T$ and $\mathbf{A}$, respectively. From a numerical point of view, given a certain numerical precision, ill-conditioned systems behave very similar to singular ones, and additional information is needed for a stable inversion. This leads to the concept of numerical singularity. Hence, although numerical null vectors may not be exact null vectors, they are effectively so by virtue of the catastrophic effect of round off errors in the numerical computations. It is therefore safely concluded that the matrix in an ill-posed problem is always ill-conditioned and that its numerical null space is spanned by vectors with many sign changes. The many sign changes is not a feature rigorously proven to exist for all ill-posed inverse problems, but is rather based on experience from a large pool of various types of inverse problems. A more detailed discussion on this topic is found in *Hansen* [1998a].

Similar to the continuous case, the SVD may be used to illustrate the smoothing of an integral operation and the amplification effect of the reciprocal differential operation. Using (4.10), one obtains the spectral forms

$$\mathbf{x} = \sum_{i=1}^{n} \langle \mathbf{v}_i^T \mathbf{x} \rangle \mathbf{v}_i \quad \text{and} \quad \mathbf{Ax} = \sum_{i=1}^{n} \sigma_i \langle \mathbf{v}_i^T \mathbf{x} \rangle \mathbf{u}_i \qquad (4.13)$$

Obviously, the high-frequency components of $\mathbf{x}$ are damped, by the multiplication with a smaller σ_i, than the low-frequents components. The inverse problem, recovering the cause given the output, consequently has the opposite effect, and amplifies the high-frequency constituents of the right-hand side $\mathbf{y}$. This leads to the definition of a generalised inverse in terms of the SVD, as well as a means to characterise the instability of the inverse problem at hand.

If $\mathbf{A}$ is invertible, its inverse is given by $\mathbf{A}^{-1} = \sum_{i=1}^{n} \mathbf{v}_i \sigma_i^{-1} \mathbf{u}_i^T$ and, straightforwardly, the least squares solution is $\hat{\mathbf{x}} = \sum_{i=1}^{n} \sigma_i^{-1} \langle \mathbf{u}_i^T \mathbf{y} \rangle \mathbf{v}_i$, cf. (4.6). However, to accommodate a possible rank-deficiency of the matrix $\mathbf{A}$, the *generalised inverse* matrix operator (or Moore-Penrose inverse) $\mathbf{A}^\dagger$ is introduced [*Björk, 1996; Golub and Van Loan, 1996*]:

$$\mathbf{A}^\dagger = \sum_{i=1}^{\text{rank}(\mathbf{A})} \mathbf{v}_i \sigma_i^{-1} \mathbf{u}_i^T \qquad (4.14)$$

The least squares solution $\hat{x}$ now reads

$$\hat{x} = A^\dagger y = \sum_{i=1}^{\text{rank}(A)} \frac{u_i^T y}{\sigma_i} v_i \tag{4.15}$$

In this sense the full-rank inverse is merely a special case of a Moore-Penrose inverse. The condition number of the matrix A, which signifies the sensitivity of the solutions $\hat{x}$ to perturbations in A or in y, is in this context measured by the 2-norm

$$\text{cond}(A) = \|A\|_2 \|A^\dagger\|_2 = \sigma_1/\sigma_{\text{rank}(A)} \tag{4.16}$$

The GSVD plays a role when, in addition to a measurement-derived system of equations, some kind of regularisation matrix is introduced. This is a generalisation of the SVD algorithm to two matrices, the pair (A, L) [*Hansen*, 1989, 1998a]. The GSVD plays a role in the development of theories for the regularisation of ill-posed problems, but since it is always possible to transform an ill-posed problem into a *standard form* (see Sect. 4.3.1), and, hence, the GSVD is usually not explicitly computed, its discussion is deferred to Appendix B.

4.1.3 Regularisation and filtering

Regularisation implies the incorporation of further information about the solution, in order to stabilise the problem as well as to single out a useful (and in some sense best) solution. For this purpose, a range of regularisation methods are possible. Surveys of such methods have shown up in the literature over the past decades, as inverse problems are plentiful enough to warrant the development of a structural framework for their treatment. A summary of regularisation methods for geodetic applications is given by *Bouman* [1998]. The dominating approach to regularisation is to allow a certain residual norm associated with the regularised solution,

$$r(f) = \left\| \int_a^b K(x, y) f(y) dy - g(x) \right\|_2 \tag{4.17}$$

The regularised solution is then typically found by minimisation of [*Hansen*, 1998a]:

- $r(f)$ subject to the constraint that f belongs to a specified subset of solutions; or,
- $r(f)$ subject to the constraint that a measure $s(f)$ of the size of f is smaller than some upper bound α_1; or,
- $s(f)$ subject to the constraint that $r(f)$ is smaller than some given upper bound, $r(f) \leqslant \alpha_2$; or,
- a linear combination of $r(f)^2$ and $s(f)^2$, i.e.,

$$\min \left\{ r(f)^2 + \alpha_3^2 s(f)^2 \right\}$$

where α_3 specifies the relative weighting of the two elements.

The positive parameters α_1, α_2 and α_3 are the so-called *regularisation parameters*, prescribed by the specific regularisation method.

In the spectral framework of the SVD such regularisation replaces the SVD solution (4.15) by a filtered counterpart

$$\hat{\mathbf{x}}_\alpha = \mathbf{A}_\alpha^\dagger \mathbf{y} = \sum_{i=1}^{\mathrm{rank}\,(\mathbf{A}_\alpha)} \delta_i \frac{\mathbf{u}_i^T \mathbf{y}}{\sigma_i} \mathbf{v}_i \qquad (4.18)$$

where the *filter factors* $\delta_i \in [0, 1]$. The filtered residual vector corresponding to the filtered solution now reads

$$\mathbf{y} - \mathbf{A}\hat{\mathbf{x}}_\alpha = \sum_{i=1}^{n} (1 - \delta_i)\, \mathbf{u}_i^T \mathbf{y} \mathbf{u}_i + \mathbf{y}_0 \qquad (4.19)$$

where the vector $\mathbf{y}_0 = \mathbf{y} - \sum_{i=1}^{n} \mathbf{u}_i \mathbf{u}_i^T \mathbf{y}$ is the component of $\mathbf{y}$ orthogonal to the vectors $\mathbf{u}_i$, and therefore outside the range of $\mathbf{A}$. In the case that all filter factors are equal to one, the solution is the standard least squares solution; in the general case filter factors characterise the dampening of the erroneous SVD components. Regularisation methods differ only in how they choose the filter factors. For some methods there exist explicit formulas for the filter factors, while for others they are not known, but rather determined by some iterative procedure. The amount of filtering, or, in other words, the choice of optimal regularisation, is another matter of prime importance which will be discussed in Sect. 4.5.

In practice, all inverse problems are affected by various types of errors in both $\mathbf{A}$ and $\mathbf{y}$. Although, in causation problems, the physical model is given and $\mathbf{A}$ is known exactly, whereas the right-hand side may be contaminated with errors, this is nothing more than an approximation of the actual situation. The real-life error sources include measurement errors, *i.e.* $\mathbf{y}^\varepsilon = \mathbf{y} + \varepsilon$, rounding errors involved in the numerical computation of both $\mathbf{A}$ and $\mathbf{y}$ and also discretisation errors involved in setting up the linear system. In the following only measurement errors shall be considered, satisfying $E\{\mathbf{y}^\varepsilon\} = \mathbf{y}$. In other words, systematic effects are disregarded.

4.1.4 The discrete Picard condition

From a strictly numerical point of view, a finite-dimensional problem in the absence of data errors always satisfies the Picard condition (4.7), as the discretisation in itself causes the minimum norm solution to be bounded. This is trivial, however, and does not counteract the instability caused by the decay in the singular values σ_i as $i \to \infty$ or some upper index number. Given the appearance of the discrete Fourier coefficients $\langle \mathbf{u}_i^T \mathbf{y} \rangle$ and the singular values σ_i in the regularised solution (4.18), it is clear that a discrete version of the Picard condition (4.7) plays an important role in understanding the numerical behaviour of the problem. Intuitively, it may be expected that if $|\mathbf{u}_i^T \mathbf{y}|$ exhibit a slower decay than σ_i, then filtered solutions cannot produce a useful regularised solution. Moreover, the faster the decay of the

Fourier coefficients, the better the regularised solution approximates the exact solution. The *discrete Picard condition* is satisfied by the unperturbed right-hand side **y** if, for all numerical non-zero generalised singular values, the Fourier coefficients $|\mathbf{u}_i^T \mathbf{y}|$ on average decay to zero faster than the singular values [*Hansen*, 1990].

Figure 4.1 depicts the Picard plots for the unregularised normal equation of GLGM–2 and, *idem*, for the square root information filter of LP75G. While a direct comparison of the two systems is complicated by the fact that the GLGM–2 system is a normal matrix equation and the LP75G system is a triangular QR-factorised equation, which causes the singular values and Fourier coefficients to be of a different magnitude, the fact remains that the two systems describe the same fundamental thing: the final linear and unregularised equation relating selenopotential coefficient corrections to the measurement residuals. In other words, systematic trend differences are related to the relative numerical performance of the two equations.

The very fact that the right-hand side is not error-free complicates the picture, since one can never expect or guarantee that the errors ε satisfy the discrete Picard condition. Rather, in problems where under-sampling is not an issue and only measurement errors are of concern, the singular values σ_i tend to decrease monotonically until they level off at some level determined by the errors in the matrix **A**. *Vice versa*, the Fourier coefficients $|\mathbf{u}_i^T \mathbf{y}^\varepsilon|$ decay until they settle at a level determined by the error ε in **y** [*Hansen*, 1998a]. In other words, if the singular values level off from some index $i \geq i_\mathbf{A}$ and $|\mathbf{u}_i^T \mathbf{y}^\varepsilon|$ level off beyond index $i \geq i_\mathbf{y}$, then one can only hope to recover those singular value components of the solution for which the errors in σ_i and $|\mathbf{u}_i^T \mathbf{y}^\varepsilon|$ are not dominant, *i.e.* components $|\mathbf{u}_i^T \mathbf{y}^\varepsilon|$ for $i \leq \min(i_\mathbf{A}, i_\mathbf{y})$.

From the singular values one cannot only compute rather catastrophic condition numbers for the 70×70 GLGM–2 normal matrix and 75×75 LP75G SRIF of 2.098×10^{17} and 6.520×10^7, respectively[1], but also deduce that while the Fourier coefficient of LP75G level off at approximately $10^1 - 10^2$, those of GLGM–2 continue to decay right up to the last index. In other words, beyond $i \approx 1500$ (this number is of course not rigorously determined by a qualitative plot only) errors in the LP75G SRIF matrix start to play a role, whereas for GLGM–2 such model errors are inferior for the entire spectrum. This is likely to be a result of the difference in available tracking data going into the final linear equation system from which the models are derived, in favour of the superior tracking data from the Lunar Prospector mission, combined with the data weighting schemes. Secondly, the numerical properties of the QR-factorisation outperform the normal equation formulation, making the LP75G system less sensitive to a build-up of rounding errors. However, the levelling-off of the Fourier coefficients in the case of LP75G directly indicates the number of spectral components that could possibly be recov-

---

[1]Forming a normal matrix from the LP75G SRIF yields a condition number of 7.652×10^{15}, which is, for direct comparison, about a factor ~ 27 better than GLGM–2. The fact that $\left(6.520 \times 10^7\right)^2 < 7.652 \times 10^{15}$ shows that squaring of near-singular matrices is a non-trivial operation, and illustrates the potential problem of the normal equation formulation of the lunar gravimetric problem

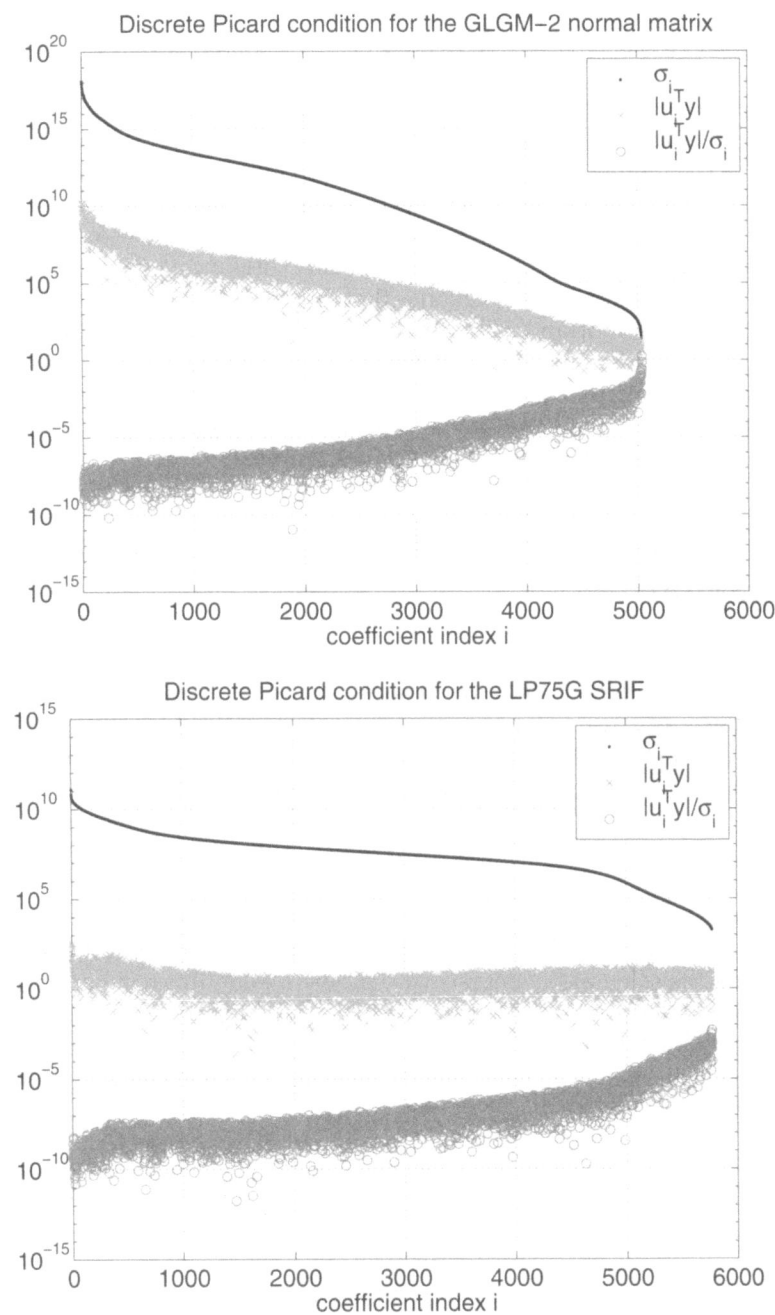

Figure 4.1   Picard plots for the unregularised normal equation of GLGM–2 and the
unregularised square root information filter equation of LP75G

ered directly from the SRIF to be smaller than 1500, the slow decay could be an indication that the real number is smaller. This is verified later using contribution measures in Sect. 4.4.2.

Furthermore, both matrix systems generally satisfy the discrete Picard condition, as the ratio $|\mathbf{u}_i^T \mathbf{y}^\varepsilon|/\sigma_i$ stays below one. Notice, however, the numerical instability in the computation of the very last spectral component for GLGM–2, which shows up as a single index with a Picard coefficient greater than one. The matrix equation right-hand sides are therefore smooth enough to allow inversion. However, the truly large condition numbers evidently discourage such an approach. A perhaps more enigmatic phenomenon is that for GLGM–2 errors in $\mathbf{A}$ and in $\mathbf{y}^\varepsilon$ are in fact of inferior importance. This is a proof that regularisation of lunar gravity field solutions requires more than compensation, by means of structural *a priori* information, for incorrect data collection, but some additional countermeasure for the severe under-sampling. This leads directly to the multiple and sometimes contradictory roles of regularisation in lunar gravimetry.

4.2   The multiple roles of regularisation in lunar gravimetry

A general principle behind all work on ill-posed inverse problems is that exact data should give the exact solution. This is, of course, where trouble starts in lunar gravimetry, because independent of the quality of the near-side sampling, the global spherical harmonic basis functions will remain incompletely sampled. On the other hand, applying a regularisation technique, which induces a significant bias where the bias may be parasitic, is also sub-optimal. Hence, the type and level of regularisation may be expected to affect the solution in a multitude of ways. Clearly, such considerations are not exclusive to the lunar gravimetric problem. Rather, they are common to all inverse problems suffering from incomplete or inadequate sampling. However, as will become clear in the following, as it is posed today, the selenopotential recovery problem takes this to an extreme.

Kress [1999] defines the key requirements for any regularisation scheme. First, the regularisation scheme must be *point-wise convergent*, *i.e.* replacing the ill-posed least squares solution $\hat{f} = A^\dagger g^\varepsilon$ (or $\hat{\mathbf{x}} = \mathbf{A}^\dagger \mathbf{y}^\varepsilon$ for the discrete case) with the approximation

$$\hat{f}_\alpha = A_\alpha^\dagger g^\varepsilon \quad \text{such that} \quad \lim_{\alpha \to 0} \hat{f}_\alpha = f$$

it must hold

$$\lim_{\alpha \to 0} A_\alpha^\dagger A f = f \tag{4.20}$$

The positive parameter α is called the regularisation parameter. It must be seen in a general sense, and may also be an entire set of positive constants. Second, the regularisation scheme must be *regular*, that is, if the error level $\varepsilon \to 0$ the regularisation parameter $\alpha(\varepsilon)$ should also vanish

$$\lim_{\varepsilon \to 0} A^{\dagger}_{\alpha(\varepsilon)} g^{\varepsilon} = A^{\dagger} g \qquad (4.21)$$

Regularity implies that the regularisation parameter is a monotonic function of the error level. The two conditions together assure that exact data yield the exact solution. In order to comply with the mathematical requirements, the consequence is that the filter factors $\delta_i(\alpha, \sigma)$ may not be arbitrarily chosen.

The combination of high-quality data sampling of the lunar near-side, *e.g.* the extensive DSN tracking of Lunar Prospector at altitudes down to $\sim$15 km with a precision level $\sim$0.25 mm/s, and no data collection at all over the lunar far-side, represents the challenge for a lunar gravity model developer. Obviously, the data do not specify anything for the lunar far-side, and, hence, large excursions and non-physical variations might be expected for this part of the Moon. It is only through the integrated orbit behaviour (as the satellite again becomes visible for the tracking devices) that one can actually deduce some overall and bulk properties of the far-side gravity field. That is, one should be very careful to interpret far-side features appearing in lunar gravity field models as real, since the evidence is circumstantial, at best.

Since the data specify a non-unique and unstable global lunar gravity field solution, lunar physics is usually entered into the equation. Such *a priori* physical information (in many cases merely assumptions) effectively limits the coefficient degree variations by biasing the solution towards zero in some way, and at the same time enforces the error measures to be consistent with the applied constraint, *e.g.* Fig. 3.1. Hence, selenopotential models may well be internally consistent. The solution to such problems is, however, not unique since tuning of the constraint yields a solution with different characteristics, and therefore carries an intrinsic twist towards the physical thinking behind the constraint choice, frequently related to the intended application of the model. For example, models to be applied in orbit determination may be better served by a smoothness constraint that retains the high-degree information, similar to models that focus on local or regional selenophysical studies, since valuable information is found exactly in the high-degree variations, whereas global-scale selenophysical problems may encourage more smoothing.

Practical examples of such considerations are the different choices of data weighting schemes going into GLGM–2 and LP75G, as well as the debate on whether or not to include the lunar topography when choosing the constraint relations [*Konopliv*, private communication]. In other words, a key issue is how to deal with the scattered high-frequency information that current data sets undeniably contain. These data sets only specify certain scattered orbital frequencies, and certainly not the global range of the gravitational model. Even more important is the fact that they lead to different constraint choices, in the form of a Kaula rule, for GLGM–2 and LP75G, as shown in Table 4.1. Another issue is also - again - the quest for some sort of verification of the results, *i.e.* one would like to infer the

	GLGM–2	LP75G
$l = 2 - 25$	15	25
$l = 26 - 30$	15	20
$l = 31 - l_{max}$	15	15

Table 4.1   Scaling parameters β of the Kaula-type regularisation applied in the development of GLGM–2 and LP75G. The Kaula rule is given by $\{\overline{C}_{lm}, \overline{S}_{lm}\} = \beta \times 10^{-5}/l^2$. For GLGM–2, a single parameter is used for the entire range of harmonic degrees, whereas for LP75G, three different β values are used. The higher value of β for the lower degrees of LP75G indicates a relaxation in the constraints with respect to GLGM–2. Sources: *Lemoine et al.* [1997] and *Konopliv* [private communication]

results as independently as possible from other (previous and related) findings. Obviously, this is complicated if only marginally improved data sets are available.

All in all, it may therefore be concluded that there is a contradiction between the numerical stabilisation which replaces the original ill-posed problem by a *nearby* well-posed problem and which would be ideal for purely near-side applications, and arguments stemming from lunar physics that are largely based on empirical data, assumptions or perhaps orbit fit, and in many cases lack verification. In other words, the independence of the estimation is lost, and in fact the actual problem solved may not be so "close" to the original one specified by the data. Again, in terms of filter factors, this means that from a mathematical point of view the optimal regularisation parameter α should be as small as possible, *i.e.* $\delta_i \to 1$, while in order to account for errors and under-sampling the regularisation parameter α should be large, *i.e.* $\delta_i \to 0$. Finally, if the constraints themselves are not verified, and it is *e.g.* only through continued iterative fits of the satellite orbit data that some Kaula-type constraint (or other) is chosen, the gravity field solutions may only be internally verified, and certainly not so by independent measurements. In other words, one should be careful to view the model as correct simply because it fits some *a priori* empirical function and the satellite orbit data fit well globally when using multi-day arcs.

The conclusion is therefore that the criteria for the selection of the regularisation parameter in lunar gravimetric problems highly depends on the application. A large range of solutions is actually possible due to the strong influence of the regularisation, both in terms of the choice of method and in terms of amount. For orbit analysis and models in terms of global basis functions a certain global smoothness is required which makes the model development dependent on assumptions stemming from lunar physics, *e.g.* a Kaula rule. On the other hand, if the problem could be posed differently in terms of compactly based functions, and solved only for the sampled areas of the Moon, the optimal regularisation might be chosen independently of any prior information, directly from the properties of the linear equation system. It might even be that regularisation in itself practically becomes a non-issue, since the sampling of the near-side is truly excellent. Such regional gravity field solutions are presently under study at Delft Institute for Earth-Oriented Space Research [*Goossens*, private communication].

4.3 Regularisation methods and the view on the estimation process

Having demonstrated the qualitative and quantitative effects of regularisation in selenodesy in a variety of ways in the previous, it is only appropriate to discuss regularisation methods in a more mathematical framework. In the following, focus is on regularisation methods that play a role in satellite selenodesy and geodesy, and on the consequences the choice of the estimator has for the lunar gravity field solution and associated quality description. It will be shown that the usual collocation framework of Bayesian estimation may be unacceptable due to implicit assumptions of unbiasedness.

From a general perspective one can essentially distinguish between two classes of regularisation methods: methods that are based on some kind of "canonical decomposition", like the (G)SVD, QR-factorisations, bi-diagonal reductions and the similar, and methods that avoid such decompositions. Direct methods, which will be investigated for the lunar problem in this book, are a member of the first class, while iterative methods are a member of the second class. Since direct methods are based on standard matrix operations and decompositions, the computational effort in solving a regularised problem may be estimated *a priori*. This is not the case for the iterative types of methods, which evidently rely on some stop criterion to indicate convergence. Iterative regularisation schemes are preferable to direct methods when the coefficient matrix $\mathbf{A}$ is so large that it is too time-consuming and memory-demanding to work with explicit decompositions of $\mathbf{A}$. Instead, the iterative methods access the matrix $\mathbf{A}$ only through matrix-vector multiplications involving $\mathbf{A}$ or $\mathbf{A}^T$ and producing a sequence of iterate solutions that converge to the desired solution. At the most basic level, the desired intrinsic property of such schemes is that they pin-point the singular value components $\left(\mathbf{u}_i^T \mathbf{y}^\varepsilon / \sigma_i\right) \mathbf{v}_i$ that correspond to the larger singular values, and, hence, the components most desired in a regularised solution. The iteration number here plays the role of the regularisation parameter.

In the sequel, the lunar gravimetric problem is investigated and interpreted in the framework of Tikhonov regularisation and biased estimation. Other direct methods for the regularisation include truncated (G)SVD methods [*Hansen*, 1989, 1998*a*, 1998*b*]; for a gravity field example involving satellite gravity gradiometry see _Xu_ [1998]. Such methods are not investigated here due to the desire to solve the full problem as posed by developers of lunar gravity field models, and therefore avoid spectral truncation of singular vectors that are numerically close to the null-space of the matrix. In other words, the posing of the problem in terms of a fixed-size spherical harmonic expansion is not altered. Furthermore, if such a truncation were to be performed, one is confronted with the problem of deciding which components to truncate, as, obviously, in ill-posed problems, on the contrary to truly rank-deficient problems, the singular values are slowly decaying towards zero. In other words, there is no clear boundary between "interesting" or "useful" spectral components, and "uninteresting" ones, eligible for truncation.

Over the past decade, iterative methods have also become popular in gravity field recovery applications [_e.g., Schuh et al._, 1996; _Schuh_, 1996; _ESA_, 2000; _Van Geemert_, 2000; _Van Geemert et al._, 2000], in particular in connection with the upcoming dedicated spaceborne gradiometry experiment for the determination of the high-resolution and high-accuracy geopotential field [_e.g., ESA_, 1996, 1999; _Balmino et al._, 1998]. Due to the shear size of the inverse problem, involving the determination of a spherical harmonic model up to at least degree and order 200, direct methods become less attractive, and methods like (preconditioned) conjugate gradients (PCCG) [_Shewchuk_, 1994; _Golub and Van Loan_, 1996; _Hansen_, 1998a] are preferred. However, such methods are still to find their way into planetary gravity field modelling, where the size of the expansions thus far is manageable in a direct regularisation scheme, and shall therefore not be considered in the present context. Nevertheless, it deserves mentioning that the problem size is not the only criterium for choosing PCCG methods. Perhaps equally important is the decay rate of the singular values, where a poor conditioning of the normal matrix might make such methods unattractive, since it hampers the optimal extraction of the spectral components. So far, however, no efforts have been made to solve the lunar gravimetric problem using PCCG, and it would be premature to make in-depth predictions about their possible role in future lunar gravimetric research.

It is also possible to consider a third class of regularisation methods, namely those that lead to non-linear optimisation problems, but also such methods are outside the scope of the present work. For a discussion on the role of non-linearity in inverse problems the reader is referred to _Snieder_ [1998].

4.3.1 Tikhonov-Phillips regularisation

The most famous member of the class of direct methods is undoubtedly the Tikhonov-Phillips regularisation method. It was first proposed in a special form by _Phillips_ [1962], and later formulated in a more general framework by _Tikhonov_ [1963a, 1963b]. Their key idea was to incorporate _a priori_ information about the solution (signal) amplitude and/or smoothness. In the continuous case the Tikhonov-regularised solution is the solution obtained by the minimisation of

$$J_\alpha(f) = \|Af - g^\varepsilon\|_G^2 + \alpha \|Lf\|_F^2 \tag{4.22}$$

where L is some operator applied to the solution f. Frequently, but not always, it is a differential operator, or a series of successive differentials. The regularisation parameter α controls the weight given to the minimisation of the regularisation terms, relative to the residual norm. In the discrete case, the function to minimise reads, in the standard L_2 norm,

$$J_\alpha(\mathbf{x}) = \|\mathbf{Ax} - \mathbf{y}^\varepsilon\|_2^2 + \alpha\,\|\mathbf{Lx}\|_2^2 \tag{4.23}$$

and $\mathbf{L}$ is now typically either the identity matrix $\mathbf{I}_n$, a diagonal weighting matrix, or some discrete approximation of a derivative operator. If an *a priori* estimate $\mathbf{x}_{ap}$ of the regularised solution is also available, this may be incorporated by replacing the second term in (4.23) by $\alpha\|\mathbf{L}\left(\mathbf{x} - \mathbf{x}_{ap}\right)\|_2^2$. Consequently, the regularised solution will be biased towards the *a priori* estimate, for gravity field applications frequently zero (for other applications not necessarily so) .

A regularisation problem is said to be of *standard form* if the matrix $\mathbf{L}$ is the identity matrix. From a numerical point of view standard form problems are simpler to analyse than the general form, mainly because they involve only one matrix and therefore make superfluous all matrix operations involving the full pair $(\mathbf{A}, \mathbf{L})$, *e.g.* the explicit GSVD. *Eldén* [1982] proposed and gave the algorithm to transform (4.23) into

$$\bar{J}_\alpha(\mathbf{x}) = \|\bar{\mathbf{A}}\bar{\mathbf{x}} - \bar{\mathbf{y}}^\varepsilon\|_2^2 + \alpha\,\|\bar{\mathbf{x}}\|_2^2 \tag{4.24}$$

which for the simple case that $\mathbf{L}$ is square and invertible (like it *e.g.* is for regularisation based on a Kaula rule) simply reads $\bar{\mathbf{A}} = \mathbf{AL}^{-1}$, $\bar{\mathbf{y}}^\varepsilon = \mathbf{y}^\varepsilon$, and the backward transformation becomes $\mathbf{x}_\alpha = \mathbf{L}^{-1}\bar{\mathbf{x}}_\alpha$. For $\mathbf{L} \neq \mathbf{I}_n$ the algorithm is based on the GSDV and is given by *Eldén* [1982] and *Hansen* [1998a].

Alternative formulations to (4.23) are

$$\left(\mathbf{A}^T\mathbf{A} + \alpha\mathbf{L}^T\mathbf{L}\right)\mathbf{x} = \mathbf{A}^T\mathbf{y}^\varepsilon \quad \text{and} \quad \min_{\mathbf{x}}\left\|\begin{pmatrix}\mathbf{A}\\\sqrt{\alpha}\mathbf{L}\end{pmatrix}\mathbf{x} - \begin{pmatrix}\mathbf{y}^\varepsilon\\\mathbf{0}\end{pmatrix}\right\|_2 \tag{4.25}$$

Hence, if $\mathcal{N}(\mathbf{A}) \cap \mathcal{N}(\mathbf{L}) = \{0\}$, *i.e.* if the matrix $\mathbf{A}$ has full rank, $\mathrm{rank}(\mathbf{A}) = n$, such that the null spaces of $\mathbf{A}$ and $\mathbf{L}$ intersect trivially, the Tikhonov solution $\hat{\mathbf{x}}_\alpha$ is unique and satisfies

$$\hat{\mathbf{x}}_\alpha = \left(\mathbf{A}^T\mathbf{A} + \alpha\mathbf{L}^T\mathbf{L}\right)^{-1}\mathbf{A}^T\mathbf{y}^\varepsilon \tag{4.26}$$

By inserting the (G)SVD into the above equation it is straightforward to show that the filter factors for Tikhonov regularisation in standard and generalised form are given by [*Hansen*, 1998a]:

$$\delta_i = \frac{\sigma_i^2}{\sigma_i^2 + \alpha}, \; \mathbf{L} = \mathbf{I}_n \quad \text{and} \quad \delta_i = \frac{\gamma_i^2}{\gamma_i^2 + \alpha}, \; \mathbf{L} \neq \mathbf{I}_n \tag{4.27}$$

where γ_i are the generalised singular values of the matrix pair $(\mathbf{A}, \mathbf{L})$, see Appendix B.

The solution to (4.23) could be found by forming the normal matrix $\mathbf{A}^T\mathbf{A} + \alpha\mathbf{L}^T\mathbf{L}$ and compute its Cholesky factorisation. This is, *e.g.*, the approach in the

GEODYN II software [*Eddy et al.*, 1990; *Pavlis et al.*, 1998], which has been used to develop the GLGM–2 model. It is emphasised that, in principal, this approach should preferably be avoided, since forming $\mathbf{A}^T\mathbf{A}$ and its inverse may lead to loss of information in finite-precision arithmetic. An example of such behaviour is the previously discussed computation of the condition number of LP75G. Also, it is not ideal for studies of various regularisation parameters α, which would require a new Cholesky factorisation for each investigated value of α. The advantage of normal equations, however, is that for $m \gg n$ it involves about half the arithmetic compared to other methods, as well as less storage requirements. One alternative is to compute QR-factorisations, *e.g.* by means of *Householder transformations* or *Givens rotations* [*Golub and Van Loan*, 1996; *Montenbruck and Gill*, 2000]. For less than moderate residual signatures, such methods are known to perform better than normal equations, and, moreover, the regularisation may be seen as additional observations, and hence require a limited number of additional orthogonal transformations. The most efficient and numerically stable way to compute the solution to the Tikhonov problem is the bi-diagonalisation algorithm due to *Eldén* [1977]. Such a method is, however, not applied in present-day gravitational potential solutions, and is not further elaborated here.

The least squares collocation and Bayesian interpretations

The least squares estimator (4.26) is a maximum likelihood estimator under the assumption that the probability density functions of the model and the observations are Gaussian (normal) [*Bierman*, 1977; *Press et al.*, 1992; *Teunissen*, 2000]. In the Tikhonov framework it may be seen as a maximum likelihood estimator in the presence of norm and smoothness constraints. There are nevertheless several ways to interpret the Tikhonov solution. *Moritz* [1989] interpreted the regularised solution as a kind of least squares collocation, and it may also be seen as a Bayesian type of estimator [*e.g., Backus*, 1988]. In the next section another view on the estimation process shall be advocated, namely that of a biased estimator. This may seem like a trivial consideration, but, as will become clear, in fact it may have a profound influence on both the estimation strategy and, hence, the solution, as well as on the quality measures of the end result.

In least squares collocation, prior information may be available in the form of the first and second moments of the estimation parameters, *i.e.* the prior expectation and dispersion of the solution,

$$E\{\mathbf{x}\} = \boldsymbol{\mu}_x \quad \text{and} \quad D(\mathbf{x}) = \mathbf{Q}_x$$

where $\boldsymbol{\mu}_x$ is the expected value of $\mathbf{x}$ and $\mathbf{Q}_x$ its *a priori* variance-covariance matrix. A weighted least squares collocation solution is obtained under the optimality criterion [*Rummel et al.*, 1979; *Rummel*, 1997; *Xu*, 1992b]

$$(\mathbf{Ax} - \mathbf{y})^T\mathbf{Q}_y^{-1}(\mathbf{Ax} - \mathbf{y}) + (\mathbf{x} - \boldsymbol{\mu}_x)^T\mathbf{Q}_x^{-1}(\mathbf{x} - \boldsymbol{\mu}_x) = \min \qquad (4.28)$$

where $\mathbf{Q}_y$ is the *a priori* covariance matrix of the observations. For uncorrelated measurements, $\mathbf{Q}_y$ is simply a diagonal inverse weight matrix. Although this is a frequent assumption, it is of no concern for the present development. Hence, the solution reads

$$\hat{x}_\alpha^{col} = \left(A^T Q_y^{-1} A + Q_x^{-1}\right)^{-1} \left(A^T Q_y^{-1} y + Q_x^{-1} \mu_x\right) \qquad (4.29)$$

Using $E\{\varepsilon\} = 0$ and, hence, $E\{y^\varepsilon\} = E\{y + \varepsilon\} = E\{y\} = Ax$, where x is the straightforward and unregularised least squares solution, this may be rewritten as

$$\hat{x}_\alpha^{col} = \left(A^T Q_y^{-1} A + Q_x^{-1}\right)^{-1} \left(A^T Q_y^{-1} Ax + Q_x^{-1} \mu_x\right) \qquad (4.30)$$

$$= \left[\left(A^T Q_y^{-1} A + Q_x^{-1}\right)^{-1} \left(A^T Q_y^{-1} A\right)\right] x + \left[\left(A^T Q_y^{-1} A + Q_x^{-1}\right)^{-1} Q_x^{-1}\right] \mu_x$$

from which it can be seen that the collocation solution is a weighted mean of the least squares estimate x and the *a priori* expectation μ_x.

Moritz [1989] proves that the least squares collocation solution yields the best approximation (minimum variance) of any linear functional of the gravitational potential, given a set of observations being linear functionals of the unknowns, provided the original model and a priori information are correct. This is immediately also the weakness of the method, because the label "best" solely depends on the quality of the *a priori* information. In other words, in lunar gravity field determination, where the verification of empirical degree variations of the selenopotential at present is impossible, there is a good chance the *a priori* information will be incorrect, and, hence, the estimator is actually biased, and the collocation framework will consequently fail to produce the optimal solution and associated error description.

A similar line of arguments applies to the interpretation of Tikhonov regularisation as a variant of Bayesian estimation, where, given a prior probability density function of the model and the observations, the goal is to maximise the *a posteriori* probability of the model [*e.g., Berger*, 1985]. Bayesian inference is different from the least squares collocation approach if the underlying probability density function is non-Gaussian. However, since the framework of least squares solutions is so profoundly connected to maximum likelihood under Gaussian assumptions, the Bayesian interpretation leads to the same solution and *a posteriori* error as the collocation approach.

In gravity field recovery, as in most other geosciences, collocation or, alternatively, Bayesian inference arguments have a near-exclusive position, and even more so for so-called satellite-only models, which are stabilised by some empirical spectra of the model, *e.g.* a Kaula constraint, [*e.g., Schwintzer et al.,* 1992, 1997; *Nerem et al.,* 1994; *Tapley et al.,* 1996; *Lemoine et al.,* 1998]. Not surprisingly, lunar and planetary gravity field solutions derived over the past decade are developed along the same lines, inheriting the views from the terrestrial gravity field modelling [*e.g., Smith et al.,* 1993, 1999b; *Lemoine et al.,* 1997; *Konopliv and Sjogren*, 1996; *Konopliv et al.,* 1998, 1999, 2001]. Some exceptions do exist, see *e.g. Tscherning* [2001].

However, in the following, it is argued that the collocation and Bayesian views exhibit some severe difficulties, which actually invites a different perspective of the regularised least squares estimator.

4.3.2 Biased estimation

One such alternative is offered by so-called *biased estimation* [*Xu*, 1992*a*, 1992*b*; *Xu and Rummel*, 1994*b*, 1994*a*; *Rummel*, 1997], in the statistical literature also known as *ridge regression* [*e.g.*, *Hoerl and Kennard*, 1970*a*, 1970*b*; *Vinod and Ullah*, 1981; *Golub and Van Loan*, 1996]. While the basic framework for Tikhonov regularisation is the integral equation of the first kind and its transformation into a nearby well-posed (second kind) problem by including prior information, the core idea of biased estimation is to add some (semi-)positive definite matrix to the system of normal equations for the sole purpose of increasing the numerical stability of the system, at the expense of some bias. It is called the *frequentist* approach to the problem, where the purpose is to derive as much physical information as possible from the data [*Xu and Rummel*, 1994*b*]. The (semi-)positive definite matrix to add to the normal equations is frequently chosen such that the total error, consisting of a data error term and the bias introduced by the regularisation, is minimal, given some minimisation criterion. Hence, the biased estimation solution reads

$$\hat{x}_\alpha^{\text{be}} = \left(A^T Q_y^{-1} A + \alpha K\right)^{-1} A^T Q_y^{-1} y \tag{4.31}$$

where K is now the regularisation matrix and α the regularisation parameter. Obviously, it is possible to extend the regression scheme, by introducing more regularisation parameters, α_j, $j = \{1, 2, \ldots, j_{\text{max}}\}$. Although this generally will lead to smaller errors [*Xu and Rummel*, 1994*b*], this approach shall not be pursued in this work, for the very reason that in lunar gravimetric applications so far K is some Kaula rule scaled to the Moon with the factor α, and, hence, the regularisation scheme is rooted in lunar physics. An example of several scaling parameters, each describing separate parts of the degree scale is LP75G, cf. Table 4.1, where the tighter constraints for the upper degree domains are a straightforward result of the decrease of the data information content with l. The choice of such a multiple set of parameters is purely empirical, based on orbit fit optimisation, as there is no direct proof for a physical change in the scaling factor α with l. Furthermore, the use of multiple regularisation parameters in a generalised bias estimation scheme is known to, in some case, induce larger errors in the long-wavelength part of the gravity field [*Xu and Rummel*, 1994*a*]; for an example in satellite gradiometry see *Bouman and Koop* [1998]. Probably the first application of ridge-type estimators for planetary gravity fields is found in *Minter and Cicci* [1998], where it is shown that viewing the estimation process as biased leads to quality improvement of an existing Mars model.

One fundamental cost of regularisation is that the regularised solution is actually the solution of a nearby and related problem. It is therefore obvious that regularisation will introduce a regularisation error, in addition to the error sources already contained in the right-hand side. The difference between the regularised solution $\hat{f}_\alpha^\varepsilon$ based on real, and hence error contaminated data, and the exact (and unknown) solution f derived from exact data is [*Bouman*, 1998, 2000]

$$\hat{f}_\alpha^\varepsilon - f = A_\alpha^\dagger (g^\varepsilon - g) + \left(A_\alpha^\dagger - A^\dagger \right) g \qquad (4.32)$$

where the first term on the right-hand side is the *data error* and the second term is called the *regularisation error* [*Louis*, 1989], the solution *bias* [*Xu*, 1992a, 1992b] or *perturbation error* [*Hansen*, 1998a].

In the finite-dimensional case it holds for the total error

$$\hat{x}_\alpha^{be} - x = A_\alpha^\dagger (y^\varepsilon - y) + \left(A_\alpha^\dagger - A^\dagger \right) y \qquad (4.33)$$

$$= \left(A^T Q_y^{-1} A + \alpha K \right)^{-1} A^T Q_y^{-1} \varepsilon$$

$$+ \left[\left(A^T Q_y^{-1} A + \alpha K \right)^{-1} - \left(A^T Q_y^{-1} A \right)^{-1} \right] A^T Q_y^{-1} y$$

which has a non-zero expectation even in the absence of data errors. That is, for $E\{\varepsilon\} = 0$, the bias yields

$$E\{\hat{x}_\alpha^{be} - x\} = 0 + \left[\left(A^T Q_y^{-1} A + \alpha K \right)^{-1} - \left(A^T Q_y^{-1} A \right)^{-1} \right] A^T Q_y^{-1} A x \qquad (4.34)$$

$$= \left(A^T Q_y^{-1} A + \alpha K \right)^{-1} \left(A^T Q_y^{-1} A + \alpha K - \alpha K \right) x - Ix$$

$$= - \left(A^T Q_y^{-1} A + \alpha K \right)^{-1} \alpha K x$$

$$\neq 0$$

Obviously, such considerations also apply to Tikhonov regularisation, but the fact remains that the estimator in that case is frequently viewed as unbiased, in the collocation or Bayesian estimation.

As the form of the solution (4.31) is identical to the Tikhonov solution, it of course yields the same filter factors as long as the regularisation matrices are the same, cf. (4.27). However, the selection criteria for the regularisation parameter might be different, and therefore result in a different gravity field solution. Moreover, in biased estimation the bias will be accounted for in the error measures.

4.3.3 Error measures and error propagation

In terms of the propagated error the collocation and biased estimation views on the regularised inverse problem behave differently. Consider first the collocation solution (4.29) and its propagated error $\hat{Q}_x^{col}$, or dispersion $\hat{D}\{\hat{x}_\alpha^{col}\}$. If and only if

$E\{\mu_x\} = 0$ and $D\{\mu_x\} = \mathbf{Q}_x$ will error propagation yield the well-known covariance matrix of the unbiased collocation estimation error:

$$\hat{\mathbf{Q}}_x^{col} = \hat{D}\{\hat{\mathbf{x}}_\alpha^{col}\} = E\left\{\left(\hat{\mathbf{x}}_\alpha^{col} - \mathbf{x}\right)\left(\hat{\mathbf{x}}_\alpha^{col} - \mathbf{x}\right)^T\right\} \tag{4.35}$$

$$= \left(\mathbf{A}^T\mathbf{Q}_y^{-1}\mathbf{A} + \mathbf{Q}_x^{-1}\right)^{-1}$$

Likewise, for an expected prior value of $\mathbf{x}$ equal to zero, *i.e.* $\mu_x = 0$ the *unbiased* collocation solution is

$$\hat{\mathbf{x}}_\alpha^{col} = \left(\mathbf{A}^T\mathbf{Q}_y^{-1}\mathbf{A} + \mathbf{Q}_x^{-1}\right)^{-1}\mathbf{A}^T\mathbf{Q}_y^{-1}\mathbf{y}^\varepsilon \tag{4.36}$$

An essential ingredient in error propagation in the collocation framework is therefore the treatment of the regularisation as *zero observations* $\mathbf{z}$ with a dispersion matrix $\mathbf{Q}_x$. That is, the required linear model must read

$$E\left\{\begin{bmatrix}\mathbf{y}^\varepsilon\\\mathbf{z}\end{bmatrix}\right\} = \begin{bmatrix}\mathbf{A}\\\mathbf{I}\end{bmatrix}\mathbf{x}, \quad D\left\{\begin{bmatrix}\mathbf{y}^\varepsilon\\\mathbf{z}\end{bmatrix}\right\} = \begin{bmatrix}\mathbf{Q}_y & 0\\0 & \mathbf{Q}_x\end{bmatrix} \tag{4.37}$$

which yields the unbiased solution

$$\hat{\mathbf{x}}_\alpha^{col} = \left(\mathbf{A}^T\mathbf{Q}_y^{-1}\mathbf{A} + \mathbf{I}\mathbf{Q}_x^{-1}\mathbf{I}\right)^{-1}\left(\mathbf{A}^T\mathbf{Q}_y^{-1}\mathbf{y}^\varepsilon + \mathbf{I}\mathbf{Q}_x^{-1}\mathbf{z}\right) \tag{4.38}$$

and, hence, by error propagation, (4.35). For more information on the interpretation of prior information as zero observations, the reader is referred to *Bierman* [1977] and *Björk* [1996].

In the view of the author, this interpretation is problematic for gravity field determination applications, for the following reasons, cf. [*Xu*, 1992b; *Xu and Rummel*, 1994b, 1994a; *Rummel*, 1997]. In satellite-derived gravity field modelling, regularisation is performed using some sort of Kaula rule, or a derivative thereof. In the collocation interpretation, the *a priori* values of the selenopotential coefficients are considered to be zero, while their variances are described by the square of Kaula's rule, given in Table 4.1. Also, there are no *a priori* correlations, but this is irrelevant for the present discussion. In other words, $\mathbf{Q}_x^{-1} = \alpha\mathbf{K}$, with $\mathbf{K}$ the diagonal matrix specifying Kaula's rule, and α the regularisation parameter (the regularisation matrix for biased estimation was therefore by no means arbitrarily selected). However, Kaula's rule of thumb actually describes the degree-wise variations of the gravity field coefficients, based on empirical analysis of gravitational potential models [*Kaula*, 1966], and certainly not their variance, cf. Fig. 3.1. In other words, the prior information contained in the Kaula rule is applied incorrectly. The companion assumption that the a priori value of the coefficients (or in the case of the linearised problem the corrections to the selenopotential coefficients) equals zero is equally difficult to accept. The sole purpose of data collection is to reveal more physical information about the selenopotential model. The individual model coefficients (and their corrections) are certainly different from zero, otherwise it makes little sense to send spacecraft into space at high cost to collect the tracking data. In other words, relying on zero *a priori* values of the unknowns appears difficult to tolerate.

One interesting aspect of these considerations is that they apply to the solution of any linear combination of spherical harmonic coefficients, *e.g.* also in the space domain in terms of gravity anomalies or selenoid undulations, provided the estimation is seen as a deterministic process. In this case, the collection of satellite (or other) data is seen as one sampling in the space domain of a given linear combination of the spherical harmonic of the gravity field. The estimation process is then in principle non-stochastic, as the possible inclusion of more measurements in the sense of a higher measurement density, *i.e.* observations at more locations, basically yields additional information, rather than a second sample of the same quantity of the gravity field. For such an interpretation of the estimation process the assumption of zero expectation is dubious.

Finally, it deserves mentioning that the search for an optimal regularisation parameter α, in this case the scaling parameter for the Kaula rule, actually accentuates the fact that some uncertainty exists concerning the value of the prior information. In other words, iterative approaches to derive a final gravity field model may also face a change in the regularisation as the iterations over the available data proceed. For example, *Konopliv et al.* [1998] use several regularisation parameters to describe the prior information. This in fact implies discontinuities in the degree-wise spectrum, something which is of course unacceptable from a physical point of view. At the very least, this should illustrate the fact that rigid prior assumptions are disputable in planetary gravity field model development. Perhaps the only rigorous way to avoid this problem is to literally interpret the "optimal" regularisation parameter in the sense of the "best" solution quality. Once a clear quality assessment framework has been established, based on the view of the estimation scheme, *i.e.* biased or unbiased, pragmatic algorithms may be put to play to directly search for the optimal condition. This is the topic of Sect. 4.5.1

Furthermore, it is clear that error propagation results, *e.g.* in terms of predicted orbit errors (see Chap. 3), or for any other linear functional of the estimate, largely depends on the understanding and choice of the estimator. In a situation where little may be confirmed by direct verification using independent data sources, this is of course not an ideal situation, and calls for clarity when presenting both the solution itself as well as associated error estimates.

The interpretation of the regularised inverse problem as a biased estimation problem suffers less from the above objections, and therefore appears preferable to the collocation framework. In this context, the error measure is based on the concept of the *mean square error matrix* (*MSEM*) of the biased estimate, which contains both the propagated measurement error and the square of the bias,

$$MSEM = E\left\{\left(\hat{\mathbf{x}}_\alpha^{be} - \mathbf{x}\right)\left(\hat{\mathbf{x}}_\alpha^{be} - \mathbf{x}\right)^T\right\} \tag{4.39}$$

$$= E\left\{\left(\hat{\mathbf{x}}_\alpha^{be} - E\{\hat{\mathbf{x}}_\alpha^{be}\}\right)\left(\hat{\mathbf{x}}_\alpha^{be} - E\{\hat{\mathbf{x}}_\alpha^{be}\}\right)^T\right\} + E\left\{\hat{\mathbf{x}}_\alpha^{be} - \mathbf{x}\right\} E\left\{\hat{\mathbf{x}}_\alpha^{be} - \mathbf{x}\right\}^T$$

$$= \hat{\mathbf{Q}}_x^{be} + \Delta\mathbf{x}\Delta\mathbf{x}^T$$

$$= \left(\mathbf{A}^T\mathbf{Q}_y^{-1}\mathbf{A} + \alpha\mathbf{K}\right)^{-1}\mathbf{A}^T\mathbf{Q}_y^{-1}\mathbf{A}\left(\mathbf{A}^T\mathbf{Q}_y^{-1}\mathbf{A} + \alpha\mathbf{K}\right)^{-1}$$

$$+ \left(\mathbf{A}^T\mathbf{Q}_y^{-1} + \alpha\mathbf{K}\right)^{-1}\alpha^2\mathbf{K}\mathbf{x}\mathbf{x}^T\mathbf{K}\left(\mathbf{A}^T\mathbf{Q}_y^{-1}\mathbf{A} + \alpha\mathbf{K}\right)^{-1}$$

The first term on the right-hand side of (4.39) is the propagated error for Tikhonov regularisation, similar to (4.35). It represents an accuracy gain due from the use of biased estimation, as obviously

$$\left\|\left(\mathbf{A}^T\mathbf{Q}_y^{-1}\mathbf{A} + \alpha\mathbf{K}\right)^{-1}\right\|_2 \leqslant \left\|\left(\mathbf{A}^T\mathbf{Q}_y^{-1}\mathbf{A}\right)^{-1}\right\|_2 \tag{4.40}$$

The second term reflects the accuracy loss due to the bias. In addition, given (4.40), it holds that [*Xu*, 1992*b*]

$$\left\|E\{\hat{\mathbf{x}}_\alpha^{be}\}\right\|_2 < \|\mathbf{x}\|_2 \tag{4.41}$$

In other words, the interpretation of regularisation as a biased estimation causes the solution to have less power than the "exact" solution, which one will never know. This is a fact often observed in gravitational potential modelling [*e.g.*, *Nerem et al.*, 1993, 1994], and also in the case of GLGM–2 for the Moon [*Lemoine et al.*, 1997]. Such behaviour can obviously not be explained by an unbiased estimator from a Bayesian or collocation point of view [*Xu*, 1992*b*]. Notice that the difference between the mean square error matrix and the error variance-covariance matrix may also be expressed as

$$\Delta MSEM = MSEM - \hat{\mathbf{Q}}_x^{col} \tag{4.42}$$

$$= -\left(\mathbf{A}^T\mathbf{Q}_y^{-1}\mathbf{A} + \alpha\mathbf{K}\right)^{-1}\alpha\mathbf{K}\left(\mathbf{A}^T\mathbf{Q}_y^{-1}\mathbf{A} + \alpha\mathbf{K}\right)^{-1}$$

$$+ \left(\mathbf{A}^T\mathbf{Q}_y^{-1}\mathbf{A} + \alpha\mathbf{K}\right)^{-1}\alpha^2\mathbf{K}\mathbf{x}\mathbf{x}^T\mathbf{K}\left(\mathbf{A}^T\mathbf{Q}_y^{-1}\mathbf{A} + \alpha\mathbf{K}\right)^{-1}$$

which again emphasises the opposite effects of the accuracy gain and the bias caused by the regularisation.

A second important conclusion is that if the actual coefficients in a solution decrease faster (in a degree-wise spectrum) to zero than the Kaula rule applied in the solution, then the *a posteriori* variance-covariance matrix $\left(\mathbf{A}^T\mathbf{Q}_y^{-1}\mathbf{A} + \mathbf{Q}_x^{-1}\right)^{-1}$ or $\left(\mathbf{A}^T\mathbf{Q}_y^{-1}\mathbf{A} + \alpha\mathbf{K}\right)^{-1}$ is a conservative error measure. On the other hand, if the solution has more power than the regularisation, the actual error is underestimated by the covariance measure, cf. *Rummel* [1997]. From this point of view, and also supported by Fig. 3.1, the GLGM–2 covariance appears to be a safer estimate of the error than the covariance LP75G. In other words, independent of the orbit error information discussed in the previous chapter, or matrix calibration information, one is able to conclude that the LP75G error estimate is optimistic, while those of the later LP100n series appear more consistent with the corresponding solutions. Undeniably, any such deductions depend of the rightfulness of the Kaula rule used as prior information in the estimation process.

One characteristic of biased estimation is that the regularisation parameter is chosen according to some optimality criterion. Frequently, such criteria are based on the minimisation of the expected difference between $\hat{x}_\alpha^{be}$ and x, *i.e.* min $E\{\|\hat{x}_\alpha - x\|_2\}$. *Hoerl and Kennard* [1970a] therefore define the scalar *mean square error* (*MSE*)

$$MSE = E\left\{\left(\hat{x}_\alpha^{be} - x\right)^T \left(\hat{x}_\alpha^{be} - x\right)\right\} \qquad (4.43)$$
$$= \text{trace}\left[MSEM\right]$$
$$= \text{trace}\left[\hat{Q}_x^{be} + \Delta x \Delta x^T\right]$$
$$= \text{trace}\left[\hat{Q}_x^{be}\right] + \Delta x^T \Delta x$$

as a measure of the deviation between the biased solution and the exact solution. *Bouman* [1998] gives the *MSE* for a range of regularisation methods used in satellite geodesy in terms of the SVD and filter factors, for the case of a standard form problem and diagonal Q_y^{-1}.

4.4   Information content and effect of the bias on the quality description

The preceding section illustrates how the view of the regularised least squares estimate influences the understanding of the selenopotential coefficient errors, and argues that the idealised assumptions of least squares collocation or unbiased Bayesian estimation are unsuited for gravity field applications. Consequently, the estimator is better regarded as biased, which invites the use of an error estimate that includes both the propagated measurement error and a bias term. In turn, using the mean square error as a measure of the estimation error, it was explained that the methods for selecting the optimal regularisation parameter(s) might lead to a different solution, *e.g.* if the mean square error is minimised. This is further elaborated in Sect. 4.5. In other words, a minimum norm collocation solution might not be optimal in view of a biased estimation.

In this section, the information content of lunar gravity field models is investigated in further detail, with particular emphasis on the bias and the practical implications of the view on the estimator. The differences between biased estimation and unbiased collocation is illustrated by means of ratio measures, such as the *bias-to-signal ratio* and *signal-to-noise ratio*, as well as propagated error measures, like the errors in selenoid height.

Two immediate problems arise in bias computations. First, (4.34) shows that the exact solution is required. This a direct penalty stemming from the fact that the regularisation is no longer seen as prior information, but simply a stabilising matrix. The degree of freedom previously found in the choice of the Kaula rule (or other types of prior information) is transferred to the bias. The exact solution **x** is not known (and will never be), and the bias must therefore be approximated from the available information. Second, in gravity field estimation from interplanetary tracking data, the fundamental relationship between the observables and the gravity field coefficients is non-linear, and one is solving for first-order corrections to an a priori model. In other words, the *a priori* gravity field model strongly affects the bias computations: a well-fitting *a priori* model would lead to small corrections, and, hence, a small bias, while a poor *a priori* model would tend to yield larger biases.

The prior model itself, however, is also biased, since it is computed according to similar principles. Therefore, in the present context, two different bias computation procedures are advocated: a *coefficient bias* derived from the gravity field coefficients themselves, *i.e.* the prior model is considered to be the central term of the gravitational potential only and the vector **x** equals the solutions from degree 2 upward; and, a *synthetic bias* computed from a vector **x** given by the Kaula rule applied in the respective model development and the sign for each coefficient chosen according to the sign of the corresponding model coefficient. The latter method of synthetic biases is in many cases a more realistic measure, since the model coefficients themselves are biased towards zero, and, hence, suffer from the same bias one tries to estimate, and the coefficient bias therefore underestimates the true bias effect. This fact is illustrated for GLGM–2 in Table 4.2. The same table shows that, also in this respect, LP75G behaves differently from GLGM–2, a fact which has all to do with the degree-wise spectrum of LP75G having a higher power than the corresponding constraints. Conform the previous, the covariance matrix of LP75G may therefore be considered an optimistic error measure.

4.4.1   Ratio measures

The bias-to-signal ratio, which signifies the severity of the bias in each coefficient, is defined as

$$BSR_{lm} = \left| \frac{\Delta K_{lm}}{K_{lm}} \right| \tag{4.44}$$

where BSR_{lm} is the bias-to-signal ratio for a coefficient of degree l and order m, and ΔK_{lm} is the bias in each coefficient K_{lm}. Evidently, a bias-to-signal ratio can also be determined in an RMS sense over subsets of coefficients, *e.g.* per degree, per order or for the entire set of coefficients.

	GLGM–2	LP75G
trace $[\hat{\mathbf{Q}}_x^{col}]$:	8.035×10^{-11}	2.064×10^{-11}
trace $[\hat{\mathbf{Q}}_x^{be}]$:	1.570×10^{-11}	4.177×10^{-12}
MSE with coefficient bias:	2.325×10^{-11}	3.655×10^{-11}
MSE with synthetic bias:	4.755×10^{-11}	1.799×10^{-11}

Table 4.2   Mean square error and trace of the propagated errors $\hat{\mathbf{Q}}_x^{col}$ and $\hat{\mathbf{Q}}_x^{be}$ for GLGM–2 and LP75G. The comparison number for the propagated collocation error and the mean square error is determined by the ratio of the Kaula constraint to the coefficient power. Therefore, assuming that the prior information (constraint relations) is correct, for GLGM–2, which has a significantly lower power than the corresponding constraint, the propagated error is a conservative error measure. For LP75G the coefficients have a slightly higher power than the model, which leaves $\hat{\mathbf{Q}}_x^{col}$ on the optimistic side

Figures 4.2 and 4.3 depict the bias-to-signal ratios for GLGM–2 and LP75G, re-spectively. The left-hand side of the figures displays the coefficient-wise bias of the $\overline{C}_{lm}$ coefficients (the situation for the $\overline{S}_{lm}$ coefficients is nearly identical), while the right-hand sides of the figures show the degree-wise and order-wise biases. In the case of GLGM–2 it is readily seen that the synthetic bias by far exceeds the coefficient bias for all orders and for all degrees beyond approximately 15. In general, this may be seen as another direct proof that the satellite tracking data incorporated in the GLGM–2 solution alone do not contain enough information to determine a full 70×70 spherical harmonic expansion. Using simple coefficient biases, the bias-to-signal ratios tend towards unity for the higher degrees, which simply indicates that the coefficients are nearly fully determined by the constraint. For the order-wise plot the tendency is similar, but there is more of a downward slope. Both curves exhibit intermediate peaks, indicating significant variations of the bias throughout the harmonic spectrum. The overall tendency of, in particular for the higher degree and order terms, an extremely high bias-to-signal ratio is a clear display of the lack of (non-prior) information going into the model develop-ment.

In the case of LP75G a similar amplitude pattern is seen, but nevertheless the point at which the bias becomes dominant over the signal is found at a lower de-gree and order. Altogether, this is caused by the relative behaviour of the slightly looser constraint (Kaula rule) and a generally higher power in the coefficients, cf. Fig. 3.1. Since the coefficient amplitude of the two models are wide apart, a direct

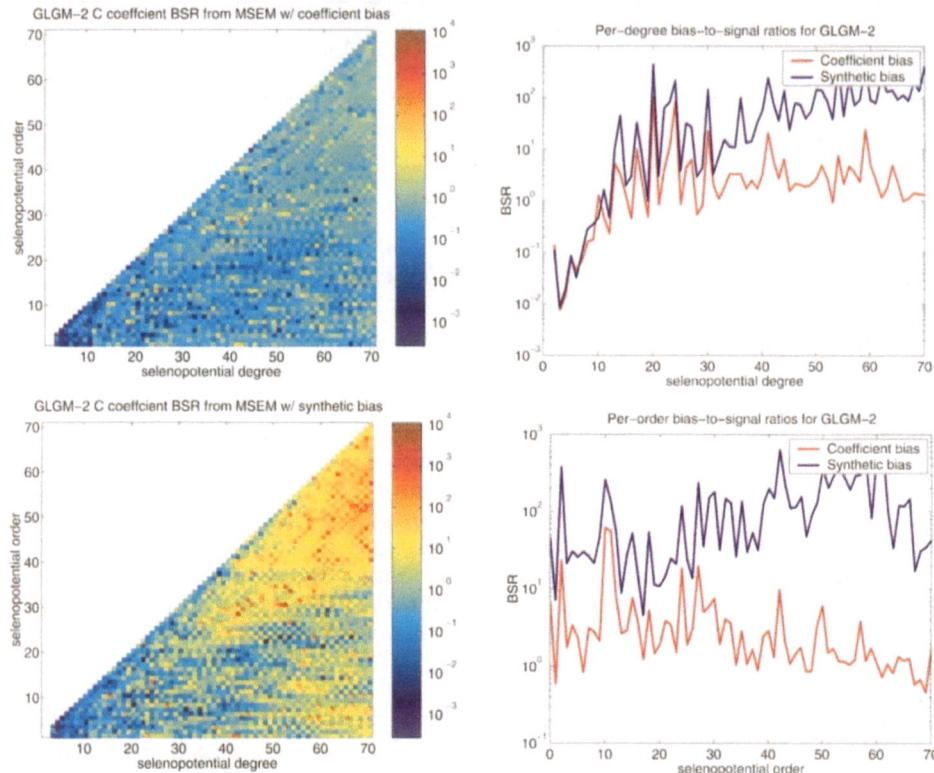

Figure 4.2   Bias-to-signal ratios for GLGM–2 for biases determined from the model
              solution (coefficient bias) as well as on the basis of the Kaula rule applied in
              the development of GLGM–2, with the sign of each coefficient chosen
              according to the sign of the model coefficient

comparison of the two models in terms of bias-to-signal ratios is not straightforward. However, the results for both models confirm that only low degree harmonics are well determined by the data, and that most higher degree and order terms
to a significant level are determined by the level of regularisation. Furthermore, it
is seen that the choice of the bias is of lesser importance for LP75G, based on the
observation that the Kaula constraint and the coefficient power are closer for this
model than for GLGM–2.

It might be rightfully argued that the biases chosen here are pessimistic, since
the entire coefficient values, or Kaula prior values, are used in their determination. In other words, no irregularity in the selenopotential is accommodated in the
a priori gravity field model, since they necessarily are biased themselves. Consequently, the actual biases as experienced in lunar gravimetry might be of smaller
magnitude than the magnitude of the biases presented here. *Xu* [1992*b*] argues
that this particular choice of the bias actually represents some upper bound mag-

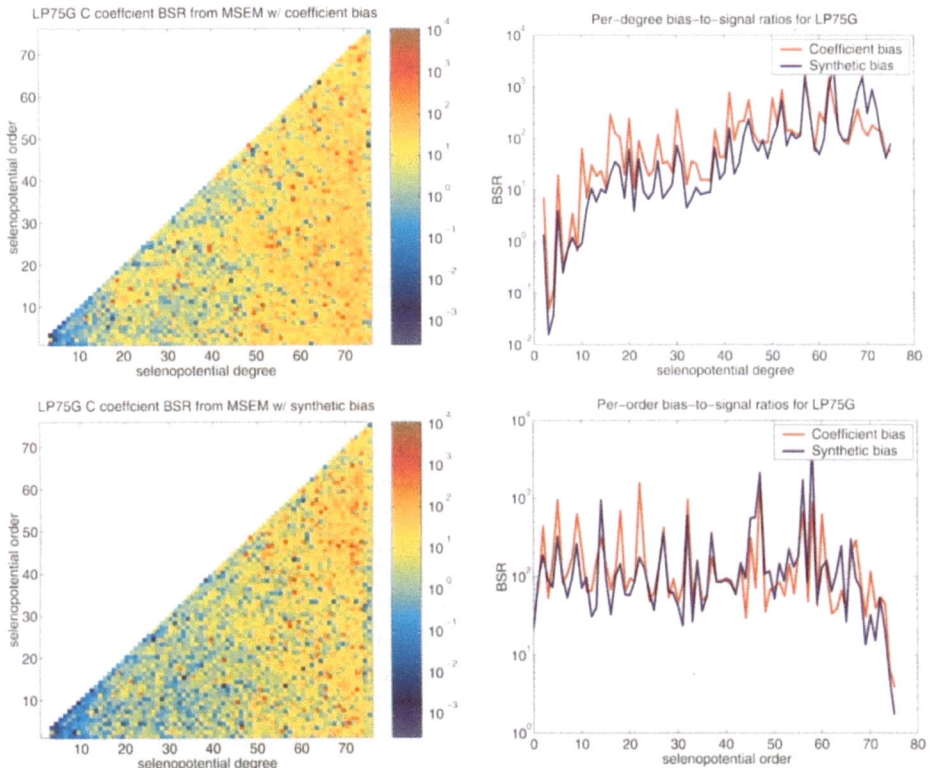

Figure 4.3   Bias-to-signal ratios for LP75G for biases determined from the model solution (coefficient bias) as well as on the basis of the Kaula rule applied in the development of LP75G, with the sign of each coefficient chosen according to the sign of the model coefficient. Notice that for LP75G the bias choice is of less influence on the *BSR* than in the case of GLGM–2

nitude. On the other hand, the large biases are also an indirect measure of the bad condition number of the unregularised normal matrix c.q. square root information filter. As the sampling of orbit perturbations, *e.g.* through SST, becomes better in the future, also a relatively pessimistic choice of the vector **x** will yield more modest bias estimates. In addition, it is considered wise to use conservative biases in order to avoid underestimation of its effect on the error measure.

Signal-to-noise ratio measures, on the other hand, express the significance of the estimated gravity field solution, and exist in several contexts in gravimetric modelling. In the sense of degree variations of the selenopotential signal and the corresponding variance, the *SNR* is obviously defined as the degree-wise ratio of the potential coefficient RMS and the corresponding standard deviations. In its coefficient-wise form the *SNR* is defined as

$$SNR_{lm} = \frac{|K_{lm}|}{\sigma_{lm}} \tag{4.45}$$

where σ_{lm} denotes the uncertainty of the coefficient estimate, either the standard deviation, or the square root of the diagonal elements of the *MSEM*. As for the BSR, both degree-wise and order-wise signal-to-noise ratios are straightforwardly computed by taking the RMS value.

Assuming that the solution is unbiased, the square root of the diagonal elements of $(\mathbf{A}^T\mathbf{Q}_y^{-1}\mathbf{A} + \alpha\mathbf{K})^{-1}$ are the standard deviations of K_{lm}. Their ratio is plotted in the top left parts of Figs. 4.4 and 4.5. It is readily seen that most of the coefficients, in the collocation case, are well below the measurement noise. For the coefficient and synthetic bias cases, respectively, it is found that the synthetic bias yields results similar to the unbiased case, while the coefficient bias case appears to be slightly optimistic. Therefore, it is concluded that using the unbiased assumption does not influence the SNR compared to the unbiased case, as long as the bias estimate is "correct". The true bias is unknown, however, which is a fact one should realise when interpreting any statement concerning the quality. Degree-wise and order-wise it is clear that the higher signal-to-noise ratios are all found at the lower end of the range, the per-degree SNR having a slightly higher amplitude than the per-order measure. These differences are, however, much more apparent for GLGM–2, where obviously the choice of the bias matters more, than for LP75G, where actually all plots and curves largely resemble each other and it is only for the highest degrees and orders that any significant dispersion of the signal-to-noise ratios is found. Finally, it deserves mentioning that the relative peaks found at the very highest degrees and orders using the LP75G covariance matrix all stem from the fact that orbit perturbation information beyond 75 cycles per revolution have aliased into the solution, cf. Fig. 3.1.

If desired, it is also possible to define a bias-to-noise ratio, which evidently measures the relative importance of the bias and the propagated measurement errors $\hat{\mathbf{Q}}_x$. Such plots are, however, not shown here, for the simple fact that the importance of the bias is adequately illustrated by other means, and therefore their additional value is limited.

4.4.2 Contribution measures

Another type of quality assessment tool is contribution measures. The key idea is that if the underlying observational model is correct, the unknowns are in part determined by the observations and in part by the regularisation, or prior information. For unbiased estimators the contribution of the observations to the solution of the unknowns is defined as

$$\text{contr}_y := \hat{\mathbf{Q}}_x\mathbf{Q}_{x,y}^{-1} = (\mathbf{A}^T\mathbf{Q}_y^{-1}\mathbf{A} + \alpha\mathbf{K})^{-1}\mathbf{A}^T\mathbf{Q}_y^{-1}\mathbf{A} \tag{4.46}$$

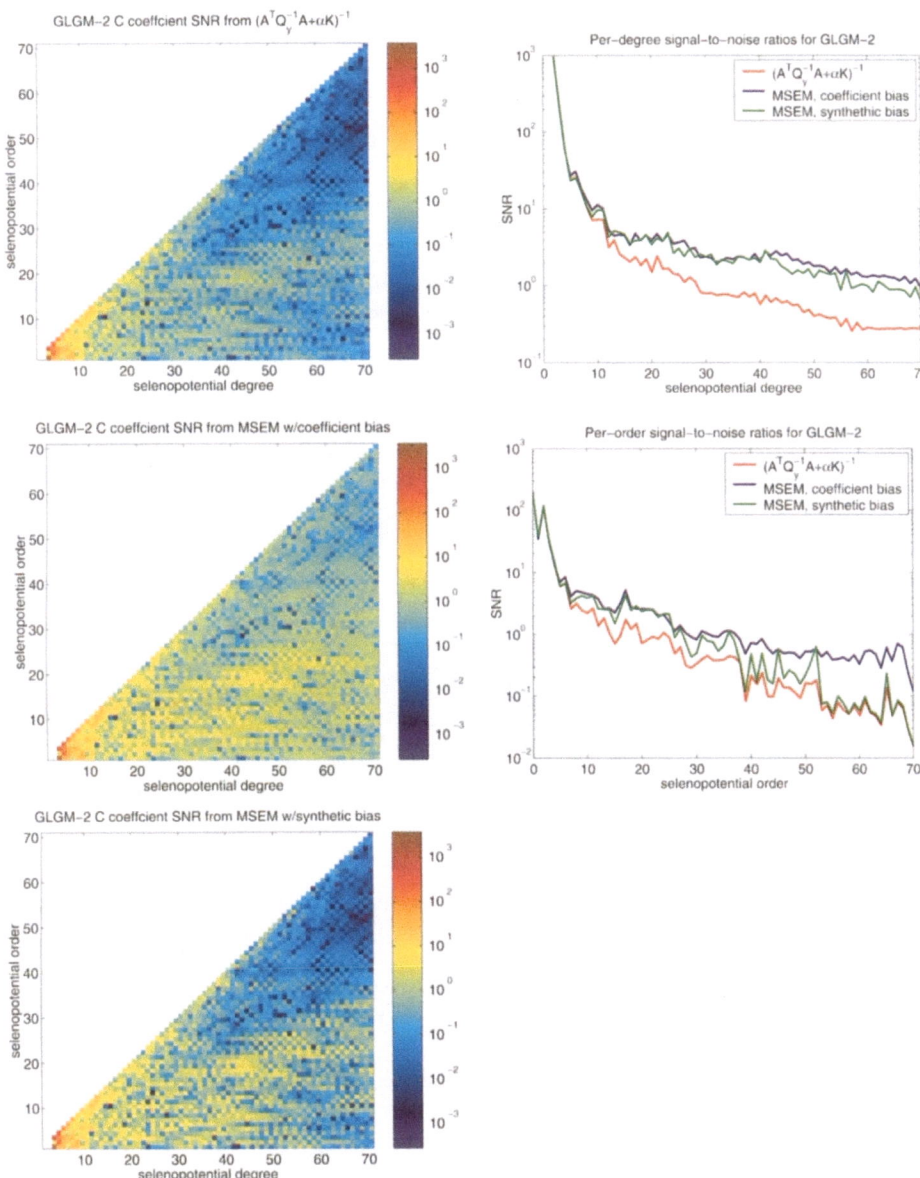

Figure 4.4   Signal-to-noise ratios for GLGM–2 using both the propagated error $\hat{\mathbf{Q}}_x^{col}$ as well as two cases of the mean square error matrix applying respectively a simple coefficient bias and a synthetic bias based on the Kaula rule used in the derivation of GLGM–2

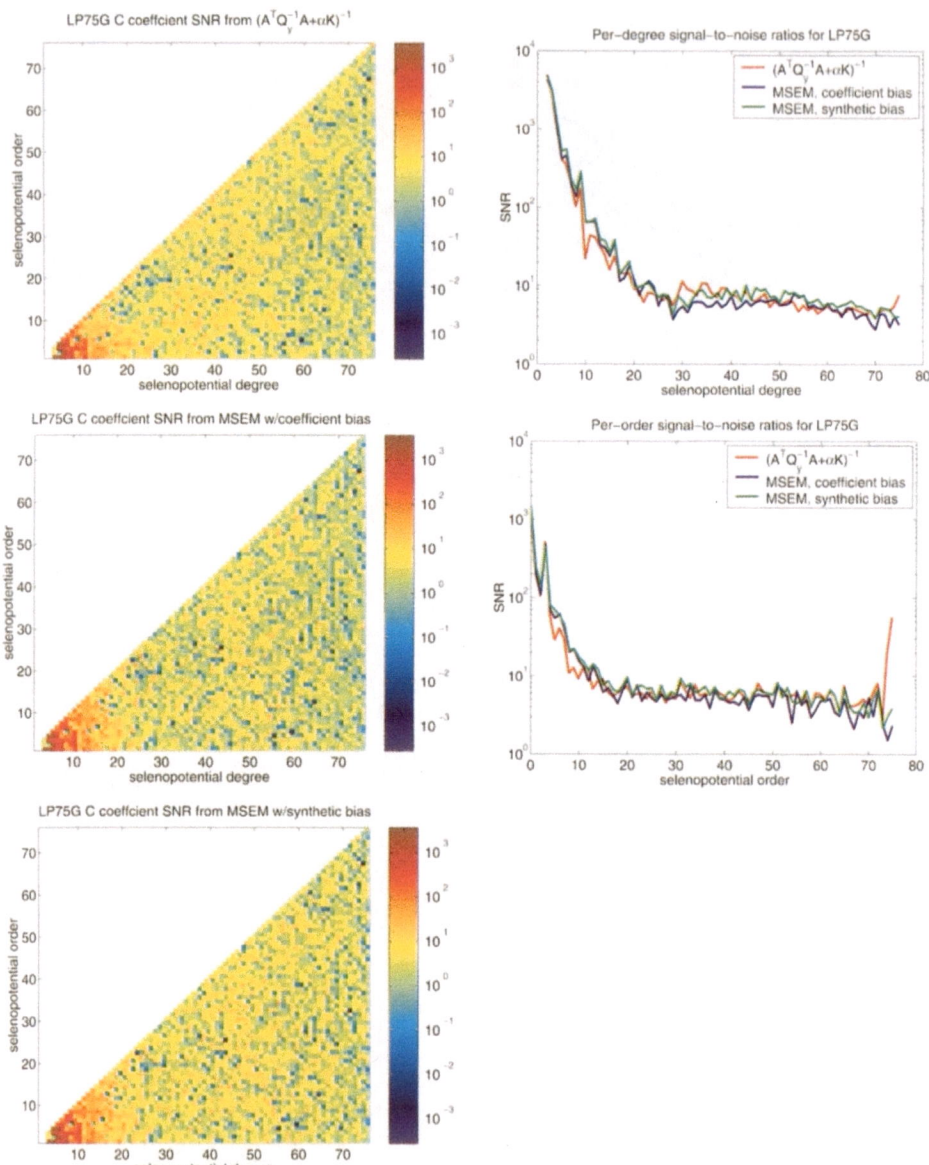

Figure 4.5   Signal-to-noise ratios for LP75G using both the propagated error $\hat{Q}_x^{col}$ as well
               as two cases of the mean square error matrix applying respectively a simple
               coefficient bias and a synthetic bias based on the Kaula rule used in the
               derivation of LP75G

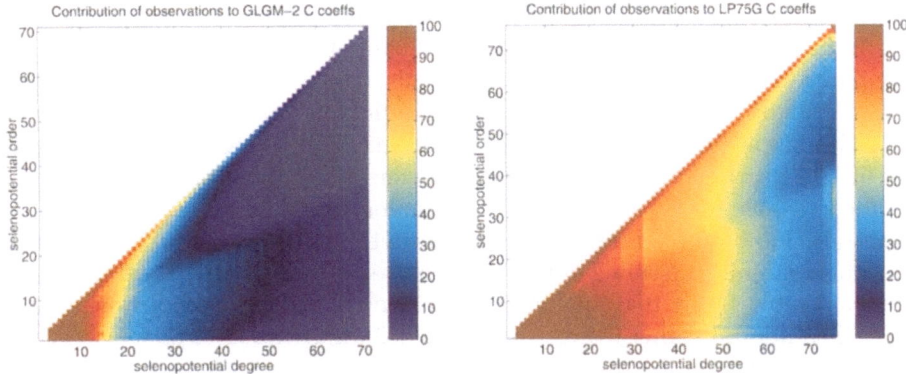

Figure 4.6   Percentage of the contribution of the observations to the individual GLGM–2
and LP75G gravity field coefficients. While LP75G suffers from two
discontinuities in the Kaula rule used in the development of the model as well
as optimistic data weighting schemes, the model still outperforms GLGM–2 in
terms of actual tracking data information. Notice also the indications of a
relatively solid determination of high-order sectorial terms for LP75G

where $\mathbf{Q}_{x,y}^{-1}$ is the unregularised least-squares normal matrix. The diagonal elements of $\text{contr}_y \in \{0, 1\}$. Zero contribution indicates that the observations do not contribute to the determination of the unknown, i.e. all information comes from the constraint, whereas a unit value implies that the observations completely determine that specific unknown. Equation 4.46 equals the gain matrix in Kalman filtering [*Bierman*, 1977]. Such a comparison makes sense, since the larger the weight of the observations relative to the *a priori* information, the larger the gain, and the larger the contribution of the data to the coefficient solution. *Schwintzer* [1990] associates the contribution measure with the *redundancy number* $r = m - n$, which appears in statistical testing of measurements. Testing is, however, not an issue here, and, moreover, the interpretation in terms of the change in covariance is engineering-wise quite intuitive. It is straightforwardly shown that the diagonal elements of (4.46) equal

$$\text{contr}_{y_i} = 1 - [\hat{\mathbf{Q}}_x]_{ii} \cdot [\alpha \mathbf{K}]_{ii}$$

which is a simple and numerically stable computation.

   The contribution of the observations to the GLGM–2 and LP75G solutions is depicted in Fig. 4.6. Again, only $\bar{C}_{lm}$ coefficients are shown, since the situation for the $\bar{S}_{lm}$ coefficients is near-identical. For GLGM–2 it is clear that only the coefficients up to degree and order ~ 15 have an observation contribution larger than 50%. With the exception of the sectorial terms, the contribution quickly drops to low levels, indicating that these coefficients can hardly be recovered from the measurements. Consequently, given the unverified situation on the value of the *a priori* information, it is likely that these coefficients are biased. In a direct comparison with GLGM–2, LP75G shows clear signs of improvement and, moreover, the figure illustrates the benefit of the high-quality Lunar Prospector tracking data compared to the previous data sets. Despite the clear signs of optimistic variances, which evidently falsely improves the contribution measure, there is no doubt that LP75G is outperforming GLGM–2. In particular the low degree and order part of the figures clearly indicate that more parameters in this range are well-determined by the available data, compared to the situation for GLGM–2. Nevertheless, the figure also indicates some of the weaknesses of LP75G, *e.g.* that the change in regularisation parameter (scaling of Kaula's rule) clearly affects the contribution value, compare Table 4.1, or that the LP75G data alone absolutely do not warrant a full 75×75 solution. In other words, also the contribution measures directly reveal the intrinsic weaknesses in present-day models.

The above considerations all apply to unbiased estimators, *e.g.* the collocation interpretation of a regularised least squares solution. However logical, a straightforward extension for biased estimators using the *MSEM* instead of the propagated error, as suggested by *Bouman and Koop* [1998], unfortunately does not provide the required contribution information. Considering the contribution measure

$$\text{contr}_y := MSEM \times MSEM^{-1}\big|_{\alpha=0} \tag{4.47}$$
$$= (\mathbf{A}^T\mathbf{Q}_y^{-1}\mathbf{A} + \alpha\mathbf{K})^{-1}(\mathbf{A}^T\mathbf{Q}_y^{-1}\mathbf{A} + \alpha^2\mathbf{Kxx}^T\mathbf{K})(\mathbf{A}^T\mathbf{Q}_y^{-1}\mathbf{A} + \alpha\mathbf{K})^{-1}\mathbf{A}^T\mathbf{Q}_y^{-1}\mathbf{A}$$

it is obvious that for $\alpha = 0$ the contribution is complete, and one has the same measure as in the unbiased case. However, for $\alpha \to \infty$ the contribution should vanish, which it does not:

$$\lim_{\alpha\to\infty} \text{contr}_y = \lim_{\alpha\to\infty} \left(\frac{1}{\alpha}\mathbf{A}^T\mathbf{Q}_y^{-1}\mathbf{A} + \mathbf{K}\right)^{-1} \left(\frac{1}{\alpha^2}\mathbf{A}^T\mathbf{Q}_y^{-1}\mathbf{A} + \mathbf{Kxx}^T\mathbf{K}\right). \tag{4.48}$$
$$\left(\frac{1}{\alpha}\mathbf{A}^T\mathbf{Q}_y^{-1}\mathbf{A} + \mathbf{K}\right)^{-1} \mathbf{A}^T\mathbf{Q}_y^{-1}\mathbf{A}$$
$$= \mathbf{K}^{-1}\mathbf{Kxx}^T\mathbf{KK}^{-1}\mathbf{A}^T\mathbf{Q}_y^{-1}\mathbf{A} \neq 0$$

In other words, a bias term remains, and destroys the usefulness of the measure. Moreover, since the bias amplitude may exceed the solution, the above measure is not confined to the $[0, 1]$ range. For these two reasons a contribution measure for biased gravity field solutions is not applied here.

4.4.3   Selenoid height errors

In lunar gravity field modelling, error propagation appears at several levels. First, one is interested in the direct propagation of observation errors into an error in the selenopotential solution. In the view of biased estimation such considerations led to the definition of the mean square error matrix. In Chap. 3, Sect. 3.3.2, the covariance matrices, that is the published unbiased error measures, of GLGM–2 and LP75G were studied in terms of their orbit errors, both in the frequency domain and in the spatial domain. In a similar fashion, the error measure of the coefficient solution may be mapped into selenophysically meaningful quantities, such as selenoid height errors or free air gravity anomaly errors.

Error propagation formulae yield, for any linear function $\mathbf{y} = \mathbf{Ax}$ [*Moritz*, 1989; *Leick*, 1995],

$$\mathbf{E}_y = \mathbf{A}\mathbf{E}_x\mathbf{A}^T \tag{4.49}$$

where in the present context $\mathbf{y}$ is a linear function of the harmonic coefficients $\mathbf{x}$, $\mathbf{A}$ is the linear matrix operator that connects $\mathbf{x}$ and $\mathbf{y}$, and $\mathbf{E}_x$ is the associated error matrix, i.e. $\mathbf{E}_x = \hat{\mathbf{Q}}_x^{col}$ (collocation) or $\mathbf{E}_x = MSEM$ (biased estimation). Limiting the linear function $\mathbf{y}$ to selenoid heights and gravity anomalies (or any other linear function not depending on the selenopotential order m), one may write:

$$y(\theta, \lambda) = \sum_{m=0}^{L} \left[\left(\sum_{l=m}^{L} \beta_l \overline{C}_{lm} \overline{P}_{lm}(\sin\phi) \right) \cos m\lambda + \left(\sum_{l=m}^{L} \beta_l \overline{S}_{lm} \overline{P}_{lm}(\sin\phi) \right) \sin m\lambda \right] \tag{4.50}$$

where β_l are the eigenvalues, which may depend on the degree only. For selenoid heights, $\beta_l = a_e$, and for gravity anomalies, $\beta_l = GM/a_e^2 \times (l-1)$. Applying the propagation law one gets a two-dimensional Fourier expansion for the error matrix (propagated error or *MSEM*) of the function y relating two points P and Q on the lunar sphere $\mathbf{E}_y(P, Q)$ [*Haagmans and Van Gelderen*, 1991; *Visser*, 1992] :

$$\mathbf{E}_y(P, Q) = \sum_{m=0}^{L} \sum_{k=0}^{L} [A_{mk} \cos m\lambda_P \cos k\lambda_Q + B_{mk} \sin m\lambda_P \cos k\lambda_Q \tag{4.51}$$
$$+ C_{mk} \cos m\lambda_P \sin k\lambda_Q + D_{mk} \sin m\lambda_P \sin k\lambda_Q]$$

where

$$A_{mk} = \sum_{l=m}^{L} \sum_{n=k}^{L} \beta_l \beta_n \mathbf{E}_x(\overline{C}_{lm}, \overline{C}_{nk}) \overline{P}_{lm}(\sin\phi_P) \overline{P}_{nk}(\sin\phi_Q) \tag{4.52}$$

$$B_{mk} = \sum_{l=m}^{L} \sum_{n=k}^{L} \beta_l \beta_n \mathbf{E}_x(\overline{S}_{lm}, \overline{C}_{nk}) \overline{P}_{lm}(\sin\phi_P) \overline{P}_{nk}(\sin\phi_Q)$$

$$C_{mk} = \sum_{l=m}^{L} \sum_{n=k}^{L} \beta_l \beta_n \mathbf{E}_x(\overline{C}_{lm}, \overline{S}_{nk}) \overline{P}_{lm}(\sin\phi_P) \overline{P}_{nk}(\sin\phi_Q)$$

$$D_{mk} = \sum_{l=m}^{L} \sum_{n=k}^{L} \beta_l \beta_n \mathbf{E}_x(\overline{S}_{lm}, \overline{S}_{nk}) \overline{P}_{lm}(\sin\phi_P) \overline{P}_{nk}(\sin\phi_Q)$$

and A_{mk}, B_{mk}, C_{mk} and D_{mk} are the Fourier coefficients of a two-dimensional series, $\mathbf{E}_x(\overline{C}_{lm}, \overline{C}_{nk})$ etc. are the coefficient error covariances or mean square error matrix components. In the sequel only point error variances are considered, i.e. $P = Q$.

Figures 4.7 and 4.8 depict the error propagation results for GLGM–2 and LP75G, with a summary of the minimum, maximum and RMS values given in

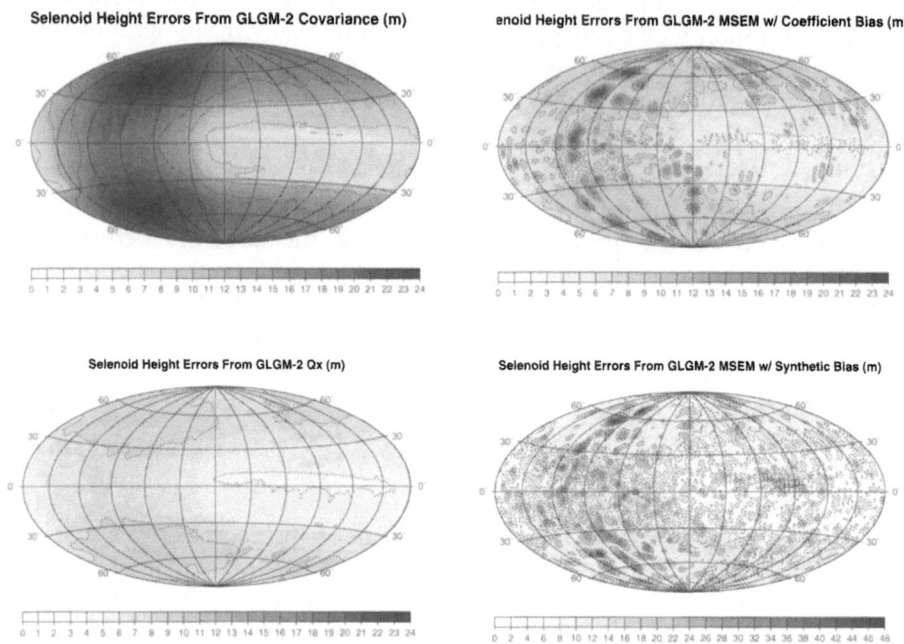

Figure 4.7　　Selenoid height errors from error measures of GLGM–2. In the left part the
results from the propagation unbiased collocation error $\hat{Q}_x^{col}$ (top) and from
the propagated biased estimation error $\hat{Q}_x^{be}$ (bottom) and on the right side the
results from the propagation of the *MSEM* with respectively a coefficient bias
(top) and a synthetic bias (bottom)

Table 4.3. The plots and table data are limited to selenoid height errors. The propagated error covariance for GLGM–2 yields an overall RMS value of 16.81 m, with
a clear relationship between the areas of good tracking, *i.e.* the near-side, and above
all its low-inclination region, and on the other side a quite smooth lunar far-side,
where both the GLGM–2 solution and error are largely determined by the regularisation. For the coefficient bias case, the overall RMS error amounts to 8.75 m,
notably smaller than what results for the pure covariance propagation, while the
synthetic case yields an RMS value of 12.18 m. On the other hand, the synthetic
bias yields larger error extremes in the selenoid than does the error covariance,
with differences with respect to the coefficient bias case, or simple collocation up
to a factor 3.5. The overall RMS value remains lower, however, due to the fact
that a large portion of the far-side exhibits smaller errors. Such behaviour is explained by the bias term. For some coefficients the bias is positive, while for others
it is negative. It is therefore not evident that the use of the *MSEM* instead of the
error covariance necessarily leads to larger formal selenoid errors. *Xu* [1992*b*] reports similar results for geopotential estimation from satellite gravity gradiometry,

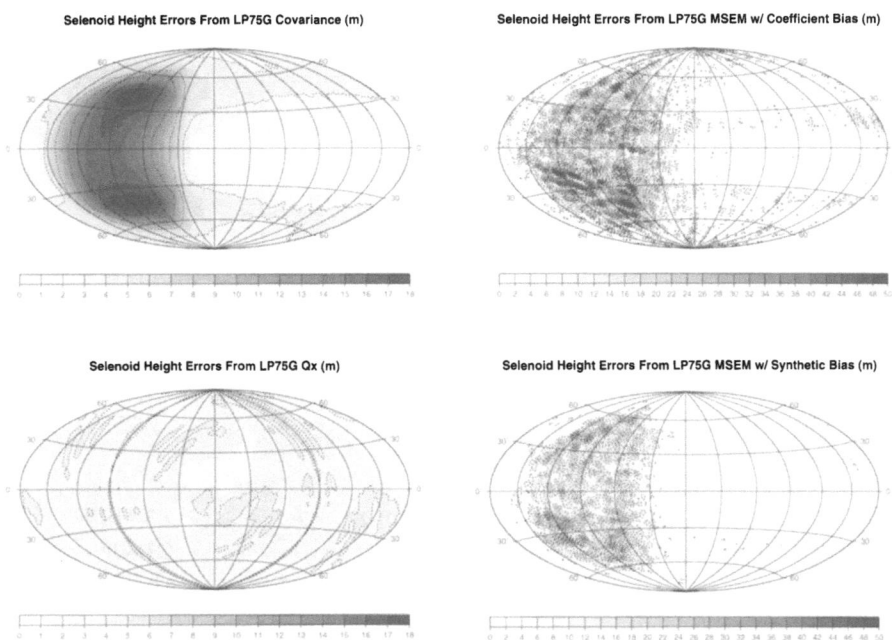

Figure 4.8   Selenoid height errors from error measures of LP75G. In the left part the
results from the propagation unbiased collocation error $\hat{Q}_x^{col}$ (top) and from
the propagated biased estimation error $\hat{Q}_x^{be}$ (bottom) and on the right side the
results from the propagation of the *MSEM* with respectively a coefficient bias
(top) and a synthetic bias (bottom)

and also shows that the degree-wise contribution to the error may increase or decrease depending on the choice of the bias. Furthermore, the similarity between
covariance-based and *MSEM*-based selenoid errors is less evident when using the
synthetic bias. In this case, the error appears more scattered, which may be explained from the use of sign according to the model, but amplitudes coming from
the Kaula rule. In some sense this is also expected from the data, which contain
orbit errors distributed over a range of orbital frequencies, and which therefore
directly contain the evidence for the spatial variations in the near-side selenopotential. By application of a global smoothing constraint in the sense of a Kaula rule
it is not realistic that all spurious components be removed from the far-side. This
is exactly what is reflected by the larger variations seen in the synthetic bias.

The case of LP75G is different in the sense that while the covariance matrix
yields an RMS selenoid error of 7.15 m, the *MSEM* with coefficient biases and
synthetic biases, yield 9.78 m and 6.71 m, respectively. In other words, the highest overall value is found for the coefficient bias-based *MSEM*, and not, like for
GLGM–2, for the covariance. This has to do with the power of the model, relative

	GLGM–2	LP75G
$\hat{Q}_x^{col} = \left(A^T Q_y^{-1} A + \alpha K \right)^{-1}$:		
min	3.866	0.709
max	24.396	17.838
rms	16.810	7.154
$\hat{Q}_x^{be} = \left(A^T Q_y^{-1} A + \alpha K \right)^{-1} A^T Q_y^{-1} A \left(A^T Q_y^{-1} A + \alpha K \right)^{-1}$:		
min	3.087	2.316
max	9.727	7.300
rms	7.255	3.657
MSEM with coefficient bias:		
min	3.118	0.675
max	23.807	67.461
rms	8.746	9.778
MSEM with synthetic bias:		
min	3.145	0.643
max	48.534	42.447
rms	12.179	6.708

Table 4.3   Selenoid height errors from the GLGM–2 and LP75G covariances, propagated biased estimation error and mean square error matrices. Units are in metres

to the Kaula power law being used in the model development, as discussed earlier. In cases where the models exhibit more degree-wise power than the Kaula rule, the covariance matrix coming from the least squares collocation view of the estimator is actually an underestimate of the error. Likewise, it is for the same reason that the covariance of GLGM–2 can be said to be on the conservative side. Of course, any such considerations inevitably hang on the rightfulness of the constraints. On the other hand, LP75G shows error propagation similarities with GLGM–2. The covariance-based results are smooth over the far-side, and the narrow-banded near-side, low-inclination error-low seen in GLGM–2 has disappeared due to the high-quality Lunar Prospector tracking over the entire near-side. Therefore, the se-lenoid height error distribution is in both cases obviously directly correlated with the measurement distribution. A further similarity is also the fact that the *MSEM*-based error propagation results are less smooth than those based on the covariance. An overall conclusion remains that the error propagation results for present-day lunar gravity field solutions vary by up to 50%, depending on the view of the estimator and the choice of the bias. Put otherwise, the estimation accuracy of the available lunar gravity field solutions is by no means a trivially solved problem.

4.5 Searching for the optimal regularisation parameter

The final section of this chapter deals with the selection of the regularisation parameter. Whereas previous sections have shown that the lunar gravimetric problem, as it is posed today, is best characterised as an extreme case of an ill-posed inverse problem, and that the regularisation is more of a compensation for a severe under-sampling of the global basis functions than a countermeasure for data errors, it should be no surprise that numerical methods are actually available for the *deterministic* selection of the regularisation parameter. Rather than relying on some prior empirical spectra of the gravitational field, based on terrestrial measurements and frequently scaled to other planets based on selenophysical assumptions, it is actually possible to deduce regularisation parameters(s) in some optimal sense directly from the matrix equations which describe the selenopotential estimation problem. In general the parameter choice rules seek to find regularisation parameters which optimally compensate for indirect or incorrect sampling of the potential field, but are only marginally able to counteract incomplete sampling. It may therefore not be expected that such methods yield regularisation parameters similar to those stemming from lunar physics.

In the present situation this appears disadvantageous, since the methods are likely to yield models with large far-side excursions, and of limited value for orbit determination or global selenophysical studies. However, such methods possess two fundamental advantages:

- they yield numerically optimal solutions to the gravimetric inverse problem as it is presently posed.

- they enable the selection of regularisation parameter(s) independent of lunar physics, and, hence, substantiate a framework for optimal regularisation analysis as soon as a (near-)global data set becomes available

The former helps illustrate the fact that the selenodetic inverse problem is, in fact, unsolvable, if the target is a fully global solution, and that the excellent near-side sampling is only of value for near-side gravity modelling, with a limited extension to some areas near the limbs and poles. Second, it also shows that, at present, optimality in the constraint choice stemming from lunar physics is vastly different from optimality predicted by the numerical properties of the least squares matrix equations. The latter is of importance for future lunar missions which will finally provide a set of measurements adequate for independent and definite inference of the selenopotential at an accuracy level sufficient for global, regional as well as local geophysical interpretation.

Summaries and performance analysis of parameter choice methods for discrete ill-posed problems are in fact not abundant in the literature. Notable exceptions are *Hansen* [1998a] who advocates the use of methods not relying on empirical error level information and exemplifies their performance for two non-geodetic problems, and *Bouman* [1998] who studies the parameter choice methods within the framework of Tikhonov regularisation for an airborne gravimetry example. For most ill-posed problems in science and engineering there are, however, many examples of the application of specific regularisation methods. In general, most methods have both *pros* and *cons*, and it is considered a wise approach to monitor several strategies and base the final choice of the regularisation parameter on the outcome of all these strategies. Consequently, no parameter choice method is guaranteed to work for all problems, and the ultimate choice of parameter is rarely the straightforward output of a single such method.

In terms of filtering, as discussed before, the proper choice of the regularisation parameter is a matter of selecting the right cut-off for the filter factors, *i.e.* the break point in the (G)SVD spectrum where damping starts to play a role. Obviously, parameter choice methods should therefore ideally determine the point at which filtering becomes desirable. It is for this reason that discrete ill-posed problems, like those found in satellite geodesy, are more complicated than truly rank-deficient problems, where the spectral components are clearly separated, and the truncation level is obvious from the spectrum alone.

In this section, several methods for the selection of the regularisation parameter are therefore investigated, notably the L-curve method and the quasi-optimality criterion. The current analysis is limited to single regularisation parameters. Minimisation of the *MSE* within the framework of generalised biased estimation is therefore not discussed. The results of the L-curve analysis are furthermore applied in the determination of fictitious GLGM–2 and LP75G solutions, where the regularisation is determined solely by the numerics and without any use of physical assumptions.

4.5.1 Pragmatic parameter choice methods for single-parameter regularisation

Three aspects are of paramount importance during the study and selection of the optimal regularisation parameter:

- the quality of the final solution, *i.e.*, given some optimality criterion, is the parameter choice method capable of finding the optimal solution?

- the convergence and rate of convergence of the solution, *i.e.* given a regularisation scheme for computing $\hat{x}_\alpha$ and a parameter choice method for computing α, how fast does $\hat{x}_\alpha$ converge to the exact solution x as the norm of the error $\|\varepsilon\| \to 0$ and $\alpha \to 0$? In particular, fast-converging parameter choice methods are attractive since they respond well to changes in the error level, thereby reducing the numerical effort.

- given the fact that experiments frequently cannot be repeated, *e.g.* due to the costs, is it possible to deduce all necessary information from the available data rather than relying on external information?

In particular the last item is frequently a problem, since it may be difficult to extract any information on the error level in the right-hand side. Generally speaking, one therefore distinguishes between two classes of parameter choice methods: *a posteriori methods* which require knowledge on either the signal norm $\|x\|_2$ or the norm of the data error $\|\varepsilon\|_2$, and *heuristic methods* that seek to extract an estimate of $\|\varepsilon\|_2$ from the given right-hand side. When *e.g.* good information on $\|\varepsilon\|_2$ is available, it should evidently be used. However, in gravity field applications, as in many other applications, one of the key issues is to avoid the use of empirical parameters or prior information, since what one actually needs is hard evidence, based on data alone, that will break part of the impasse that exists due to data scarcity. In other words, if for *a posteriori* methods, the choice of the regularisation parameter simply shifts to the choice of the prior bounds on the solution or the data error, one has actually not made much progress.

A posteriori methods

At least two *a posteriori* methods are in use today, being the method of *quasi-solutions* [*Ivanov*, 1962] and the *discrepancy principle*, usually attributed to *Morozov* [1984]. The methods are cross-linked by the following relationship [*Kress*, 1999]: while, for a given $c > 0$, the method of quasi-solutions minimises the defect $\|Af - g\|_G$ subject to the constraint that the solution norm is bounded by $\|f\|_F \leqslant c$, the discrepancy principle minimises the norm $\|f\|_F$ subject to the constraint that the defect is bounded by $\|Af_\alpha - g\|_G \leqslant \epsilon$, for a given $\epsilon > 0$. In practice, given either c or ϵ, the optimal α for discrete cases may be found by Newton-Raphson iterations seeking the zero crossing of respectively $\|\hat{x}_\alpha\|^2 - c^2 = 0$ and $\|A\hat{x}_\alpha - g^\epsilon\|^2 - \epsilon^2 = 0$.

A test of such methods for GLGM–2 clearly demonstrated that *a posteriori* methods require tight bounds on the choice of c and ϵ. Choosing the bounds on the solution norm according to some scaled Kaula rule, actually proved to be a very sensitive problem. As will be shown in Sect. 4.5.2, the residual norm corresponding to a given solution norm is ill-determined, cf. Fig. 4.10. Convergence of the zero crossing was therefore impossible to achieve. Besides, such approaches obviously do not circumvent the problem of verification of the prior information. The further use of such methods is therefore abandoned, and in the following only heuristic methods are studied.

Heuristic methods

Recalling from (4.32) and (4.33) that the total error consists of a regularisation error part and data error part, the optimal regularisation parameter should minimise the distance between $\hat{x}_\alpha$ and the exact solution x, based on certain optimisation criteria. This pragmatic point of view evidently fits well in the interpretation of the lunar gravimetric least squares problem as a case of biased estimation, where one includes both the data error and the regularisation error in the error propagation. Moreover, when the discrete Picard condition is satisfied, then, on average, the regularisation error decreases and the data error increases, as less regularisation is introduced. Obviously, the opposite is true if tighter regularisation is applied. Although several optimisation criteria exist, one common aspect of all these criteria is that the optimal solution for α is found near the point at which the two error components balance each other. The optimal α is therefore always in some sense an approximation of the minimum *MSE*. At this point, the resolution limits i_y and i_A will match, and determine the number of spectral components inferable from the linear system at hand. This is the simple and pragmatic fundament of the heuristic parameter choice methods, of which there are several in use. The most widespread has been the method of *generalised cross-validation* [*Golub et al.*, 1979; *Wahba*, 1990]; two other methods are the *quasi-optimality criterion* [*Morozov*, 1984] and the recently popular *L-curve* method [*Hansen*, 1992; *Hansen and O'Leary*, 1993; *Hansen*, 1998a, 1999], originally proposed by *Lawson and Hanson* [1974]. In this work, the focus is on the two latter methods, while the former is briefly discussed for completeness.

The L-curve

The L-curve is basically a graphical representation of the compromise between the minimisation of the discrete smoothing (semi-)norm of the regularised solution $\|\mathbf{L}\hat{\mathbf{x}}_\alpha\|_2$ versus the residual norm $\|\mathbf{A}\hat{\mathbf{x}}_\alpha - \mathbf{y}^\varepsilon\|_2$, which is the core of any regularisation method, as a function of the regularisation parameter. For discrete ill-posed problems, it turns out that the L-curve, when plotted on a *log-log* scale, very often has an L-shaped appearance (hence its name) with a distinct corner separating the vertical and horizontal parts of the curve; see Fig. 4.9. The vertical part of the curve corresponds to the smaller values of α. The emphasis is on minimising the residual $\|\mathbf{A}\hat{\mathbf{x}}_\alpha - \mathbf{y}^\varepsilon\|_2$, allowing $\|\mathbf{L}\hat{\mathbf{x}}_\alpha\|_2$ to become large. The horizontal part of the L-curve, on the other hand, corresponds to solutions where the residual norm $\|\mathbf{A}\hat{\mathbf{x}}_\alpha - \mathbf{y}^\varepsilon\|_2$ is most sensitive to the scaling parameter because the solution is dominated by the regularisation error. In terms of the filter factors, when little regularisation is introduced most of the filter factors δ_i are close to 1, and the error $\hat{\mathbf{x}}_\alpha - \mathbf{x}$ is dominated by the data error $\mathbf{A}^\dagger (\mathbf{y}^\varepsilon - \mathbf{y})$, which corresponds to under-smoothing. *Vice versa*, when a large amount of regularisation is introduced, most filter factors $\delta_i \ll 1$ and over-smoothing takes place. In between, there is a region where both error components contribute to similar extent, and this region defines the convex corner of the L-curve. Generally speaking, the faster the decay of the Fourier coefficients $\mathbf{u}_i^T\mathbf{y}$ (and hence the better the discrete Picard condition is met), the smaller the crossover region, and the sharper the L-curve corner. As such, the L-curve describes in an intuitive and graphical way the fundamental behaviour of the regularised least squares minimisation problem as a function of α.

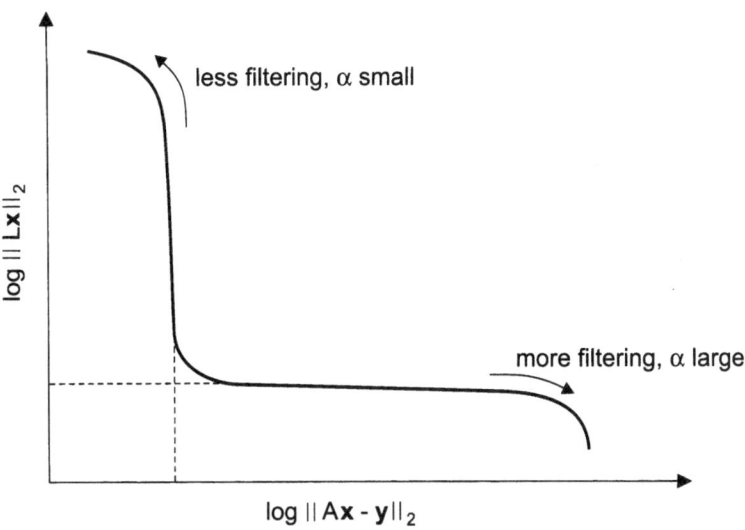

Figure 4.9   The generic form of the L-curve, from *Bouman* [1998]

The L-curve furthermore plays an essential and illustrative role in Tikhonov-type regularisation for discrete ill-posed problems, because it divides the first quadrant into two regions [*Hansen*, 1992, 1998*a*]. It is impossible to construct any solution that corresponds to a point below the Tikhonov L-curve, and any regularised solution of the inverse problem must lie on or above the curve. In causation, for a given right-hand side y^ε, α is therefore an optimal regularisation parameter that balances the two error components and which will never be far away from the corner of the L-curve. By locating the corner of the L-curve one can compute an approximation to the optimal regularisation parameter in terms of the mean square error. This is the *L-curve criterion* for optimal regularisation parameter selection. A strong feature of this criterion is that it is based on a fundamental characteristic of all discrete ill-posed inverse problems, and that it is independent of any statistical properties of the errors or any additional properties of the problems at hand. The L-curve criterion is therefore robust and capable of handling correlated errors.

It is important to plot the L-curve on a *log-log* scale in order to emphasise the two different parts of the curve. There is both strong intuitive and theoretical justification for this [*Hansen and O'Leary*, 1993; *Hansen*, 1998*a*, 1999]. Intuitively, since the singular values typically span many orders of magnitude, the behaviour of the problem is better illustrated on a logarithmic scale. Furthermore, the "horizontal" and/or "vertical" part(s) are emphasised on a log-log scale when the variations in either part is small compared to the other, which is often the case. Most importantly, *Hansen* [1998*a*] proves that, on a linear scale, the L-curve is always convex, independent of the right-hand side. For similar reasons it is important that the "operational" definition of the corner point is based on the curvature of the L-curve, rather than simply the minimum distance from the origin [*Hansen and*

O'Leary, 1993]. The curvature κ is a purely geometric quantity. Therefore, if the regularisation parameter is continuous, the point of maximum curvature may be found straightforwardly by maximisation of

$$\kappa(\alpha) = \frac{\xi'\eta'' - \xi''\eta'}{[(\xi')^2 + (\eta')^2]^{3/2}} \tag{4.53}$$

where $\xi = \log_{10} \|A\hat{x}_\alpha - y^\varepsilon\|$, $\eta = \log_{10} \|L\hat{x}_\alpha\|$ and the primes denote differentiation with respect to α. In most cases, however, the L-curve is given in only a few points, in which the curve may be approximated by a 2D spline, and the corner of the L-curve is defined as the point closest to the corner of the spline curve. An alternative is to consider the point C where the L-curve is concave and the tangent has a slope -1. The concave condition is required, since the slope may be -1 near the endpoints of the curve, in addition to the corner. *Regińska* [1996] and *Engl et al.* [1996] show that point C is a corner of the L-curve if and only if the function

$$\psi(\alpha) = \|\hat{x}_\alpha\|_2 \|A\hat{x}_\alpha - y^\varepsilon\|_2 \tag{4.54}$$

has a local minimum at $\alpha = \alpha_C$.

Another interesting aspect of the L-curve is how it relates to the discrepancy principle and the method of quasi-solutions. Since, for the discrepancy principle the residual norm is given, the discrepancy principle-based solutions correspond to the interception point of the L-curve with the vertical line $\|A\hat{x}_\alpha - y^\varepsilon\|_2 = \epsilon$. *Vice versa*, all quasi-solutions lie on an interception point given by the horizontal line $\|\hat{x}_\alpha\|_2 = c$ and the same L-curve. Since the L-curve method by definition seeks to locate the corner point, it is generally a more robust parameter choice method.

Note that L-curve regularisation requires the design matrix A, which in gravity field applications frequently is not available. The final linear equation system for gravity field model adjustment is usually the combination of such systems for many individual satellite arcs, or even the combination of a multitude of observation types. Second, in satellite applications the $m \times n$ design matrix per arc is generally huge, and any explicit computation (and storage) is therefore impractical. Hence, the individual design matrices are usually not stored, and one needs to estimate the residual norm. Since the data errors behave like σ^{-1}, cf. (4.15), the actual observations y^ε may, however, be approximated by $Ax^\varepsilon = A(x + e)$, where e is an error component. An obvious choice for the exact solution x is the gravity field solution itself, *e.g.* the coefficients of GLGM–2 or LP75G. In order to accommodate errors up to a 3σ level, the error e can be approximated by

$$\frac{3N(0,1)}{\sigma_i} \times \begin{bmatrix} \sigma_{\bar{C}_{lm}} \\ \sigma_{\bar{S}_{lm}} \end{bmatrix}$$

where $N(0,1)$ is the standard Gaussian distribution with zero mean and a standard deviation of 1 and $\left\{ \sigma_{\bar{C}_{lm}}, \sigma_{\bar{S}_{lm}} \right\}$ are the standard deviations or square root of the diagonal elements of the *MSEM* for harmonic coefficients of degree l and order m. In this case an upper bound for the residual norm is given by

$$\|\mathbf{A}\hat{\mathbf{x}}_\alpha^e - \mathbf{y}^e\|_2 = \|\mathbf{A}\hat{\mathbf{x}}_\alpha^e - \mathbf{A}\mathbf{x}^e\|_2 \leqslant \|\mathbf{A}\| \, \|\hat{\mathbf{x}}_\alpha^e - \mathbf{x}^e\|_2 \tag{4.55}$$
$$= \sqrt{\lambda_1} \|\hat{\mathbf{x}}_\alpha^e - \mathbf{x}^e\|_2$$

where λ_1 is the largest eigenvalue of the unregularised normal matrix $\mathbf{A}^T\mathbf{A}$ which can easily be obtained with the power method [*Kreyszig*, 1999]. For square root information filters, the $\mathbf{A}^T\mathbf{A}$ matrix is simply the square of the filters.

Whereas the advantages of the L-curve are the intuitive and direct view on the regularisation problem, its robustness and independence of statistical information, it evidently also has its limitations. The first one involves the computational cost. Since the computation requires both the (semi-)norm of the regularised solution and the norm of the residuals to be available, the determination of the finer scale of the curve is a costly operation, in particular for large-sized problems. *Calvetti et al.* [1999] developed a method for the determination of the optimal α based on only a few points. Their method is based on *L-ribbons* that contain the L-curve in their interior, and they show that such ribbons may be computed fairly inexpensively by partial Lanzcos bi-diagonalisation of the matrix $\mathbf{A}$. Lanzcos methods are described in detail in *Björk* [1996] and *Golub and Van Loan* [1996]. Evidently, this is show-stopper if the design matrix is not available, as discussed before.

Another, perhaps more academic, limitation is the lack of an exact understanding of the relationship between the optimal L-curve based α and the minimum *MSE*. Since the rate of change of the two error components are equal in the corner point of the L-curve, with opposite sign, it should represent the minimum norm distance between the regularised and exact solutions. Such limitations are common for other parameter choice methods as well.

Further limitations of the L-curve concern its use in the reconstruction of very smooth problems and its asymptotic behaviour as the problem size n increases. In cases where the SVD coefficients $|\mathbf{v}_i^T\mathbf{x}|$ decay very fast, in other words if the solution is dominated by the first few SVD components, *Hanke* [1996] shows that the L-curve criterion will fail, as there in such cases is a dilemma between residual fit and reconstruction of the solution. As a general rule it holds that, the smoother the solution, the faster the decay of the SVD spectrum, and, therefore, the worse the performance of the L-curve method will be. *Vogel* [1996] furthermore illustrates the problem of the asymptotic behaviour of the L-curve, and pointed out that for very large problems the optimal α computed by the L-curve may not be optimal in a minimal error sense, but rather tend to yield over-regularised solutions. Again, the degree of over-smoothing is dependent on the decay rate of the singular values: the faster the decay, the less severe the over-smoothing. Given the recent popularity of the method, however, further research is likely to lead to improvements in the near future.

Quasi-optimality

The quasi-optimality criterion is another parameter choice method that seeks to balance the data error and the regularisation error. Next to being a method in its own, by virtue of its nature it also provides an opportunity to verify the L-curve results. In the literature, at least two lines of thought are said to be behind the quasi-optimality principle. A first line of thought assumes that the error is given by

$$\|\hat{\mathbf{x}}_\alpha - \mathbf{x}\|_2 \approx \left(\mathbf{y}^T \left(\mathbf{A}\mathbf{A}^T + \alpha\mathbf{K}\right)^{-4} \mathbf{A}\mathbf{A}^T\mathbf{y}\right)^{1/2},$$

the minimisation of which leads to [*Hansen*, 1998*a*]

$$\min_\alpha \frac{1}{2} \left\|\alpha\frac{d\hat{\mathbf{x}}_\alpha}{d\alpha}\right\|_2 \tag{4.56}$$

Morozov [1984] and *Engl et al.* [1996], on the other hand, derive the quasi-optimality criterion as the minimum distance between two successive regularised solutions,

$$\min_\alpha \|\hat{\mathbf{x}}_{\alpha,j} - \hat{\mathbf{x}}_{\alpha,j-1}\|_2 \tag{4.57}$$

Given the minimisation problem

$$\min_\alpha \|\mathbf{A}\mathbf{x} - \mathbf{y}^\epsilon\|_2^2 + \alpha \|\mathbf{x} - \mathbf{x}_0\|_2^2$$

where usually $\mathbf{x}_o = 0$, such a two-fold of regularised solutions is related as

$$\hat{\mathbf{x}}_{\alpha,j-1} = \left(\mathbf{A}^T\mathbf{Q}_y^{-1}\mathbf{A} + \alpha\mathbf{K}\right)^{-1} \left(\mathbf{A}^T\mathbf{Q}_y^{-1}\mathbf{y}^\epsilon + \alpha\hat{\mathbf{x}}_{\alpha,j-2}\right) \tag{4.58}$$

$$\hat{\mathbf{x}}_{\alpha,j} = \left(\mathbf{A}^T\mathbf{Q}_y^{-1}\mathbf{A} + \alpha\mathbf{K}\right)^{-1} \left(\mathbf{A}^T\mathbf{Q}_y^{-1}\mathbf{y}^\epsilon + \alpha\hat{\mathbf{x}}_{\alpha,j-1}\right)$$

$$= \hat{\mathbf{x}}_{\alpha,j-1} - \alpha \left(\mathbf{A}^T\mathbf{Q}_y^{-1}\mathbf{A} + \alpha\mathbf{K}\right)^{-1} \left(\hat{\mathbf{x}}_{\alpha,j-1} - \hat{\mathbf{x}}_{\alpha,j-2}\right)$$

$$= \hat{\mathbf{x}}_{\alpha,j-1} - \alpha \frac{d\hat{\mathbf{x}}_{\alpha,j-1}}{d\alpha}$$

and (4.57) therefore leads to the norm minimisation

$$\min_\alpha \left\|\alpha\frac{d\hat{\mathbf{x}}_\alpha}{d\alpha}\right\|_2$$

cf. (4.56). It is therefore readily seen that the given error estimate and the minimisation of the distance between two successive solutions yield the same regularised solution.

One problem of the quasi-optimality criterion is that the function to optimise may tend to have several local minima, and that the global minimum therefore might be difficult to trace. As such, the method may be less robust than the L-curve criterion, although the underlying thought is similar.

Generalised cross-validation

For the sake of completeness and due to its popularity, the method of generalised cross-validation (GCV) [*Golub et al.*, 1979; *Wahba*, 1990] is discussed in brevity. This method is based on statistical considerations, and, in particular, the idea that a well-chosen regularisation parameter should be able to predict missing data values. In other words, if a particular element of the right-hand side y_k^ε is left out, the corresponding regularised solution should predict this observation well. Moreover, the GCV method is developed to be independent of orthogonal transformations of the problem, including permutations of the elements of y^ε [*Hansen*, 1998a].

In its simplest form, GCV is a predictive scheme for the minimisation of the mean square error. However, since error-free data are not available, the method instead works with the minimisation of the GCV function

$$G(\alpha) = \frac{\left\| A\hat{x}_\alpha - y^\varepsilon \right\|_2^2}{\left[\text{trace}\left(I_m - AA_\alpha^\dagger \right) \right]^2} \tag{4.59}$$

Wahba [1990] shows that if the discrete Picard condition is satisfied, and the data error (noise) is white, then the regularisation parameter α_{GCV} leads to a regularised solution that corresponds closely to the minimum *MSE*. However, the assumption of white noise is crucial, and for coloured noise, generally no minimum of $G(\alpha)$ is found [*Hansen*, 1998b, 1999]. Intrinsically, this governs the strong and weak features of the method: when the GCV method works, it generally works very well, but the method is unable to handle correlated data errors, and is in this sense less robust than, *e.g.*, the L-curve criterion.

4.5.2 L-curve and quasi-optimality regularisation parameters for GLGM–2 and LP75G

The L-curve and quasi-optimality parameter choice methods have been applied to the GLGM–2 normal equation and to the square root information filter of LP75G. As already discussed, the purpose of this is not to derive lunar gravity solutions in an optimal orbit quality or global lunar physics sense, but rather to illustrate the true value of the satellite measurements available today. In other words, since the parameter choice methods yield, given certain criteria, numerically optimal solutions, these solutions are expected to demonstrate what kind of solutions one would get if a prior degree variations scheme, like Kaula's rule, were to be disregarded. Nevertheless, the solutions here are based on regularisation with a Kaula like function, *i.e.* it is assumed that the $\{\overline{C}_{lm}, \overline{S}_{lm}\}$ coefficients behave like $1/l^2$. Obviously, as better information becomes available about the selenopotential in the future, other choices of coefficient decay may be chosen. As will be clear, this tends to yield optimal solutions vastly different from those published in the literature, given the lack of sampling of the lunar far-side. A remarkable fact, however, is that, as will become clear, the gravity field over the near-side can still be

	GLGM–2	LP75G
L-curve	0.545×10^{-8}	0.329×10^{-8}
Quasi-optimality	0.540×10^{-8}	-

Table 4.4   Optimal regularisation parameter α as determined by means of the quasi-optimality method and the L-curve criterion for GLGM–2 and LP75G. The quasi-optimality criterion produces a range of local minima for LP75G, but fails to yield a global minimum. The scaling parameter for the corresponding degree-wise variations of the gravity field coefficients is given by $\beta = 1/\sqrt{\alpha}$

well-recovered from both sets of equations. The effect of under-sampling therefore, even in a global basis function framework, remains largely restricted to the unsampled areas.

The optimal α solutions are given for both methods and models in Table 4.4, and the corresponding L-curve plot for GLGM–2 only is depicted in Fig. 4.10. For GLGM–2 the two parameter choice methods yield very similar results, with an optimal α of 0.545×10^{-8} and 0.540×10^{-8}, respectively. The corresponding Kaula rules are, since $\beta = 1/\sqrt{\alpha}$, for the L-curve criterion, $\{\overline{C}_{lm}, \overline{S}_{lm}\} = 13454 \times 10^{-5}/l^2$, and for quasi-optimality $\{\overline{C}_{lm}, \overline{S}_{lm}\} = 13608 \times 10^{-5}/l^2$. Clearly, these are much more relaxed constraints, by a factor ~ 900, than those applied in the development of GLGM–2. From the GLGM–2 L-curve it is furthermore readily seen that while the corner point is easily located by the min $\psi(\alpha)$ criterion, the L-shape of the curve is not pronounced. The solution norm never becomes dominating, as obviously the data error for small values of α remains inferior to the regularisation error.

For LP75G the optimal L-curve α is 40% smaller than for GLGM–2, which reflects the increased value of the tracking data information in the LP75G SRIF. This evidently predicts that, from a numerical point of view, the LP75G SRIF constraints should be relaxed with respect to those applied in GLGM–2. The quasi-optimality criterion, however, yields a number of local minima, with no real indication of a global minimum. Therefore, it is concluded that the norm of the differences between successive fictitious, quasi-optimality regularised LP75G solutions may be close for a range of values of α, and that quasi-optimality in this respect is not the preferred method.

Generally speaking, it may be concluded from both these parameter choice method results that the methods tend to yield solutions close to the exact inverse. In terms of filtering, the damping which might be desired from a physical point of view is nowhere seen, and the filter factors remain close to one.

The optimal α values as determined by the L-curve criterion have furthermore

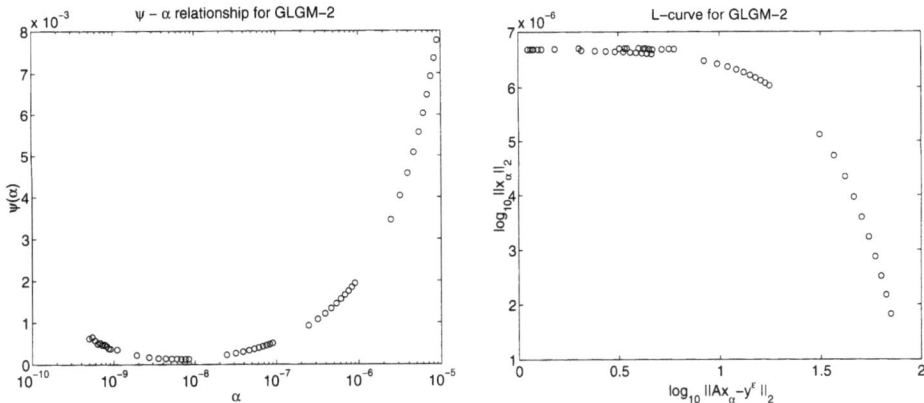

Figure 4.10   The L-curve of the GLGM–2 normal equation. The left figure depicts the
minimisation criterion for the determination of the corner of the L-curve seen
in the right figure. The optimal alpha is 0.545×10^{-8}, corresponding to a
Kaula rule of approximately $\{\overline{C}_{lm}, \overline{S}_{lm}\} = 13454 \times 10^{-5}/l^2$. While the corner
is easily located, the L-shape is much less evident, as obviously the data
error never dominates that change in residual to the regularisation. Basically,
only the right leg of the L is therefore seen, corresponding to the larger
values of α. Similar curves are found for LP75G, which yields a slightly
smaller α_{opt} of 0.329×10^{-8}

been applied in the determination of two fictitious gravity field solutions, depicted
in terms of selenoid heights in Figs. 4.11 and 4.12. The reference field is the best-
fitting ellipsoid, *i.e.* the central and $\overline{C}_{20}$ terms are excluded, and the projection is -
as before - a Hammer projection centred at $270°$ eastern longitude, such that the
entire far-side is found to the left of the centre line and the near-side to its right.

A striking feature for the fictitious GLGM–2 solution is that basically only low-
inclination near-side features are well-resolved. For these areas, the pre-Lunar
Prospector tracking data was qualitatively speaking adequate for a sound gravity
field recovery, cf. Chap. 2. The major near-side gravity highs (lunar lowlands) and
lows (lunar highlands) as well as the major ring basins are clearly distinguishable.
The overall expected negative near-side correlation with topography is therefore
present. However, the poorer sampling of the polar areas and the lack of far-side
data are the primary reasons for the catastrophic conditioning of the unregularised
GLGM–2 normal equation, which is directly visualised in Fig. 4.11. As outlined in
Sect. 4.1.2, the numerical null-space of the GLGM–2 normal matrix is spanned by
vectors with many sign changes and with small corresponding singular values σ_i.
This is exactly what is seen in the figure, where the far-side oscillations, and the
oscillations near the poles and limbs are severe. Obviously, the propagated errors
of the solution will be huge in these areas, precisely as predicted by the inverse
modelling theory.

The merit of the Lunar Prospector data is, of course, the much improved sam-

Selenoid Heights From L-curve Regularised GLGM-2

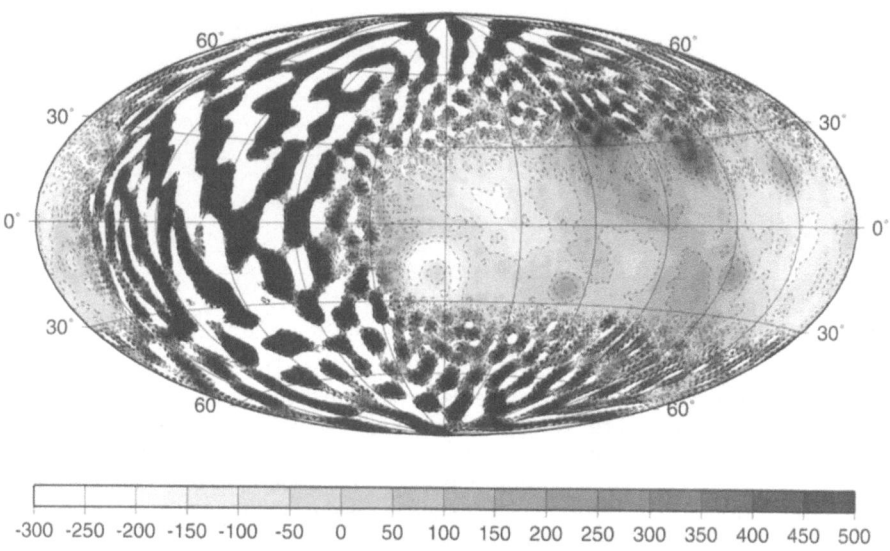

Figure 4.11   Selenoid heights in metres from a fictitious GLGM–2 solution derived using
the regularisation parameter according to the L-curve criterion. Given the
unavailability of far-side measurements, this solution reflects the true
numerical information content of the GLGM–2 lunar gravity field
determination problem, as it is posed in terms of global spherical harmonic
functions and without the use of prior information stemming from lunar
physics. The degree-wise decay of the $\{\overline{C}_{lm}, \overline{S}_{lm}\}$ coefficients is nevertheless
assumed to behave like $1/l^2$

pling over the entire near-side, the polar areas and also the limbs. By direct comparison of Figs. 4.12 and 4.11, these improvements are directly visible. Whereas the
oscillations in unsampled areas basically remain, the solution is much improved
where data is available. For example, the distinct near-side features are resolved
at accuracy levels higher than with GLGM–2, as are the features near the limbs,
e.g. Mare Orientale, or the poles. The two plots therefore, in terms of the equipotential surface elevation above the best-fitting ellipsoid, directly illustrate the increased knowledge of the lunar gravity field over the past five-six years, going
from Clementine in 1994 to the end of the Lunar Prospector mission in mid-1999.
However, in both cases, GLGM–2 and LP75G, the overall conditioning of the L-
curve regularised linear equation systems remains poor, as the L-curve method, or
any other search algorithm for the optimal choice of the regularisation parameter,
does not compensate for the severe degree of under-sampling.

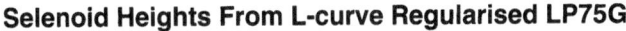

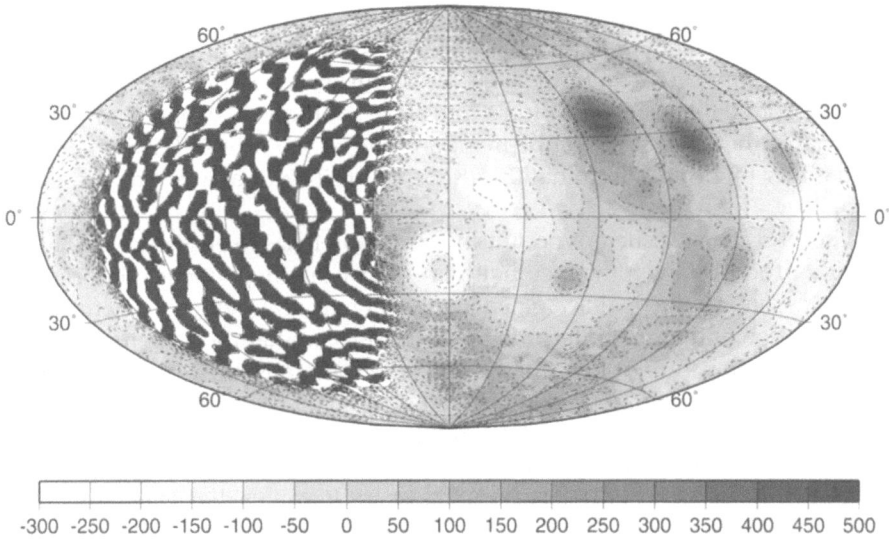

Selenoid Heights From L-curve Regularised LP75G

-300 -250 -200 -150 -100 -50 0   50 100 150 200 250 300 350 400 450 500

Figure 4.12   Selenoid heights in metres from a fictitious LP75G solution derived using the
regularisation parameter according to the L-curve criterion. Similar to the
case of GLGM–2 such a plot demonstrates the actual value of the
information available, without the use of prior information stemming from
lunar physics. As for GLGM–2, the degree-wise decay of the $\{\overline{C}_{lm}, \overline{S}_{lm}\}$
coefficients is nevertheless assumed to behave like $1/l^2$. Notice the
improvements over the poles and in areas just beyond the limbs with respect
to GLGM–2, which are the strong characteristics of the Lunar Prospector
tracking data compared to previous data sets

4.5.3   Discussion and general remarks on error assessment and calibration of selenopotential models

This chapter has been devoted to a detailed description of the ill-conditioning of
discrete inverse problems as met in selenopotential modelling. Following a general
introduction on the tools and underlying mathematics, the gravity models GLGM–
2 and LP75G have been analysed in terms of their singular value decomposition
and their compliance with the discrete Picard condition. A fundamental result is
that regularisation in global lunar gravity field recovery - at present - frequently is
seen as more than regularisation in the usual sense of the word, as the schemes are
also required to compensate for the severe under-sampling of the problem. This is
not only a result of the lack of far-side observations, but also of the desire to solve
the problem in an optimal sense for orbit analysis and/or global selenophysical

applications.

The two models have further been analysed in terms of their information content, using ratio measures, contribution measures and error propagation tools, which provide a non-orbit related way of assessing the selenopotential model quality. Such tools not only reveal the weaknesses of present-day models, but obviously also substantiate a framework for rigourous assessment of future models. In other words, the use of such tools for future solutions is strongly advocated, as they enable the analyst to complement and verify the orbit-related analysis provided either by orbit determination results or orbit error analysis as presented in Chap. 3.

A second major issue advocated in this chapter has been the view on the estimation process. Given the fact that there is a strong dependence on prior information, assuming the goal is a global spherical harmonic solution useful for orbit determination and global interpretation purposes, and at the same time the same prior information cannot be rigorously verified, there is a legitimate chance that the constraint relations are wrong. Therefore, it is not obvious that the collocation framework for lunar gravity models provides the correct solution, and, moreover, the bias that the erroneous regularisation introduces is in such cases not included in the error measures. For this reason, this chapter argues that selenopotential recovery is better regarded as a biased estimation process, and the aforementioned models GLGM–2 and LP75G are investigated within this framework. The role of the bias is a crucial point, since it influences both the estimation of the true information content of the models, as well as the error measures of the solutions. It is found that up to a factor 2 difference may be found in terms of RMS selenoid height errors between a collocation definition of the estimator and biased estimation. In other words, the quality description of selenopotential solutions is strongly influenced by the understanding of the estimator as well as the choice of the regularisation scheme.

Third, this chapter has zoomed in on deterministic methods for the selection of the regularisation scheme. The obvious aim is in the first place to avoid "circular arguments" in selenopotential modelling by rejecting assumptions and instead pragmatically determine the optimal level of regularisation directly from the available tracking data information. It has been shown that numerically optimal gravity field solutions at present yield solutions vastly different from the models presented in the literature, and for which the constraint relations are usually chosen on the basis of orbit fit or some selenophysical assumption. However, the strong points of such methods are i) that the methods reveal optimal solutions given the current data sets and disregard any kind of assumptions regarding what the model should optimally look like (in other words, the methods are not biased in any kind of way); and ii) such methods are directly applicable to future, near-global data sets, when they become available in the near future, *e.g.* through the SELENE mission.

To conclude the chapter, some remarks are given on the current practice of data calibration and subsequent error assessment. In Chap. 3 the calibration as it was applied in GLGM–2 was presented, and it was concluded that, based on the differences between GLGM–2 and LP75G, this calibration effect has been largely successful in deriving a single error measure that describes, to a reasonable level of confidence, the intrinsic error in GLGM–2. However, this type of calibration also has a parasitic effect. First, the error propagation is no longer simply the propagation of the measurement error, but also contains systematic effects. Or, phrased in another way, the sum of all error sources is lumped into a single measure, and the effect of individual components is completely hidden. Second, by means of a down-weighting of the observations, the true quality of the observations is minimised, while in fact there might be very little wrong with the measurements. Instead, it is the poor knowledge of other parameters that affect the estimation problem that is the limiting factor. In fact, this is important in situations where the target is to derive regional or even local variations in the gravity field, where obviously the high-frequency information is of crucial importance.

For these reasons, it is tempting to advocate the use of data weights according to the best estimate of the measurement precision. The propagated error in such cases contains only the data error, as is also understood in the theory of discrete ill-posed problems. Additionally, if information on the correlations of measurement errors is present, it could be incorporated in the estimation process accordingly. In a second instance, based on off-line estimates of other error sources, consider-covariance tools [*Montenbruck and Gill*, 2000] may be used to include the effect of other error sources. Combined with the bias estimates, one finally arrives at a three-component mean square error matrix as the overall error measure, all of which are well-understood. Of course, the danger of this approach is that unless a reasonable estimate may be found for a given error source, it is neglected in the mean square error, and, hence, all other propagation results, like orbit or selenoid height errors.

Chapter 5

Lunar gravity field modelling experiments with European data sets

'*For my part, the most unfeeling thing I know of is the law of gravitation: it breaks the neck of the best and most amiable person without scruple, if he forgets for a single moment to give heed to it.*'

J. S. Mill, economist, 1867

5.1 Introduction

In different ways, the previous chapters present critical views on the present-day understanding and modelling of the lunar gravity field. Starting from the satellite selenodetic standpoint of orbit quality using linear perturbation theories, passing via a thorough assessment of the true selenopotential information provided by the available tracking data, and ending in an in-depth discussion on the role of regularisation for the lunar gravimetric problem, it has been demonstrated that the data alone do not warrant the recovery of an extended global model, even if the near-side data contain valuable high-frequency orbit perturbation information. It has, however, also been shown that the effects of under-sampling remain largely restricted to the same unsampled areas. In other words, based on present models, care should be taken not to draw premature conclusions for unsampled regions of the Moon. In addition, it has been pointed out that the choice of constraint relations, and in particular the choice of one or more regularisation parameters, may influence the solution by "tuning" it toward a particular type of application. Obviously, in such a situation, unambiguous recovery of the selenopotential is impossible, since the constraints cannot be verified; and, even if they could, they would be incapable of adequately decorrelating the estimated parameters. Spectral leakage is therefore a significant problem. A third and related conclusion of the previous chapters was that there might be a "datum shift" between the gravity field model best-suited for orbit modelling (excellent fit of orbit data) and a model that is well-suited for local, regional or global selenophysical studies. Such problems have previously been long debated in the satellite geodetic community for the terrestrial gravity field determination problem, where it was found that optimal orbit performance in many cases also implied reduction of parameters not related to the gravity field, but still affected the satellite orbit. In other words, it may well be the case that models starting from a selenophysical perspective, rather than that of orbit-derived measurements, *e.g.* when *in situ* gravimetric surface measurements become available, will behave differently from a model derived with optimal orbit determination performance in mind.

Given the impetus of the present book to illustrate the nature of the lunar gravimetric problem, the previous chapters may appear to be a criticism of current practise in the research field. The contrary is, however, true. While the main driver for devoting research efforts to existing models is the understanding of the foundations of present interpretations (be it orbit-related or selenophysical), it is recognised that in lack of sufficient information, one has little choice but to use whatever information is available. This is where assumptions and "educated guesses" in the form of prior information come into play, and this fact is not to be disqualified simply because the estimation problem does no longer fit some well-defined mathematical formulation. Rather, the realisation that the problem to a large extent depends on prior information (or assumptions) simply underlines the fact that the lunar gravimetric problem poses a great challenge to the satellite data analyst, and that skill is needed to recover the information from the data in the best way possible. It is therefore rather a compliment to model developers that their models work remarkably well in respective areas, despite the significant lack of information.

Along the same line, it follows that it is all too simplistic to deliver criticism of the present-day practise without really getting involved with the actual data reduction process. It is in dealing with the "nuts and bolts" of the reduction of the satellite tracking data that one is confronted with the choices of data weights and regularisation parameters, to name but two of the decisions that need to be taken in the process. Second, it is only by analysis of actual data that experience is gained in how to compute realistic and in some sense optimal gravity field solutions, particularly under sub-optimal data circumstances. Combined with the desire to prove technology readiness for future lunar gravimetry experiments, these two reasons formed the direct incentive to organise a tracking campaign of a low lunar orbiter, and subsequently attempt to recover the gravity field.

This chapter deals with the first experiments in Europe directed towards improvement of existing lunar gravity field models. To this end, during the summer of 1998, Doppler data of the Lunar Prospector spacecraft were collected by the 30 m deep space antenna at Weilheim/Lichtenau, Germany. After an introduction of the tracking campaign, the hardware and the resulting data quality, the chapter will focus on orbit computation and gravity field modelling aspects, and in particular on the parameter choices and data weighting schemes stressed in the previous chapters. The resulting Weilheim-adjusted solutions are compared with the *a priori* model LP75G, using several of the tools introduced in Chapters 3 and 4.

5.2   The Lunar Prospector tracking campaign at Weilheim

In a joint venture between the German Space Operations Center (GSOC), a subsidiary of the German Aerospace Establishment (DLR), and Delft Institute for Earth-Oriented Space Research (DEOS) of Delft University of Technology, and by special permission from the Lunar Prospector mission management, a six weeks tracking campaign of Lunar Prospector was organised in the summer of 1998. During the period from 6 July (day of year 187) through 15 August (day of year 227), passive 3-way Doppler data were collected by DLR's 30 m deep space antenna at Weilheim, located about 50 km southwest of Munich, Germany, and shown in Fig. 5.1. The entire tracking period fell within the nominal mission phase, which means that the spacecraft orbited in a polar near-circular orbit at a mean altitude of 100 km, cf. Table 2.1.

As described in Chap. 2, after being launched on 7 January 1998 at 02:28 UTC, the Lunar Prospector mission came to an end on 31 July 1999 at 09:51 UTC, when the satellite performed a controlled hard landing in a crater near the south pole of the Moon, in support of on-going research into lunar volatiles and the possible detection of water ice in the polar regolith. The Lunar Prospector spacecraft, shown in Fig. 5.2, was a relatively simple, light (296.3 kg wet mass at lunar orbit insertion), drum-shaped (1.37 m diameter, 1.29 m axial length) and spin-stabilised craft designed to need only a minimum of operational effort. The nominal orientation of the spin axis was perpendicular to the ecliptic. Next to the selenodetic experiment using the on-board transponder, the spacecraft instrumentation included a gamma ray spectrometer, an alpha particle spectrometer, a neutron spectrometer, and electron reflectometer and a magnetometer [*Binder*, 1998], some of which were combined into the same unit in order to be mounted on one of the three spacecraft booms. The magnetometer was placed on a boom extension in order to be as far away from on-board sources of magnetic perturbation as possible [*Andolz et al.*, 1998]. Tight budgetary constraints (total mission cost of U.S. $ 63 million, including launch services, instruments and one year of operations) and project management according to the principles of "faster, better, cheaper" missions [*Huntress*, 1999; *Saunders*, 1999] meant that cost-effective and proven design approaches were employed, *e.g.* the choice of a spinning spacecraft. Both the attitude and the nominal spin rate were thruster-controlled. Furthermore, the spacecraft had no on-board computer. On the top panel was the 1.6 m tall antenna assembly, consisting of an omni-directional antenna (top cone) and a medium-gain helix antenna (coni-

Figure 5.1   The DLR 30 m deep space antenna at Weilheim

cal cylinder). These two antenna systems provided data down-link and command up-link communications between the satellite and the ground stations.

The measurement principle, commonly known as 3-way Doppler tracking, differs from the more usual 2-way counterpart by using separate stations for the up-link and the down-link, as illustrated by Fig. 5.3. Passive 3-way Doppler tracking furthermore relies on the up-link by other deep space stations, in this case NASA's Deep Space Network. The up-link signals, generated by the 26 m and 34 m stations at Madrid (DSS-61, -66), Canberra (DSS-42, -46) and Goldstone (DSS-16, -24, -27), were received by the Lunar Prospector spacecraft, coherently retransmitted by the on-board transponder, and finally received by the Weilheim 30 m antenna at times of common visibility. Given the longitudinal separation of the DSN stations, as well as the relative proximity of Weilheim and Madrid, it speaks for itself that the up-link for the bulk of the data was provided by the Madrid antennas. Considerable overlaps were nevertheless also found for Canberra and Goldstone. While the Weilheim station is also capable of down-linking in the X-band, all Lunar Prospector tracking occurred in the S-band, primarily due to mission constraints. At Weilheim, the down-link signal, with a nominal carrier frequency of 2273 MHz, was processed by an ESA standard Multi Purpose Tracking System (MPTS) [*Gaudenzi et al.*, 1990] after passing a low noise amplifier and a recently installed low-Earth

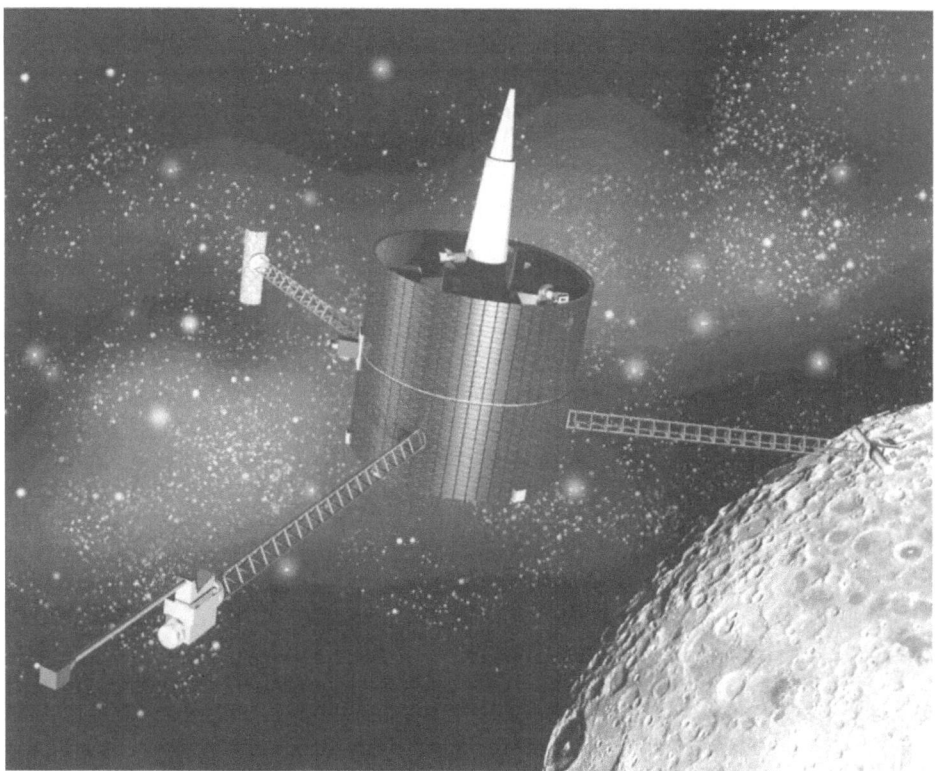

Figure 5.2   An illustration of Lunar Prospector in lunar orbit. Main features of the
spinning craft were the three booms carrying the scientific payload and the
antenna assembly on the north-facing panel. The top conical part was the
omni-directional low-gain antenna, the below conical cylinder was the
medium-gain helix antenna. Artistic impression by Boris Rabin, NASA Ames
Research Center, www.arc.nasa.gov

S-band receiver [*Weischede et al.*, 1999].

In contrast to 2-way measurements, drifts or biases in the up-link frequency
generation are not cancelled by corresponding errors in the frequency standard of
the Doppler measurement unit. A hydrogen maser atomic clock operated at the
Weilheim ground station [*Nau et al.*, 1994] was therefore connected to the MPTS
system to provide a highly accurate and stable reference for measuring the carrier
frequency of the received down-link signal. Non-destructive Doppler measurements were registered at count intervals of 1 s, but in a later stage combined into
effective 30 s count intervals for orbit determination and gravity field recovery
purposes. The measurements exhibit a 1σ noise level of about 3 mm/s at 1 s count
intervals or, equivalently, 0.5 mm/s at 30 s integration time. Besides possessing
a notably reduced data noise level, the integrated measurements have the advan-

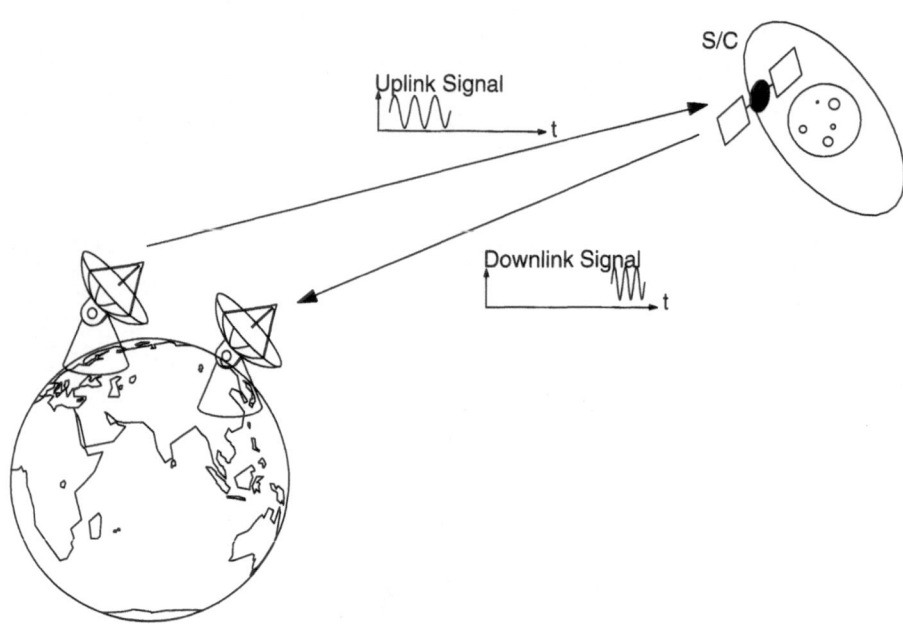

Figure 5.3   The principle of 3-way Doppler tracking. In the case of the tracking campaign
at Weilheim, the up-links were performed by stations of the Deep Space
Network and the data were collected by passively listening to the down-link
signal

tage of being essentially free of periodic Doppler variations caused by the nominal
5 s/rev (12 rpm) rotation of the spinning Lunar Prospector. Characteristics and
effects of the spacecraft rotation are further elaborated in Sect. 5.2.1. The Doppler
counts were subsequently converted to average range-rate measurements using
the nominal Lunar Prospector up-link frequency published by DSN. Furthermore,
the data preprocessing consisted of the assignment of the proper up-link station
identification to the data records, based on the forecasted DSN operations sched-
ules. Finally, the initial analysis included monitoring of data residuals, followed
by a manual data editing in order to remove records affected by up-link station
transitions or occultation phases.

The six weeks campaign delivered a total data set of 1,030,000 measurement
points at 1 s integration intervals, which by the above-described averaging process
was reduced to 34,335 observations of 30 s integration time. Effectively, the campaign therefore produced about one-third the number of measurements provided
by the Clementine spacecraft in 1994, which delivered 361,794 2-way and 3-way
Doppler measurements at 10 s integration intervals [*Lemoine et al.*, 1997], the orbit
characteristics of Lunar Prospector of course being by far superior to Clementine's

Selenographical coverage of the Weilheim measurements

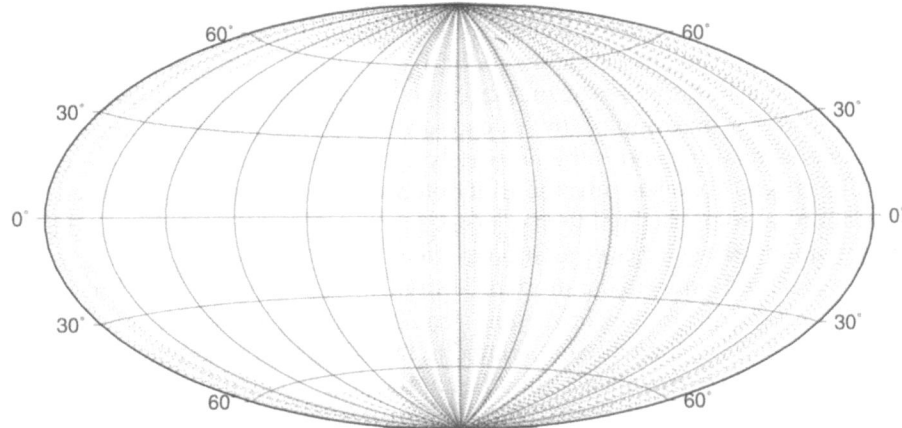

Figure 5.4   Total coverage of the 30 s 3-way Doppler data obtained through the six
               weeks tracking campaign of Lunar Prospector at Weilheim. The plot is,
               similar to previous selenographical plots, centred around 270° eastern
               longitude. Tracking was initiated at the beginning of calender week 28 in
               1998 and was discontinued in the same year at the end of calender week 33.
               A total of 1,030,000 observations at 1 s count intervals were collected; by
               averaging, this number was later reduced to 34,335 30 s observations.
               Despite a reasonably good coverage of the near-side, it is readily seen that
               sectorial gaps remain, with a maximum equatorial range of about 150 km

in terms of gravity field mapping applications. The overall coverage of the near-side, depicted in Fig. 5.4, is encouraging, given the fact that the orbiter was tracked with a single ground station. Nevertheless, longitudinal gaps are clearly visible, emphasising the fact that the campaign did not provide complete coverage of the observable part of the Moon. The maximum longitudinal gaps are in the order of 150 km at the equator.

5.2.1   Spacecraft rotation and its effect on the Doppler observations

It is known for long that spacecraft rotation introduces a constant Doppler shift [*e.g., Marini,* 1970]. This principle is straightforwardly understood by considering the emission of a circularly polarised wave from an antenna on-board the spinning spacecraft. In the antenna reference system, the field vector would, for a constant spin-rate, rotate while transmitting the nominal frequency. For an observer at distance, the frequencies received would oscillate between the nominal transmitter frequency ± the antenna (or spacecraft) spin rate. In the event that the transponder ratio is exactly one, and the same polarisation is used for reception and transmission, the effect of the satellite spin would cancel out. In the case of Lunar Prospector, however, different polarisations and a transponder ratio of approximately 240/221 result in a constant Doppler offset. Nevertheless, for passive 3-way tracking, basically caretaking only the down-link, based on limited knowledge of the up-link frequency, measurement biases are estimated for each *pass*[1]. Hence, any constant Doppler shift due to satellite spin is absorbed by the measurement bias estimation, and is therefore only a minor concern, through correlations with the epoch orbit elements, for orbit determination or subsequent gravity field estimation based on the Weilheim data set. Furthermore, after combination into 30 s observations, the spacecraft rotation is effectively averaged out.

The observed sinusoidal pattern of 3-way Doppler residuals based on 1 s measurements, *i.e.* prior to the conversion into 30 s integrated Doppler observations, shown in Fig. 5.5, is therefore not related to the nominal spin of Lunar Prospector. A more plausible explanation is an offset of the antenna phase centre with respect to the spin axis of the spacecraft, either due to misalignment of the medium-gain antenna on the spin axis, or a nutation of the spin axis itself. Lunar Prospector was equipped with helical low-gain and medium gain antennae, which were nominally aligned with the spacecraft spin axis [*Andolz et al.*, 1998]. Any radial offset between the spin axis and the symmetry axis of the receiving and/or transmitting antenna therefore results in an oscillation of the phase centre with a period equal to the spin period of 5 s. At short 1 s Doppler count intervals, the spacecraft rotation is clearly discernible from the tracking data, with an amplitude δv of about 7 mm/s. This finding is in agreement with *Beckman and Concha* [1998] of the Lunar Prospector navigation team. In the case of antenna misalignment, it would indicate an offset $\delta v / \omega$ of about 6 mm, where ω is the spacecraft's angular rotation rate.

By means of Fourier analysis of the tracking data residuals using a 30 min data arc with 1 s count interval, the best estimate of the spacecraft rotation period is $T = 5.02 \pm 0.02$ s. In addition to the main signal, a second oscillation with twice the frequency and about one-third the amplitude has also been identified. It indicates a notable asymmetry of the periodic phase centre variation, and is likely to be related to phase centre variations with azimuth angle [*Jinsong et al.*, 2000]. Such azimuth-dependencies causing cycloidal rather than circular phase centre variations have been observed for low- and medium-gain helix antennae used for the GPS satellites [*Tranquilla and Colpitts*, 1988]. Since the spin period differs slightly from an ideal value of 5 s, it can unfortunately not be fully compensated by choosing count or sampling intervals that are integer multiples of 5 s. Consequently, a pronounced beat pattern with a typical period of approximately 30 min may *e.g.* be observed by selecting every fifth measurement from a set of 1 sec count interval data. The corresponding spin period of $T = (1/5 - 1/1800)^{-1} \approx 5.014$ s

---

[1]For planetary orbiters the definition of a *pass* is not evident, since the satellite remains in the same part of the sky for a considerable time. In this book, a pass is defined to contain all tracking data separated by less than approximately 2.5 hours. Per-pass biases are estimated per up-link ground station, due to frequent variations in the DSN up-link frequency scheduling

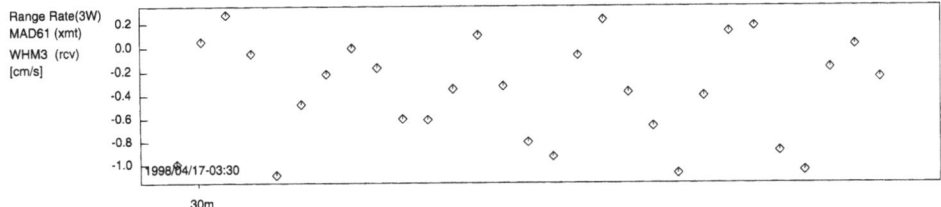

Figure 5.5   An example 30 second interval of short-arc 1 s Weilheim 3-way Doppler
             residuals illustrating the 5 s rotation period of Lunar Prospector. The up-link
             is here provided by DSS-61 at Madrid. Based on the observed periodic
             Doppler shift of ±7 mm/s, the offset of the antenna phase centre with respect
             to the spin axis of the spacecraft is estimated to be approximately 6 mm

is compatible with the value derived from the above described discrete Fourier
transformation. In case of a 30 s count interval the same beat period applies, but
the amplitude of the averaged Doppler variation amounts to a mere 0.02 mm/s,
which may be neglected in comparison with the overall data noise. In other words,
the effect averages out, and has a minuscule impact on orbit determination and
gravity field reduction based on satellite tracking data arcs of up to 3 days.

Attempts to model the spacecraft rotation parameters from the induced
Doppler variations, by application of a bi-harmonic sinusoidal rotation model,
unfortunately turned out unsuccessful, even for long data arcs. Apparently, the
rotational period was not strictly constant, but subject to minor variations. These
may be attributed to variations of the moment of inertia caused by thermal expansion of the booms of Lunar Prospector. Similar effects have been observed for
other spinning satellites [*McElrath*, Jet Propulsion Laboratory, private communi-
cation], but could not be quantified for Lunar Prospector in this study due to the
unavailability of relevant telemetry information.

5.3   Orbit determination using Weilheim data

Since the motion of the satellite is governed by a non-linear second-order ordinary
differential equation and, furthermore, the measurements are non-linear functions
of – among others – the orbit parameters, optimal orbit computation is a non-linear
parameter estimation problem. Numerically, the non-linear problem is solved by
means of iterative procedures, yielding successive first-order corrections to a prior
parameter model, until defined convergence criteria are satisfactorily met. The
orbit determination accuracy depends on a range of facets, among others: i) the
accuracy of the orbit integration procedure as implemented in the software; ii) the
knowledge of the force model; iii) the proper definition of the reference frames
in which the orbits are computed; iv) the correct modelling of tracking measurements; v) the orbit geometry, as well as satellite orbit and attitude control strate-

gies; and vi) the orbit determination strategy and the choice of a set of estimation parameters. Before describing the Lunar Prospector orbit determination results using the Weilheim data, these aspects and the corresponding choice of dynamical and measurement models are outlined. A summary of the orbit modelling parameters and characteristics is given in Table 5.1.

Table 5.1: Summary of dynamic and measurement models along with estimation parameters employed in the Lunar Prospector orbit determination based on Weilheim data using GEODYN II

Measurement models and constants

Observations	3-way passive Doppler measurements, integrated over 30 s; 6° cut-off elevation; data weight 1 mm/s; dynamic editing of spurious measurements
Station coordinates	DSN published coordinates for DSS-16,-24,-27,-42,-46,-61,-66; Weilheim coordinates according to DLR GPS-based solution
Tidal displacement	Love model, including frequency-dependent tides ($h_2 = 0.609, k_2 = 0.0852$); pole tide
Speed of light	$c = 299792.458 \text{ km s}^{-1}$

Force model

Gravity model	LP75G, complete to degree and order 75; $GM = 4,092.800269 \text{ km}^3 \text{ s}^{-2}$, $a_e = 1,738.000 \text{ km}$, $f = 1/100,000.0$
Third body attraction	Sun, Mercury, Venus, Earth (no J_2), Mars, Jupiter, Saturn, Uranus, Neptune, Pluto; positions according to JPL DE-200 ephemeris; relativistic corrections included
Non-conservative	"cannonball" model for solar radiation pressure, including umbra and penumbra and occultation by the Earth; lunar albedo not included in orbit determination
Tidal gravity	dynamical Earth tides not included
Attitude manoeuvres	a single event during the entire campaign; estimated off-line

Satellite model

Mass	187.0 kg; the loss of mass due to above-mentioned attitude manoeuvre was negligible

Cross-section	2.0 m^2; no drum, antenna or boom geometric modelling; no anisotropic surface material properties

Reference frame

Inertial frame	EME2000; orbit integration in true-of-reference date
Lunar body-fixed	Analytical fit to DE-403 lunar libration parameters; following JPL practice no use was made of the IAU system due to the desire to develop gravity models in a principle axis system
Precession	IAU 1976 (Lieske) model
Nutation	IAU 1980 (Wahr) model
Earth rotation	IERS Bulletin A

Estimation parameters per orbital arc

State vector	position and velocity estimated at epoch per 2 and 3 days
Solar radiation	one scaling parameter per 24 hours, nominal C_R value of 1.2
Empirical forces	none estimated, except for the off-line estimation of a single attitude manoeuvre on 1998/07/27 16:10:17 UTC
Biases	per-pass simple measurement biases due to incomplete knowledge of up-link frequencies

5.3.1 Computational models

The Lunar Prospector tracking campaign has previously been described as a joint DLR/GSOC and DEOS initiative. As a consequence, orbit determination efforts have been made using two different orbit determination software systems. While DLR/GSOC mainly used the tracking campaign as a test bed for a new orbit determination tool called DEEPEST [*Weischede et al.*, 1999; *Weischede*, 2000], the work presented in this book relies on the use of versions of the GEODYN II program [*Eddy et al.*, 1990; *Pavlis et al.*, 1998] tailored towards interplanetary tracking, mainly 9510 (IBM UNIX) and 9604 (Cray UNICOS), kindly provided by NASA/GSFC. The GEODYN II program is suitable for accurate orbit prediction and orbit determination, as well as the determination of other parameters related to the tracking measurements. It is capable of handling a variety of observables and multiple satellite arcs. DEOS has a long-time experience with versions of the GEODYN II software package designed for Earth satellite applications.

Obviously, any orbit determination program knows two modes: an orbit generation mode and a data reduction mode for the actual parameter estimation. The purpose of the former is the integration of the equations of motion in a suitable reference frame, in order to obtain the satellite position and velocity as a function of time, given a set of models describing the dynamics of the satellite motion. The selected GEODYN II integrator is an 11th order Cowell predictor-corrector with a fixed time step. The advantage of the relatively high order is found in the high level of accuracy for relatively large step sizes. While other choices of time-marching integration engines are commonplace in satellite orbit problems [*e.g.,* *Shampine and Gordon,* 1975; *Montenbruck,* 1992; *Montenbruck and Gill,* 2000], this particular method is well-suited for near-circular orbits typically found in Earth observation or planetary mapping.

The task of the data reduction mode of GEODYN II is to solve for a specified number of unknowns, given a set of tracking data. It generates an orbit using the first mode and subsequently fits it to the available measurement information using a Bayesian least squares estimation technique based on normal equations. Given the non-linear nature of the underlying mathematical problem, the estimation process is iterative, each iteration providing a first-order differential correction. The overall goal of the successive iteration scheme is to minimise the squared sum of the weighted differences between actual measurements and "computed" measurements derived from the best estimate of the satellite orbit, in other words the *residuals.* In the Bayesian scheme it is furthermore possible to invoke constraints in the form of prior information, a feature which may be used to tune the solution in the case that valuable *a priori* parameter information is available.

In terms of dynamic models applied in the Lunar Prospector orbit determination, the most crucial aspect concerns the choice of selenopotential model. As outlined in Chapters 2 and 3, pre-Lunar Prospector gravity models have been known to predict widely different orbit behaviour. The coming of the LP series of lunar gravity models [*Konopliv et al.,* 1998; *Konopliv and Yuan,* 1999; *Carranza et al.,* 1999; *Konopliv et al.,* 2001] has furthermore made clear that the pre-Lunar Prospector Lun60D model, developed by *Konopliv et al.* [1993], is more closely related to the LP series of models than is GLGM–2, which includes Clementine data and which is also based on a different data reduction strategy [*Lemoine et al.,* 1997]. The differences between LP75G and GLGM–2 have been thoroughly described in previous chapters. In terms of Lunar Prospector orbit determination performance, these conclusions are also supported by *Beckman and Concha* [1998] and *Lozier et al.* [1998], who performed Lunar Prospector orbit determination using 2-way DSN data. Of importance here, however, is that, in terms of Lunar Prospector orbit determination performance, the JPL-developed models appear to outperform GLGM–2. Therefore, to be able to illustrate the quality of both hardware and software modelling aspects of the orbit determination process, the LP75G model was selected for the Lunar Prospector orbit computations.

Third-body perturbations by the Sun and all nine planets are included in the orbit computations. Non-conservative forces include direct solar radiation pressure, including umbra, penumbra and evidently occultation modelling. Following standard practise, the mean direct radiation pressure at one astronomical unit is

4.5783×10^{-6} N/m^2. For the radiation force analysis, the satellite is at all times considered to be a cannonball with isotropic absorption and reflection properties, which implies that any effect of spacecraft geometry, attitude and anisotropic material properties are disregarded in the computation of the radiation pressure effects. The mass of the spacecraft is 187.0 kg, and the cross-sectional area 2.0 m^2. Relativistic corrections include general relativistic light time corrections to the measurement model, general relativistic point-mass accelerations and the relativistic Coriolis force.

In the case of planetary orbiters, the orientation of body-fixed reference frames is coupled to the instantaneous orientation of the Earth equator. While the quasi-inertial frame is given by the equinox and mean equator of J2000.0 (EME2000) , the right ascension and declination of the lunar north pole as well as the instantaneous position of the lunar prime meridian (defined by international agreement) are dependent on the lunar libration parameters. In GEODYN II, these parameters are standardly given by the current solution from the IAU/IAG/COSPAR Working Group on Cartographic Coordinates and Rotational Elements of the Planets and Satellites [*e.g., Davies et al.*, 1992, 1996]. However, these solutions for the body-fixed lunar coordinate system do not coincide with the pole position according to the principle axes of the Moon and do moreover not provide the required level of accuracy. For obvious reasons, like *e.g.* the avoidance of parasitic time-dependent gravity effects, it is desirable to use a principle axis system, which is also recent practise at JPL, among others for the LP series of gravity field models. While the numerically integrated lunar nutation and libration parameters are included in recent versions of the JPL Development Ephemerides products, like DE-403 and DE-405 [*Standish et al.*, 1995; *Standish*, 1998], the option to read planetary libration parameters from this file is not included in GEODYN II. To overcome this, an un-published analytical fit to the numerical parameters, computed by J. G. Williams of JPL, was instead implemented in GEODYN II. A comparison between the orbits derived in the thus defined body-fixed frame and a coordinate frame defined by the numerical librations of DE-403, computed using the DEEPEST software, indicated that the orbit position differences are at all times smaller than approximately 3 m. The systematic differences between the numerically integrated libration parameters and the corresponding IAU/IAG/COSPAR tables are discussed in Sect. 3.3.3. A particular feature of GEODYN II is furthermore that, whereas the inertial coordinate system is defined by EME2000, the actual integration takes place in a true-of-reference date coordinate frame, in turn defined by EME2000 and precession, nutation and libration parameters of the reference date.

The modelling of the tracking measurements as well as tracking station positions was further enhanced by including the geometric effect of tidal displacement, using standard *Love numbers* $h_2 = 0.609$ and $k_2 = 0.0852$. A pole tide solution was also applied, and five-day Earth rotation parameters were included according to IERS Bulletin A of the International Earth Rotation Service, available on the world wide web. The station coordinates of the Weilheim station were first mapped by a VLBI-campaign in 1993 [*Manabe et al.*, 1993] and later converted to the WGS84 reference frame [*NIMA*, 1997] by means of GPS-levelling [*Gill*, private communication]. Tectonic plate motion and ocean loading effects are not modelled, nor is the dynamical effect of the luni-solar tide on the Earth's gravitational potential.

5.3.2   Estimation parameters

In precise orbit determination applications, a set of estimation parameters is carefully selected, in order to maximise the orbit accuracy. Obviously, the orbit accuracy is not known as long as the exact orbit is unknown, and, in practise, the estimation parameters are therefore selected on the basis of the achieved goodness-of-fit and the adequate observability of the individual parameters. On the one hand, the inclusion of more estimation parameters is helpful in improving the fit of the measurements. On the other hand, the information on the estimation parameters must be present in the tracking data set, and, moreover, possibly be individually separable.

The typical set of estimation parameters for the Lunar Prospector orbit determination using Weilheim 3-way data includes, per arc, the orbit epoch parameters (position and velocity), scaling parameters for the solar radiation pressure effects and measurement biases of the 3-way Doppler measurements to account for incomplete knowledge of the up-link frequency and the simplifying assumption of a nominal reference frequency in the tracking data preprocessing and range rate computation. For the detailed RMS-of-fit results presented in the next section, a standard arc length of 2 days is chosen as a reasonable compromise. Long-arc orbit determination results will obviously display larger dynamical modelling errors.

Furthermore, in present-day high-precision orbit determination applications, it is common practise to estimate empirical accelerations (or velocity increments), in order to account for largely unknown effects, in particular secular and one cycle per revolution effects. As an example, zero and one cycle per revolution defects of the gravity field model are often sought to be absorbed in such a fashion. This approach is not followed here, for the simple reason that the orbits derived in this section serve – in a second stage – as starting points for a gravity field reduction process aimed at the improvement of LP75G. An inherent assumption in this process is that all non-physically explained orbit perturbations are due to incomplete knowledge of the gravitational potential, and that their prior removal by means of estimating empirical force parameters is therefore not warranted. In all fairness, this is only an approximation of the real situations, since orbit errors are due to both gravitational and non-gravitational effects, and evidently, if force model defects of non-gravitational origin are not removed by empirical means, these will enter the gravity field solution. However, finding the optimal way of dealing with unknown error sources, having spectral characteristics partially overlapping those of gravity-induced orbit errors, is a non-trivial analysis, and this is exactly why improvements in the gravity field models are inter-connected with improvements in force and measurement models.

In the tracking period of six weeks, a single attitude correction manoeuvre occurred. Due to data scarcity around the time of the manoeuvre, the parasitic effect of this manoeuvre on the satellite orbit is not estimated within a standard 2-day batch, but rather in an "off-line" manner, involving data both prior to and after the correction thrust. In general, due to the previously mentioned data scarcity, this estimation is not very robust, and it is shown later on that the orbit determination results around the manoeuvre are inferior to those achieved during the periods of no thrusting activity. A second important issue involved in the interpretation of the orbit determination results is the fact that the Weilheim campaign relies on a single station for the down-link. This causes problems for the orbit determination process during non-favourable tracking geometries, like the edge-on orbit situation (when the line-of-sight is close to the orbital plane of the satellite). Such edge-on situations obviously occur at approximately 14 days intervals. Approximate dates for such geometries affecting the Weilheim campaign are 14 July, 28 July and 11 August. For continuous DSN tracking using 2-way Doppler data, such problems have not been signalled by deterministic orbit computation results. However they have been indicated by covariance analysis [*Beckman and Concha*, 1998]. It is therefore not surprising that they appear for weaker tracking complements, such as the one used in the present analysis.

5.3.3   Orbit evaluation

A number of tools is available for the assessment of the results of the orbit computation. The orbit determination process produces a rather extended set of information regarding measurement fit, estimation parameter recovery statistics including their correlations, tracking station performance and per-arc and iteration summaries of the orbital solution. Basically, valid orbit solutions are tested against given criteria, like i) the solution must converge, ii) the estimation parameters must have physically meaningful values, iii) the residual RMS-of-fit of the data must meet given quality standards, and iv) the orbit determination program must indicate no problems in the computational process. In the following, the goodness-of-fit of the measurements and the consistency between successive orbit solutions is investigated. Other quality aspects of the solutions are explained along the way.

After the orbit computation process has converged, the RMS of the measurement residuals gives a good indication of the quality of the orbits. Strictly speaking, the orbit accuracy cannot be derived directly from the goodness-of-fit. However, residual statistics for multiple arcs do provide information about the consistency of the orbit solutions. The residual statistics evidently also signify the quality of the data (tracking complement), since large residuals, relative to the theoretical noise level, are unmistakable signs of deficiencies in the modelling or systematic data problems.

For the orbit determination, the increase of the RMS of the residuals with arc length is illustrated by Fig. 5.6. Such an increase is predominantly related to shortcomings in the dynamic force modelling. On the other hand, the low residual level for relatively short arcs is a direct proof of the quality of the Weilheim mea-

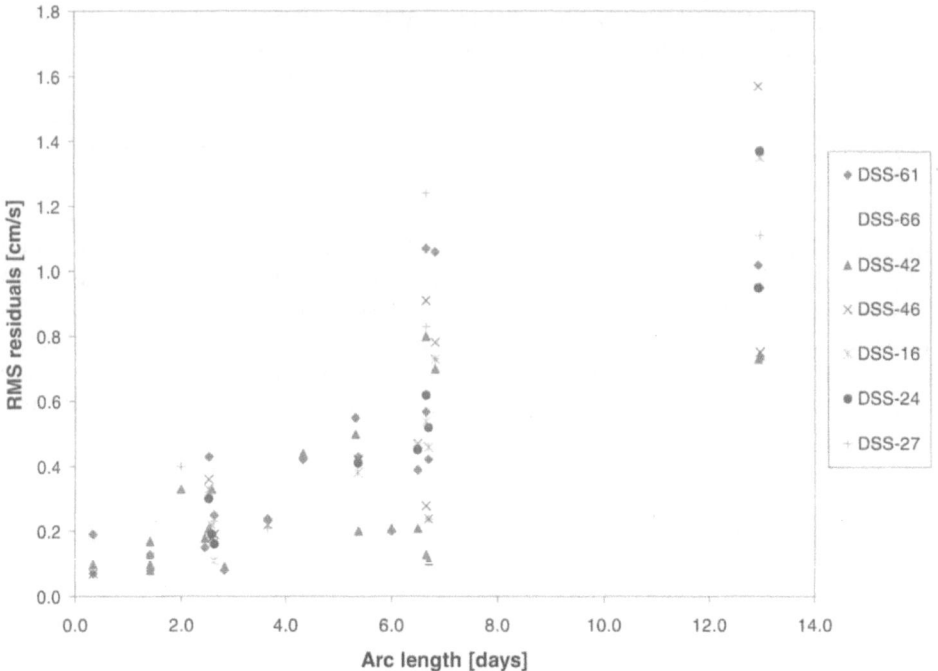

Figure 5.6   RMS of the 3-way Doppler residuals as a function of arc length and up-link
               station. Courtesy of DLR/GSOC, based on orbit computations using the
               DEEPEST program

surements, as well as the decent quality of the LP75G model for short-term and
medium-term orbit predictions.

For 2-day arcs, the RMS of the residuals of the DSS-61 and DSS-42 stations
are depicted in Fig. 5.7. Since the Madrid antennas DSS-61 and DSS-66 share the
highest common visibility with Weilheim among the DSN stations, the number of
measurements for a 2-day arc is higher than those from Goldstone or Canberra.
However, no systematic anomaly related to a specific station or antenna is apparent in any orbit solution, as similar or better residual levels are found for other
up-link stations. Under the favourable conditions of a face-on geometry, range
rate residuals of less than 1 mm/s are obtained for 30 s count intervals and 2-day
arcs, thus confirming both the accuracy of the tracking system and the good quality of the LP75G gravity model for orbit determination purposes. In other words,
the reliability of the tracking complement and the software modelling aspects are
proven. The problem caused by single-station tracking is, however, evident for
edge-on conditions, where the RMS values approach 0.35 cm/s.

The correlation between the measurement bias parameters and the state vector
components is typically lower than 0.2, while the standard deviation of the bias
estimates generally is better than 0.1 mm/s. In other words, the estimation of any

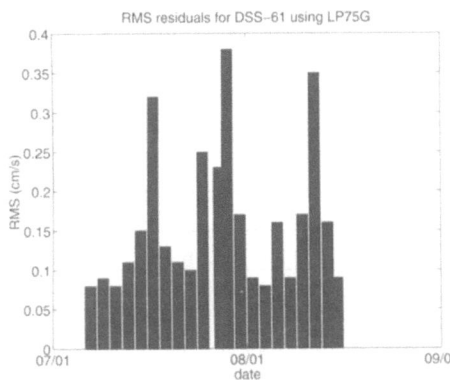
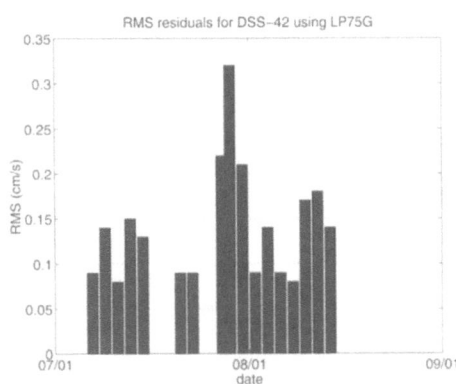

Figure 5.7   RMS value of the residuals for 2 day arcs of the Weilheim 3-way Doppler
measurements for up-links by DSS-61 at Madrid and DSS-42 at Canberra,
over a five-week period, using the LP75G model. Edge-on orbit geometries
occurring around 14 July, 29 July and 11 August, in combination with
single-station ground segment, explain the peaks around these dates

constant off-set in the Doppler measurement is robust, and poses few problems for
the overall orbit determination performance.

Further assessment of the solution quality is possible by means of a comparison
between overlapping orbits. Such analysis directly addresses the robustness of the
orbit estimate, as any inconsistency between overlapping arcs is due to the quality
of the orbit information provided by the data for the overlap period, in combination with force model deficiencies causing successive arcs to yield different orbit
solutions. For the overlap analysis, standard data arcs of 3 days (72 hours) are generated, which are extended by 12 hours in forward and backward directions, thus
creating 24 hour overlap periods at 3 day intervals.

Figure 5.8 depicts the RMS values of the radial and total position differences,
often termed *orbit consistency*, for the 24 hour overlap periods. Overall, most of the
RMS position differences in the radial direction are in the order of 5 m, while the
total position difference RMS values average to about 100–200 m. Compared to the
pre-mission navigation requirement of 1000 m in all directions [*Andolz et al.*, 1998;
Beckman and Concha, 1998; *Lozier et al.*, 1998], this is encouraging. However, problems associated with single station tracking and the attitude correction manoeuvre
are evident. Such problems are evidently of crucial influence on applications requiring precise orbits, like gravity field recovery. For edge-on geometries, there is a
significant increase in the position differences over the overlap periods, to 10 m in
radial direction and 300 m for all components root-sum-squared. Such a temporal
variation of the overlap related to the observation geometry is rooted in relatively
large gravity model errors over the far-side, cf. Chap. 4. Therefore, while the
number of tracking measurements for edge-on passes is reduced by about 30% as
compared to face-on passes, it is not the mere number of observations, but rather

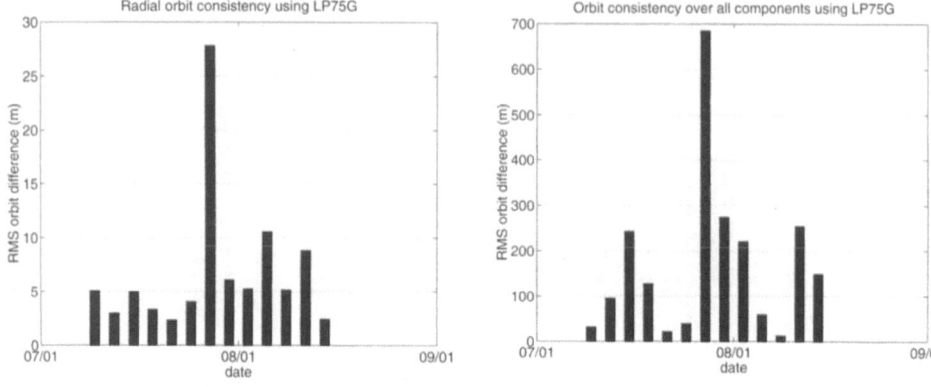

Figure 5.8   RMS values of the radial (left) and total (right) 1 day overlap orbit differences based on LP75G. Each arc has a length of 3 days, with 12 hour extensions at either end, thus creating 24 hour overlap periods between successive arcs. Notice the peak around the attitude correction manoeuvre at 27 July

the systematic force model errors, that cause the deteriorated orbit quality. Moreover, the singularity of the face-on Doppler observation geometry is somewhat removed by the extended tracking arcs of 2 days duration, which provides sufficient variation in parallax.

The effect of the attitude correction manoeuvre is clearly visible for the overlap centred around 27 July, 00:00 hours. The ΔV estimation and also the post-estimation orbit computations yield an effective velocity increment of approximately 2 cm/s. However, given the choice of the batch duration and timelining, it is evident that some unrecovered and spurious manoeuvre effects still remain. For this single overlap, the lack of robustness of the attitude manoeuvre estimation therefore amounts to an 800 m overall position difference. Nevertheless, it is safely concluded, based on the Weilheim data and the LP75G model, that a Lunar Prospector radial orbit consistency of about 5–10 m is documented and that total position errors for well-observable orbit geometries are in the order of 100 m.

Comparing such a performance with that achieved using pre-Lunar Prospector gravity field models, it is documented in the literature [*Beckman and Concha, 1998; Lozier et al., 1998*], that *e.g.* GLGM–2 produces total position difference of several kilometres, and as much as 475 m in radial direction, using continuous DSN 2-way data and based on 2 hour overlaps. Attempts to compute the Lunar Prospector orbit using the Weilheim data and the GLGM–2 model also turned out to be discouraging. In this case, face-on geometries yielded total differences of up to 9 km. Edge-on arcs were even found to be non-convergent. There is therefore little doubt that the 3 months of Lunar Prospector tracking that went into the development of LP75G, combined with an unquestionable trust in the quality of the DSN tracking data (leading to the JPL strategy for planetary orbiter data reduction), have contributed substantially to an improved knowledge of the gravitational field, and in particular the aspects of orbit modelling. Another significant result is, moreover, that Lun60D and other intermediate JPL products not including Clementine tracking data outperform GLGM–2 in terms of Lunar Prospector residual and overlap statistics.

On the other hand, recent results by *Carranza et al.* [1999] indicate that the full use of Lunar Prospector tracking is capable of reducing low lunar orbit errors by yet another order of magnitude. Using the hundredth degree and order models LP100J and LP100K, which are based on a full year of primary mission data at a mean altitude of 100 km, plus respectively 2 months and 6 months of extended mission data, where the periapsis altitude of Lunar Prospector was as low as 15 km, they report radial orbit consistencies in the order of 0.5 m, while the along-track and cross-track components are around 5 m. The overall post-Lunar Prospector orbit accuracy is therefore reportedly around 7 m. Such accuracy levels have, however, not been proven by the present campaign using the Weilheim station. Overlap analysis results using the most recent LP100J model, based on the Weilheim measurements, are further outlined in Sect. 5.4.5.

5.4 Gravity field recovery using Weilheim data

In principle, there are several ways to address the modelling of the gravitational field. Given a set of measurements independent from previously used data sets, one may opt to model the selenopotential independently from all previous modelling efforts. Or, alternatively, one may choose to add the newly gained information to the least squares equation systems of existing models, and so obtain a combined solution including both previous and newly collected data. The former approach leads to a completely independent model, which may be used to assess or verify previous models. Improvements in the knowledge of the selenopotential can in this fashion only be achieved in the event that the newly gained information is superior to previous data, where the quality improvement may be found in either measurement precision, improved sampling, or through the use of observables with improved sensitivity to the various frequency ranges of the gravity field. *Vice versa*, the end product of the latter approach is a model yielding improved data fits for the new data. The quality of such a model is, however, not necessarily better than that of existing models in terms of measurement fit or orbit consistency for data not included in the new solution. In other words, the goodness-of-fit does not automatically improve for all data sets. Hence, such an approach does not straightforwardly lead to an unambiguous improvement of existing models. Opposite to general-purpose gravity field models, this fact has led to so-called *tailored* satellite-only gravity field models, which are targeted at an optimised orbit accuracy for a particular satellite (or orbit).

The overall goal of the gravity field modelling efforts based on the Weilheim data is twofold. First, the principal target is to prove the qualification of the available equipment and tools for tracking, orbit determination and subsequent gravity field recovery. Aspects of data quality and initial orbit determination results using the LP75G model have been described in the previous section, and in this section focus is on the processing chain for gravity field model improvement. Therefore, the approach for the development of a new stand-alone lunar gravity field model is not followed. Rather, it is opted to combine the Weilheim data with all data going into LP75G, and to attempt to improve this model. Strictly speaking, such an independent procedure would also not be possible, since 2-way data collected by DSN, using the same up-link as the Weilheim 3-way data, ultimately are included in Lunar Prospector-based gravity field models. Weilheim-based solutions may therefore not be viewed as independent from the LP series of selenopotential models.

Second, and equally important, it is attempted to illustrate the effect of model design choices which face the analyst during the data reduction process. In other words, the impact of important facets, like the data weight sigma, the weighting factor of newly added least squares systems with respect to previously collected data sets and the choice of the constraint relations (in this case the scaling parameter of the Kaula rule), is analysed using the Weilheim data. No clean-cut optimal values exist for such parameters, and the model development philosophy (some would call this an art) therefore remains an important part of the process. Inevitably, all such choices affect the end result, and, hence, the basis for subsequent deterministic and statistical orbital analysis, as well as physical interpretation efforts. As such, the impact of the above-mentioned parameters is directly related to the quality assessment for existing models discussed in the previous chapters. Moreover, true experience and readiness for future gravity field modelling experiments may evidently also only be achieved by hands-on exercise with real data.

5.4.1 Approach to the gravity field estimation

The procedure for the determination of the gravity field parameters is based on the division of the first five weeks worth of data into 2-day batches (arcs), while the sixth week of the data is left for assessment purposes, for which obviously independent data are needed. For each arc, the arc-dependent parameters are determined by means of the iterative orbit determination process described in the previous section and summarised in Table 5.1. Furthermore, a per-arc normal equation for the corrections to the *a priori* selenopotential coefficients is computed, all of which are combined into a common normal equation for the lunar gravity field parameter corrections. In this process, the correlations between the arc parameters and the gravity field coefficients are taken into account, through partitioning of the arc-wise normal equations. No other parameters common to all arcs are estimated, *i.e.* only gravity field coefficients complete to degree and order 75 are included, and parameters like the lunar gravity parameter GM, Love numbers or tracking station coordinates are not estimated. The reasons for this are both pragmatic as well as scientifically motivated. First, there is little reason to hope that the estimate gravity parameter of the Moon or the Love numbers can actually be improved through a few weeks worth of Doppler tracking of a low orbiter. For station coordinates, the situation is that unconstrained coordinate estimation without some tight connection to the Earth is impossible, as too many degrees of freedom would be introduced in the overall orbit and gravity field estimation problem. Interplanetary tracking data, where the satellite remains in the same part of the sky for considerable periods of time is therefore seldom a suitable tracking type for station coordinate estimation. The combined normal equation is solved using the SOLVE program [*Ullman*, 1993], which is a companion to the GEODYN II software, suitable for the solution of large-scale linear equation systems.

As previously outlined, the Weilheim measurements are not strictly independent from the DSN 2-way data used by the Lunar Prospector radio science team in its lunar gravity field modelling efforts. Moreover, the measurements obtained by a single down-link station over a period of 6 weeks do not provide a full coverage of the lunar near-side. The Weilheim measurements are, however, independent from any data going into LP75G, since this model is based on only the first three months of the nominal Lunar Prospector mission [*Konopliv et al.*, 1998]. Given the overall goals of the gravity field modelling campaign, LP75G is therefore chosen as *a priori* gravity field model, and, moreover, the normal matrix, as derived from the square root information filter of LP75G, kindly provided by the Lunar Prospector radio science team, is added to the Weilheim-derived normal equations. That is, the combined unregularised normal equation for Weilheim-based lunar gravity field solutions reads

$$\left[\left(\mathbf{A}^T \mathbf{Q}_y^{-1} \mathbf{A} \right)_{\text{LP75G}} + \sum_{i=1}^{I} w_i \left(\mathbf{A}^T \mathbf{Q}_y^{-1} \mathbf{A} \right)_i \right] \mathbf{x} = \sum_{i=1}^{I} w_i \mathbf{A}^T \mathbf{Q}_y^{-1} \mathbf{y}_i \qquad (5.1)$$

where w_i are the weighting factors of the normal equations from each Weilheim batch $i \in [1, I]$ relative to the unit weighting factor of the LP75G normal matrix, I is the number of arcs, $\mathbf{y}_i$ are the right-hand sides of the observational equations for the same arcs, and $\mathbf{x}$ is obviously the solution of the combined problem. Notice that the right-hand side is unavailable for LP75G, since it cannot be derived from the square root information filter equation. It is also not required, since it fundamentally represents the residual information based on some (unknown) *a priori* selenopotential model, and not the intrinsic information in the data. Accordingly, the LP75G normal equation acts like a compensator for the imperfect sampling of the near-side achieved by the Weilheim campaign, thereby providing increased stability to the problem. However, it does not contribute otherwise to the corrections $\mathbf{x}$. Adding regularisation, the least squares estimation problem reads

$$\left[\left(\mathbf{A}^T \mathbf{Q}_y^{-1} \mathbf{A} \right)_{\text{LP75G}} + \sum_{i=1}^{I} w_i \left(\mathbf{A}^T \mathbf{Q}_y^{-1} \mathbf{A} \right)_i + \alpha \mathbf{K} \right] \mathbf{x} = \sum_{i=1}^{I} w_i \mathbf{A}^T \mathbf{Q}_y^{-1} \mathbf{y}_i \qquad (5.2)$$

where $\mathbf{K}$, as before, represents Kaula's rule, and α denotes the regularisation scaling parameter(s). As already shown in Table 4.1, for GLGM–2, a single parameter is used for all degrees, while for LP75G, α represents a set of three parameters, each describing distinct degree ranges.

After computing optimal orbit estimates using the procedures described in Sect. 5.3, and subsequently obtaining the arc-wise design matrices $\mathbf{A}_i$ containing the partial derivatives of the 3-way Doppler observations with respect to the gravity field coefficients, it is seen that the final lunar gravity field solution based on the combination of the LP75G data and the Weilheim measurements is fully determined by three sets of parameters:

- the weighting factors w_i assigned to the per-arc normal equations derived from the Weilheim measurements, relative to the unit weighting factor of the LP75G equation system
- the *a priori* covariance $\mathbf{Q}_y$ of the Weilheim measurements
- the scaling parameter(s) α of the regularisation scheme based on Kaula's rule

The choice of the weighting factors w_i is an empirical one. In normal situations, they should be chosen equal to one, since all the tracking information is contained in the normal (or information) matrix and the assumed precision used to weigh the individual measurements. In this sense, it is automatically ensured that each measurement gets the proper treatment during the entire data processing. On the other hand, the aim of this exercise is mainly defined as the exploration of the Weilheim data for gravity field modelling purposes, and therefore, as already mentioned, the inclusion of the LP75G normal matrix may also be rightfully seen as a compensation for the gaps in the near-side tracking, or, equivalently, the incompleteness of the Weilheim near-side data set. In this case, larger weighting factors are interesting, since they emphasise the Weilheim observations, rather than the observations on which the reference gravity field model is based. For the present purpose, weighting factors are chosen as $w_i \in \{0.1, 0.5, 1.0\}$, which in all cases effectively amounts to a significant and deliberate over-weighting of the Weilheim observations, with respect to previously collected low lunar satellite tracking data sets.

Since line-of-sight relative velocity measurements are based on consecutive readings of a Doppler counter, the measurements are usually correlated. However, this does not straightforwardly lead to correlated measurement errors. In accordance with common practice in gravity field recovery applications based on Doppler data [*Lemoine et al.*, 1997], for the Weilheim measurements, the *a priori* measurement covariance matrix $\mathbf{Q}_y^{col}$ is taken to be diagonal. In other words, $\mathbf{Q}_y^{col} = \text{diag}\{\sigma_1^2, \ldots, \sigma_m^2\}$, where σ_j is the weight sigma of the jth measurement point. Based on the data fit of the Weilheim measurements for 2-day arcs, depicted in Fig. 5.7, 1σ data weights in the range of 0.5 mm/s to 2 mm/s appear realistic. In order to avoid overly optimistic *a priori* measurement covariances in the gravity field reduction process, the cases of 1 mm/s and 2 mm/s are selected. The problems of the Weilheim-based orbit determination for edge-on orbit geometries is not accounted for in the data weighting factor selection.

The previous chapters have demonstrated that LP75G and GLGM–2 are op-

Constraint α	data σ	matrix weighting factor w_i		
as in		0.1	0.5	1.0
GLGM–2	1 mm/s	JLGM–07Bg	JLGM–07Cg	JLGM–07Dg
	2 mm/s	JLGM–08Bg	JLGM–08Cg	JLGM–08Dg
LP75G	1 mm/s	JLGM–07Bl	JLGM–07Cl	JLGM–07Dl
	2 mm/s	JLGM–08Bl	JLGM–08Cl	JLGM–08Dl

Table 5.2   Naming convention of the Weilheim-adjusted selenopotential solutions, as a function of the data weight σ, the normal matrix weighting factors w_i and the Kaula constraint scaling parameter(s) α. JLGM is an acronym for *joint lunar gravity model*, emphasising the cooperative efforts of DEOS and GSOC in the development of the solutions

posite extremes in terms of data processing strategy. Whereas GLGM–2 applies a deliberate damping of the higher degree terms, combined with an extensive calibration of the measurement weights, in support of global-scale selenophysical analysis as well as overall error analysis, the LP75G model (and all JPL-derived models in general) rely fully on the quality of the measurements, and also apply a set of regularisation parameters significantly relaxed compared to GLGM–2. This leads to better orbit prediction and orbit determination results, at the cost of an unrealistic covariance estimate and a higher degree of correlation between the gravity field coefficients. In other words, the constraint strategies of GLGM–2 and LP75G may be regarded as possible opposites, with due impact on the gravity field solution. Therefore, for the Weilheim campaign, the impact of these two different constraint schemes is investigated, keeping in mind that, possibly, a whole range of intermediate as well as more extreme choices of the scaling parameter(s) α is possible.

5.4.2   The Weilheim-adjusted selenopotential solutions

Based on the above-described solution strategy, after several iterations over subsets of the Weilheim-based observations, a total of 12 different gravity solutions has been derived, for which the naming convention is given in Table 5.2. The overall class of solutions is named *Joint Lunar Gravity Model* (JLGM), indicating that the modelling is a joint DEOS and DLR/GSOC effort, with an associated number and letter combination suffix denoting the weighting factor, data weight sigma and constraint scheme applied in the particular solution. Specifically, a "-07" solution is developed using a 1 mm/s data weight sigma, while a "-08" solution is based on a value of 2 mm/s. The uppercase characters "B", "C" and "D" denote normal equation weighting factors of 0.1, 0.5 and 1.0, respectively, and, finally, the lower case characters "g" and "l" signify regularisation schemes according to the GLGM–2 and LP75G models, cf. Table 4.1.

From (5.2), it follows that a specific JLGM solution is given by

$$\hat{x}_{\text{JLGM}} = \left[\left(A^T Q_y^{-1} A \right)_{\text{LP75G}} + \sum_{i=1}^{l} w_i \left(A^T Q_y^{-1} A \right)_i + \alpha K \right]^{-1} \sum_{i=1}^{l} w_i \left(A^T Q_y^{-1} y \right)_i$$

(5.3)

while the similar problem for the LP75G model itself would be

$$\hat{x}_{\text{LP75G}} = \left[\left(A^T Q_y^{-1} A \right)_{\text{LP75G}} + \alpha K \right]^{-1} \left(A^T Q_y^{-1} y \right)_{\text{LP75G}}$$

(5.4)

In the following, the JLGM solutions are characterised and compared to LP75G. The results include orbit-related information, both deterministic orbit computation results and results based on the LPT methods introduced in Chap. 3, as well as a selection of the methods (but not all) introduced in Chap. 4. The impact of the Weilheim data set is furthermore discussed in terms of the equipotential surface. While results in tabular form refer to the parameter choices, graphical representations use the naming convention to refer to a particular solution.

Selenoid heights and selenoid height errors

The RMS selenoid height differences of the Weilheim solutions with respect to the *a priori* model LP75G are tabulated in Table 5.3, while the selenographical distribution of the differences for four selected solutions is depicted in Fig. 5.9. Characteristically, these four solutions represent the range of selenopotential model variations that may occur by tuning the design parameters. More specifically, it is chosen i) to demonstrate the effect of the constraint relations by depicting the difference between the "-07Bg" and "-07Bl" solutions, ii) to demonstrate the effect of the data weight sigma by depicting the difference between the "-07Bl" and "-08Bl" solutions, and, finally, iii) to demonstrate the impact of the normal equation weighting factors by depicting the difference between the "-07Bl" and the "-07Dl" solutions. Obviously, other choices would also have been possible, but, based on Table 5.3, it is readily seen that these choices are characteristic, and suitable to prove the order of magnitude effect the various model development parameters have on the final selenopotential solution.

As expected, Table 5.3 clearly indicates that the difference of the Weilheim-based solutions with respect to LP75G increases with the normal matrix weighting factor w_i, or, equivalently, the weighting factor of the additional information provided by the tracking data. There is approximately a factor 2.6 – 3.2 difference between the "D" solutions and the corresponding "B" solutions, with the higher figures found for the lower 2 mm/s value of the data weight σ. Similarly, there is about a factor two variation in the difference in selenoid height compared to LP75G originating from the choice of the data weight σ. Finally, the figures indicate a distinct variation of the solutions according to the choice of constraint scheme. Clearly, the choice of the GLGM–2 regularisation parameter yields a solution closer to LP75G than the choice of regularisation parameters according to

Constraint α	data σ	matrix weighting factor w_i		
as in		0.1	0.5	1.0
GLGM–2	1 mm/s	1.51	2.94	3.95
	2 mm/s	0.77	1.74	2.36
LP75G	1 mm/s	1.67	3.48	4.84
	2 mm/s	0.84	2.01	2.79

Table 5.3   RMS values of the selenoid height differences in metres with respect to LP75G for gravity field solutions including the Weilheim 3-way Doppler measurements, as a function of the data weight σ, the normal matrix weighting factors w_i and the Kaula constraint scaling parameter(s) α. It is stressed that the figures are computed RMS-wise, and not as differences of two RMS values

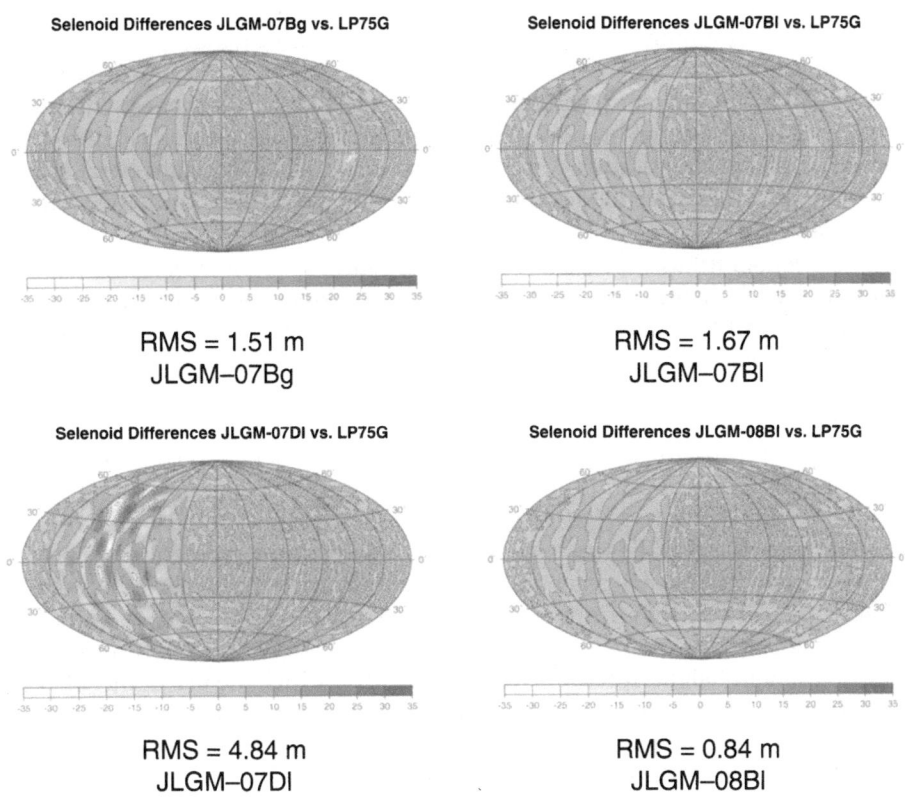

Figure 5.9   Selenographical distribution of the selenoid height differences in metres with respect to LP75G for four selected gravity field solutions including the Weilheim 3-way Doppler measurements, as a function of the data weight σ, the normal matrix weighting factors w_i and the Kaula constraint scaling parameter(s) α. As before, the plots are centred around 270° eastern longitude

Constraint α as in	data σ	matrix weighting factor w_i		
		0.1	0.5	1.0
GLGM–2	1 mm/s	2.86	3.01	3.09
	2 mm/s	2.82	2.94	3.00
LP75G	1 mm/s	2.03	2.25	2.35
	2 mm/s	1.98	2.17	2.25

Table 5.4　　RMS formal selenoid height error improvements in metres with respect to LP75G for gravity field solutions including the Weilheim 3-way Doppler measurements, as a function of the data weight σ, the normal matrix weighting factors w_i and the Kaula constraint scaling parameter(s) α

the LP75G model itself. Depending on this choice, there is a variation of up to 15% in the RMS selenoid heights relative to LP75G. Such a feature is explained by the tighter constraining of the new Weilheim-based information provided by the GLGM–2 constraint scheme. *Vice versa*, the use of the LP75G scheme leaves more freedom for the adjustment, and therefore leads to a larger difference with respect to the *a priori* gravity field model. The LP75G normal matrix is used entirely as a stabilising matrix, in addition to a Kaula scheme. Hence, the right-hand side information on which the JLGM solutions are based comes solely from the Weilheim 3-way data.

The selenographical distribution of these differences reveals that the differences are largely manifested over the far-side. Whereas the near-side differences with respect to LP75G are similar, featuring some track effects coming from the new data (as well as their gaps), the treatment of the Weilheim-based data sets may have a large impact on the far-side solution. The bulk of the global RMS selenoid height figures presented in Table 5.3 is therefore stemming from the far-side.

Considering the formal improvement in selenoid heights errors, due to the Weilheim observations and deduced from the *a posteriori* covariance matrix $\hat{Q}_x^{col}$, shown in Table 5.4 in terms of global RMS values and in Fig. 5.10 as a function of selenographical coordinates, a similar trend is apparent. Straightforwardly, the formal improvements increase with the weight assigned to the new observations, in terms of increased matrix weighting factors w_i or improvements in the data weight σ (*a priori* measurement covariance matrix). Moreover, conform the findings for selenoid height differences with respect to LP75G, the use of the GLGM–2 regularisation parameter scheme leads to larger error improvements than the corresponding LP75G scheme. On a global basis, the formal RMS improvement varies between 1.98 m and 3.09 m.

Selenographically, the selenoid height improvements are intimately related to the fact that the Weilheim tracking is non-continuous. First, the error improvements are nearly exclusively related to the lunar far-side, which supports the previous conclusion, based on selenoid height differences, that the Weilheim observations add very little to the accuracy-improvement of the near-side gravity field estimation. Second, while the error estimate from the LP75G model, which is based

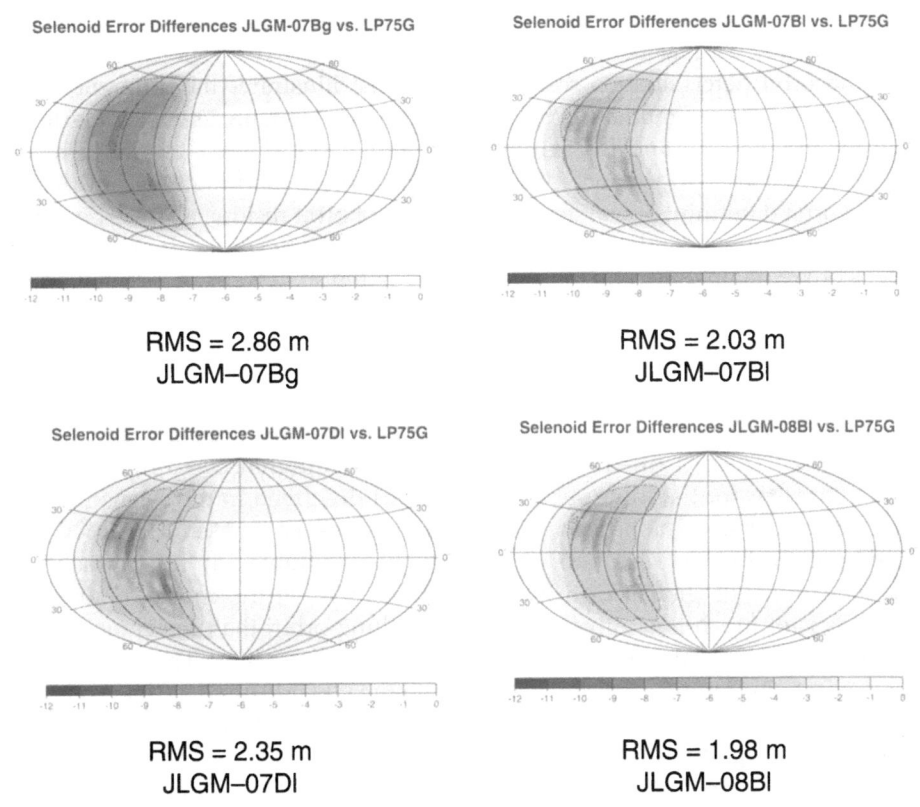

RMS = 2.86 m
JLGM–07Bg

RMS = 2.03 m
JLGM–07Bl

RMS = 2.35 m
JLGM–07Dl

RMS = 1.98 m
JLGM–08Bl

Figure 5.10   Selenographical distribution of the formal selenoid height error improvements in metres, relative to LP75G, for the four selected gravity field solutions including the Weilheim 3-way Doppler measurements, as a function of the data weight σ, the normal matrix weighting factors w_i and the Kaula constraint scaling parameter(s) α

on continuous DSN tracking, alone leads to a rather smooth far-side error, cf. Fig. 4.8, the JLGM solutions all show signs of error scattering over the far-side. This may be explained as follows: as will be shown in Sect. 5.4.3, using the JLGM solutions, all 2-day batches fit well to the available tracking data. However, due to the fact that the data stream is intermittently disrupted, the computed orbit solutions, on which the individual per-arc JLGM normal matrices are based, are found not to be very consistent. In other words, overlapping orbits do not match particularly well, as will be shown later on. A logical consequence is therefore that the far-side projection of the numerical properties of the Weilheim-based normal equations are inferior to those based on DSN arcs of continuous spacecraft tracking. In this sense, the Weilheim campaign does little to improve the ill-conditioning of the far-side gravity field solution.

5.4.3   Orbit quality

The orbit-related assessment of the JLGM solutions comprise both LPT-based analysis and orbit determination results. The former illustrates the differences with respect to the *a priori* model LP75G in the frequency-domain, and is similar to the analysis of GLGM–2 and LP75G presented in Chap. 3, while the latter reveals the deterministic performance of the models in orbit computations. In particular, the orbit determination assessment consists of orbit residuals and orbit consistency computations for the models, both for so-called *dependent data* (= the first five weeks of Weilheim data) and *independent data* (= the sixth week of Weilheim data).

Temporal orbit characteristics of the Weilheim solutions using LPT

Table 5.5 shows the RMS values of the radial orbit difference in metres with respect to LP75G, computed by means of the LPT method described in Chap. 3, while Fig. 5.11 depicts the RMS values of the orbit difference as a function of orbital frequency, for the same subset of solutions described in the selenographical plots of Sect. 5.4.2. The orbit parameters are identical to those of the 100 km altitude polar orbit used in Chap. 3 and presented in Table 3.1. It is immediately apparent from the global RMS figures of Table 5.5 that the choice of constraint relations is of much less influence on the resulting orbit, than the choice of data weighting factors σ and normal equation weighting factors w_i. There is in fact, in this respect, no significant difference between the GLGM–2 and LP75G constraint schemes, yielding about 6.2 m, 8.9 m and 9.8 m for a data weight σ of 1 mm/s and matrix weighting factors w_i of 0.1, 0.5 and 1.0, respectively. For a 2 mm/s weight σ, these numbers are significantly lower, consistent with the additional reliance on the LP75G normal matrix. For the lowest matrix weighting factor of 0.1, the difference between the 1 mm/s and 2 mm/s data weight σ amounts to some 45%, whereas for the highest matrix weighting factor of 1.0, the same difference is a mere 20%. A similar, but weaker trend was also seen for selenoid heights, where the matrix weighting factors and the data weight sigma were found to be of higher importance than the regularisation parameter(s). Generally, this supports the idea that the bulk of the difference between the JLGM and LP75G coefficient solutions is due to the *sec* information contained in the Weilheim data sets as well as how they are treated in 2-day batches, and not the constraints that are being applied to regularise the combined JLGM normal matrices.

Figure 5.11 provides insight in how the differences with respect to the a priori model are distributed over the orbital frequency range. In all cases, the major peak is found close to the 1 cpr frequency, which indicates that at least some part of the orbit difference is absorbable by means of orbit parameter reduction (orbit determination). The success of such a reduction process nevertheless would depend upon the length of the arc. Furthermore, the figure demonstrates that the overall frequency distribution of the orbit differences is similar, disregarding the choice of the constraint parameters. Generally, the effect of frequencies beyond 1 cpr are at the level of a few centimetres, and only for the largest matrix weighting factor w_i

Constraint α as in	data σ	matrix weighting factor w_i		
		0.1	0.5	1.0
GLGM–2	1 mm/s	6.23	8.90	9.86
	2 mm/s	3.47	6.69	7.86
LP75G	1 mm/s	6.22	8.87	9.82
	2 mm/s	3.47	6.68	7.84

Table 5.5   RMS values of the radial orbit difference in metres with respect to LP75G for gravity field solutions including the Weilheim 3-way Doppler measurements, as a function of the data weight σ, the normal matrix weighting factors w_i and the Kaula constraint scaling parameter(s) α, based on LPT

do such individual frequency components approach the decimetre level.

Similar to the case of the LP75G and GLGM–2 orbit differences discussed in Chap. 3, the radial orbit differences between the JLGM solutions and LP75G may be depicted in a per-degree and per-order fashion, shown in Fig. 5.12. When interpreting these plots, it is again important to realise that the error per degree l is coupled with other degrees of the same parity, while the error per order m is independent from other orders. Using any of the depicted JLGM solutions, it is readily seen that, similar to the a priori model LP75G, as demonstrated in Fig. 3.4, all JLGM solutions yield a larger radial orbit difference with respect to LP75G at a range of individual degrees than does the overall model (the overall RMS numbers from Table 5.5 are shown in the plots for reference). In other words, much like LP75G, the JLGM selenopotential solutions all suffer from negative correlations between harmonic degrees. In this respect, and fully in line with the expectation, the Weilheim-based solutions have inherited many of their characteristics from LP75G, not because they are derived in a similar fashion, but because the models are simply based on more data of the same type, which do not overcome the principle problems in lunar gravimetry.

On the contrary to the difference between LP75G and GLGM–2, cf. Fig. 3.5, which showed a dominant peak at order $m = 1$, the order-wise difference between the JLGM solutions and LP75G exhibit peaks for several orders in the low harmonic range. In fact, all odd orders $m \in \{1, 3, 5, 7\}$ are at the one metre level. This is in fact possible, because of the slow rotation of the Moon, which leads to a very small basic frequency separation. As a consequence of the small base frequency, the contribution of different orders may be close in terms of cycles per revolution.

Lunar Prospector 3-way Doppler residuals

The second part of the orbit-related assessment of the JLGM solutions concerns its performance in orbit computations. To this end, the whole range of twelve solutions has been applied in 2-day arc orbit determinations, similar to the case of LP75G presented in Fig. 5.7. For the sake of a direct comparison, all other parameters except the choice of the gravity field model are obviously identical, cf.

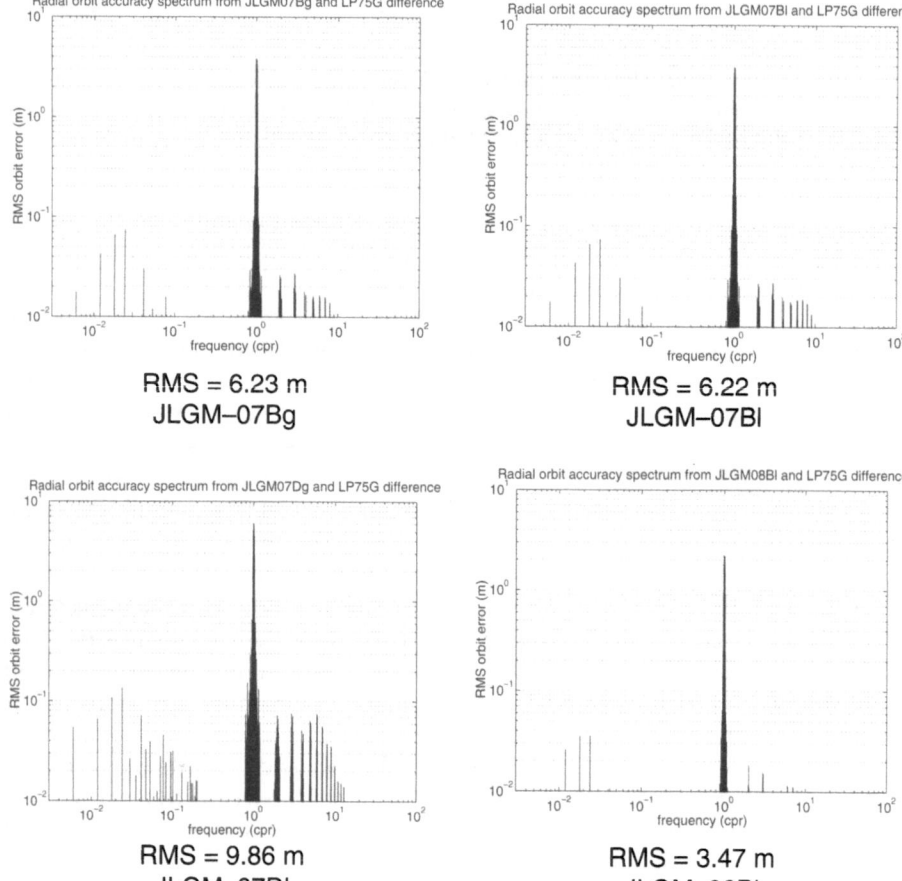

RMS = 6.23 m
JLGM–07Bg

RMS = 6.22 m
JLGM–07Bl

RMS = 9.86 m
JLGM–07Dl

RMS = 3.47 m
JLGM–08Bl

Figure 5.11   Frequency spectra of the radial orbit difference between selected gravity field
solutions including the Weilheim 3-way Doppler observations and LP75G,
based on LPT, for a 100 km altitude polar orbit, as a function of data weight
σ, matrix weighting factor w_i and regularisation parameter(s) α of the
respective Kaula scheme

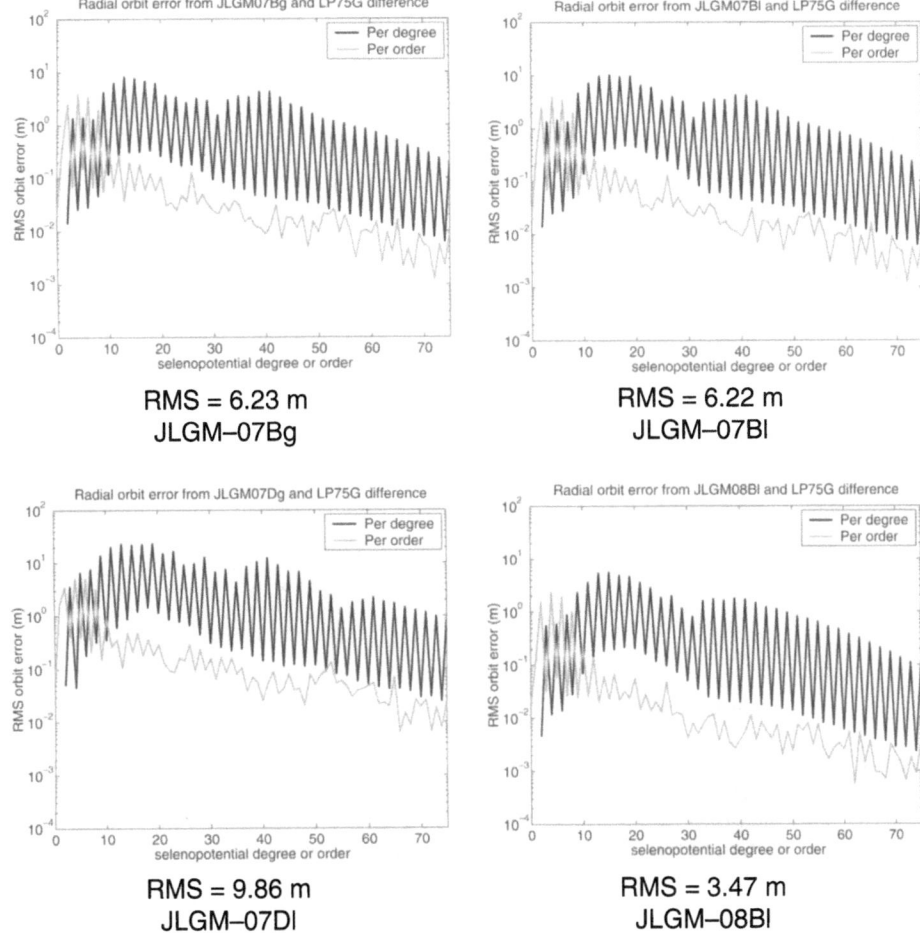

RMS = 6.23 m
JLGM–07Bg

RMS = 6.22 m
JLGM–07BI

RMS = 9.86 m
JLGM–07DI

RMS = 3.47 m
JLGM–08BI

Figure 5.12   Per-degree and per-order radial orbit differences for a 100 km altitude polar
orbit, as a function of data sigma, matrix weighting factor and regularisation
parameter(s), from the coefficient differences of selected gravity field
solutions including the Weilheim 3-way Doppler observations with respect to
LP75G. The global RMS value is the root-sum-square over all harmonic
orders, or, equivalently, all frequencies

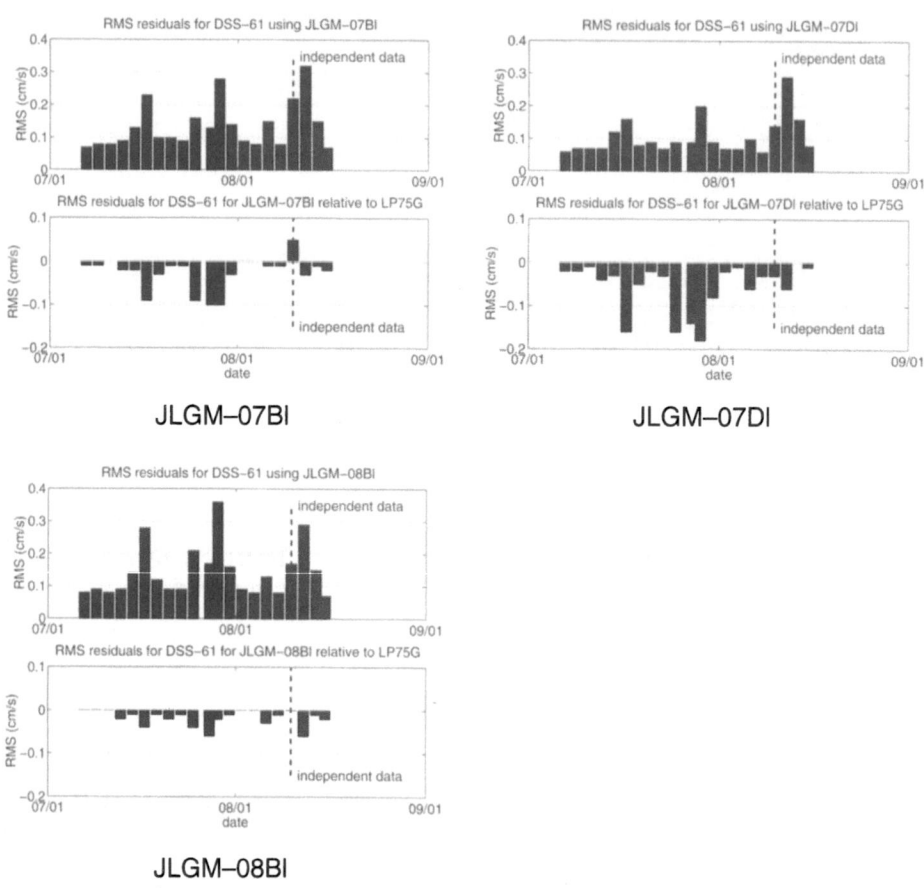

Figure 5.13   Residuals statistics for up-links by DSS-61 at Madrid, based on 2-day arcs, for three selected "I"-type JLGM solutions, as a function of the data weight σ, the normal matrix weighting factors w_i and the Kaula constraint scaling parameter(s) α. Significant residual improvements, relative to LP75G, are achieved for both dependent and independent data

Table 5.1.

Figure 5.13 depicts the residual statistics for the entire six weeks of tracking, in batches of two days, for up-links provided by DSS-61 at Madrid. In addition to the absolute residual plots, the figure also depicts the residuals relative to the same residuals achieved using LP75G, thus providing a direct comparison between all JLGM solutions and LP75G. For comparison, Fig. 5.14 depicts the same information for up-links by DSS-42 at Canberra. Overall, for preferable orbit geometries, *i.e.* no edge-on situations, the residual statistics for both up-link stations consistently yield RMS values around 0.1 cm/s, while for edge-on geometries residual RMS values up to 0.35 cm/s are obtained. A general finding is, moreover, that

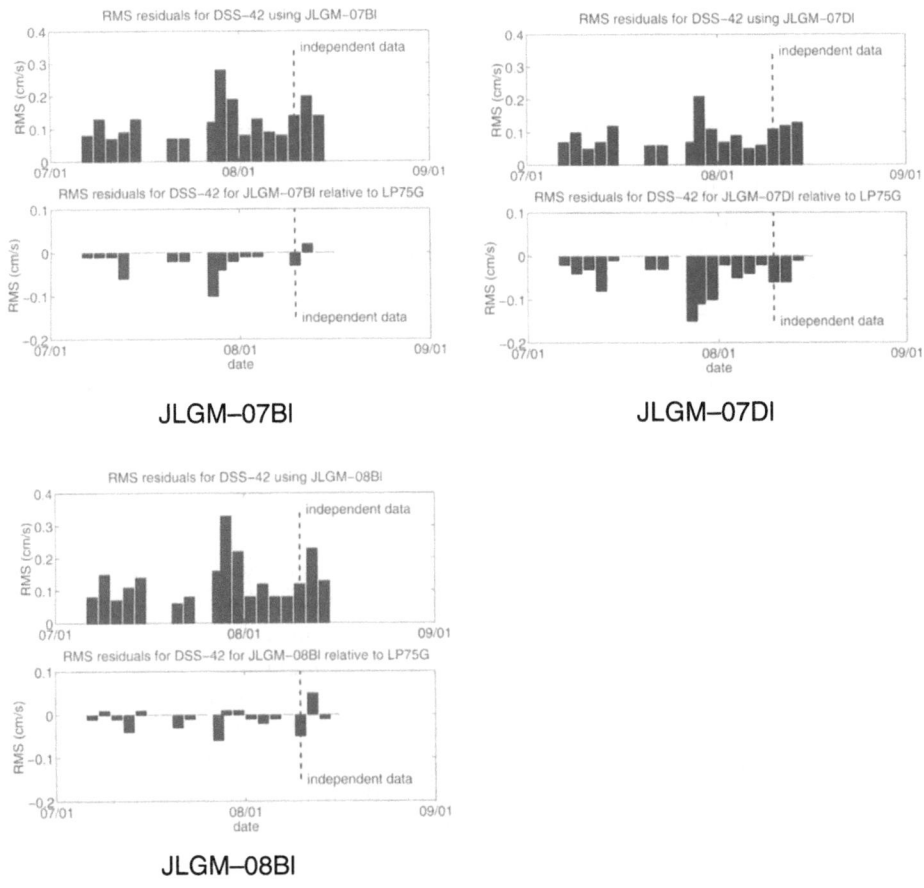

JLGM–07BI

JLGM–07DI

JLGM–08BI

Figure 5.14   Residuals statistics for up-links by DSS-42 at Canberra, based on 2-day arcs, for three selected "l"-type JLGM solutions, as a function of the data weight σ, the normal matrix weighting factors w_i and the Kaula constraint scaling parameter(s) α

the measurement fit of the 3-way Doppler measurements are largely independent of the choice of the regularisation parameter(s) α investigated here. To within 0.1 mm/s, the "g" and "l" solutions produce identical goodness-of-fit values. This is also the reason why Figs. 5.13 and 5.14 only depict type "l" solutions. On a general level, the residual RMS values are consistent with improvements, relative to LP75G, in the range from 0 to 0.10 cm/s for favourable orbit geometry situations and in the range from 0 to 0.17 cm/s for non-favourable geometries. Occasionally, a case of non-improving residuals statistics occur, but such arcs are rare and furthermore related to data scarcity as a result of the single station down-link. For comparison, the DSN operations schedule of Lunar Prospector involved up to seven DSN up-link and down-link stations during the same period.

One encouraging aspect of the residual analysis is the fact that the residual improvement also extends to data sets not included in the JLGM solutions. For so-called *dependent* data, *i.e.* data that has been used in the model development, such a residual improvement is nothing more than a natural result of the data reduction process. However, the persistent residual improvement achieved for the sixth week of Weilheim data, here termed *independent data*, is a first indicative of a more general improvement of the *a priori* model. It is stressed, however, that the pool of independent data spans no more than one week of single station tracking data, and that the data in principle is too scarce to draw conclusions about the general quality of the JLGM solutions, and in particular its performance in orbit computations compared to that of LP75G. Moreover, as will be shown later, the improved residual level for 2-day arcs is not straightforwardly translated into an improvement in orbit consistency. Nevertheless, the residual improvement is a quality verifier for the overall gravity field reduction processing chain.

Lunar Prospector orbit consistency

Using the technique described in Sect. 5.3.3, the consistency of the Lunar Prospector orbit using the series of JLGM selenopotential solutions may be computed. As for LP75G, cf. Fig. 5.8, the analysis is based on 24 hour orbit overlaps of consecutive 3-day arcs. These arcs are therefore one day longer than the arcs on which the per-arc JLGM normal equations are based. Figure 5.15 depicts the orbit consistency in the radial direction, for the same subset of JLGM solutions as used for the residual analysis, while Fig. 5.16 depicts the root-sum-square (RSS) values over all three position coordinates. As for the residuals, the RMS-wise orbit differences are given in both absolute values and relative to LP75G.

Whereas, for favourable orbit geometries, the consistency in radial direction using LP75G generally was found to be in the order of 5 m, the JLGM solutions, with the exception of a single overlap case, reduce this by a maximum of approximately 4 m for both dependent and independent data periods. However, for the overall orbit differences, root-sum-squared over all three position coordinates, the performance of the JLGM solutions is unfortunately less flattering. Still, for data included in the Weilheim solutions, a moderate improvement is seen, but for the independent sixth week of tracking, the orbit consistency is significantly deteriorated with respect to that of LP75G, with up to 100 m difference. Moreover, the severity of the reduced consistency grows with increasing reliance of the Weilheim data, both in terms of data weight σ and normal matric weighting factors w_i.

In other words, a fundamental result of the Weilheim campaign and subsequent orbit and gravity field reduction is that although a significant improvement of the measurement fit has been achieved for 2-day arcs, this does not translate into an accompanying improvement in orbit consistency. Instead, one is faced with a more difficult situation where the radial component of the orbit seems well-resolved by the Weilheim-based solutions, using independent data, whereas the overall difference is not. On the basis of these results, it is evidently not warranted

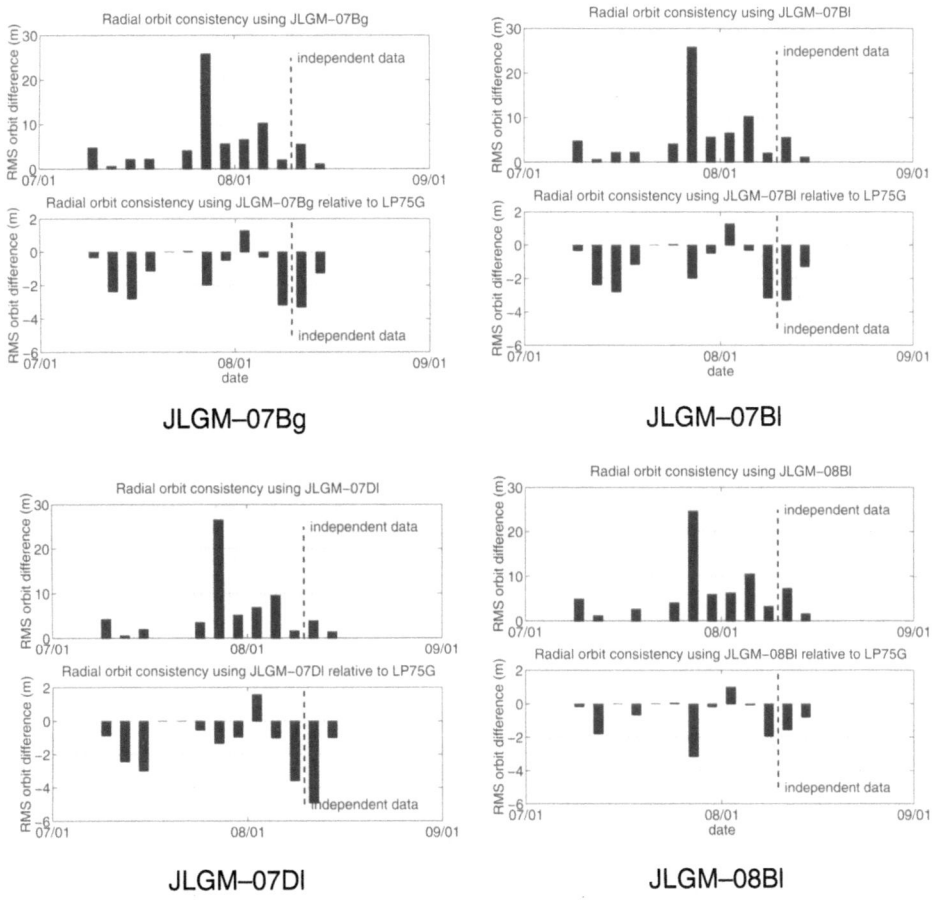

Figure 5.15   RMS values of the radial orbit differences for 1 day overlap periods, based on 3-day batches for orbit computation, for a selection of the Weilheim-based solutions, as a function of the data weight σ, the normal matrix weighting factors w_i and the Kaula constraint scaling parameter(s) α

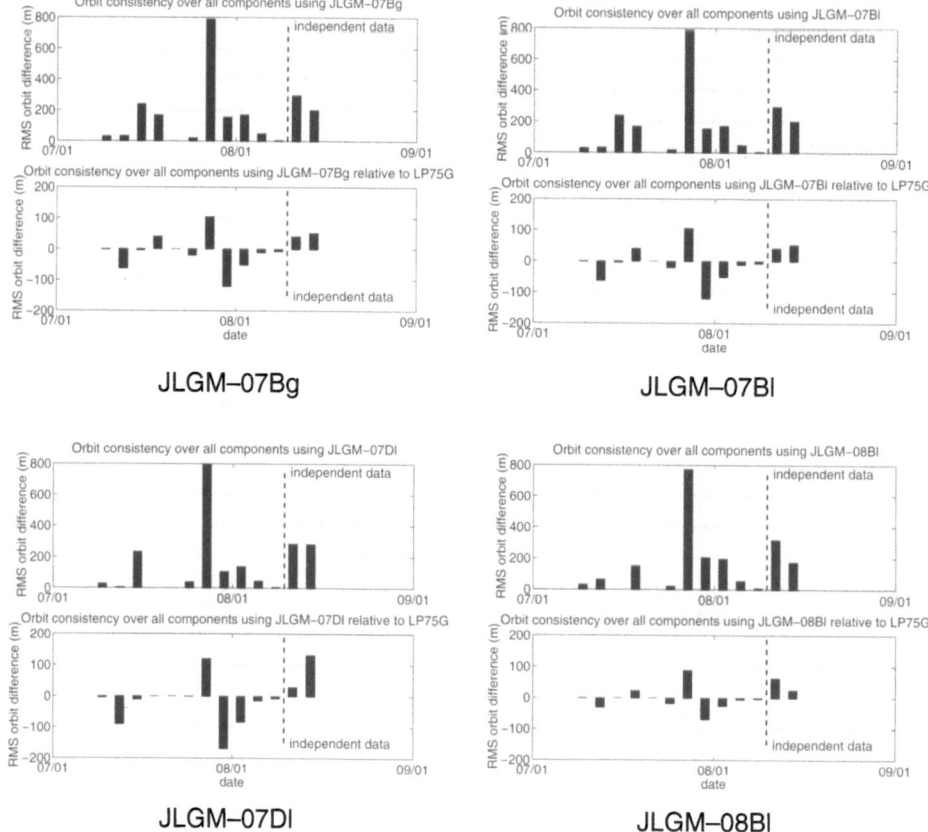

JLGM–07Bg   JLGM–07Bl

JLGM–07Dl   JLGM–08Bl

Figure 5.16   RMS values of the overall orbit differences (RSS over all three components)
              for 1 day overlap periods, based on 3-day batches for orbit computation, for a
              selection of the Weilheim-based solutions, as a function of the data weight σ,
              the normal matrix weighting factors w_i and the Kaula constraint scaling
              parameter(s) α

to view the JLGM solutions as improvements over LP75G. Yet again, however, it is an unfortunate fact that the data pool is too limited for an extended validation of the Weilheim-based selenopotential solutions over a longer period of time. Nevertheless, it is appropriate to emphasise that the JLGM selenopotential solutions are based on intermittent tracking using a single ground station. Irrespective of the fact that the 3-way Doppler observations carry somewhat less information than the 2-way counterpart, it is the fact that the JLGM solutions are based on non-continuous tracking, in an environment characterised by significant dynamical force model errors, that makes modelling of the selenopotential using these data difficult. In other words, the problems related to the consistency of the orbit solutions are somewhat to be anticipated. Within the limits of the tracking complement, and in the absence of any budget at all, the JLGM solutions are nevertheless interesting results in an exercise to put the available infrastructure and modelling tools to the test. Furthermore, these consistency results emphasise the importance of continuous tracking for a stable and consistent recovery of the satellite orbit.

Finally, for all components, in accordance with the LPT-based results, it is readily seen that the orbit performance of the Weilheim-based selenopotential solutions does not depend on the choice of the constraint relations, within the bounds of the schemes applied in GLGM–2 and LP75G. The "g" and the "l" solutions produce, to within a few centimetres, identical orbit differences for 1-day overlaps.

5.4.4   Degree variations and solution consistency

The assessment of the JLGM selenopotential solution is concluded with a discussion on the degree-wise signal power spectrum and the corresponding spectrum of the variances, plotted in Fig. 5.17. In Chap. 4, it was concluded that the LP75G model shows clear signs of optimistic error calibration, since the degreewise signal-to-noise ratio, based on the covariance matrix Q_x^{col}, remains higher than one for the entire degree range of the model. Obviously, by adding the Weilheim observations to the data sets, the corresponding formal selenopotential error of the JLGM solutions will improve accordingly. In other words, the data weight sigma of the JLGM–07 and JLGM–08 solutions, which in turn are based on the fit of the Weilheim measurements for 2-day arcs, cf. Fig. 5.7, still leads to an underestimation of the real error, much like the LP75G model.

However, the availability of orbit fit results, in combination with knowledge of the applied data weights, now enables a proper interpretation of the degree-wise signal-to-noise factors. First, the fact that the data weights are consistent with measurement fit statistics is a direct indication that the weighting schemes are reasonable, at least for orbit computation applications. The fact that the JLGM solutions have a higher power than the corresponding variances at the high end of the degree scale therefore indicates that the overall data set, including all observations used to develop LP75G as well as the Weilheim measurements, principally carries a significant amount of even higher degree information, *i.e.* beyond $l = 75$. In fact, it is exactly this type of reasoning that led to the development of the LP100n series

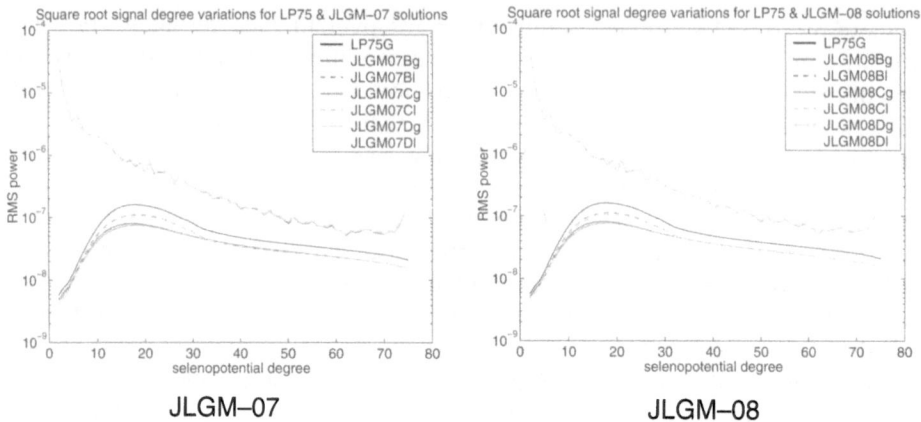

JLGM–07   JLGM–08

Figure 5.17   Degree-wise RMS amplitude spectra of the JLGM selenopotential solutions
(thick lines) and the corresponding error variances (thin lines of same colour).
Noticeable deviations of the JLGM selenopotential degree variations from
LP75G are only seen for the $l = \{15, \ldots, 25\}$ degree range, *i.e.* where,
according to the analysis in Chap. 4, the constraint relations for LP75G start
to play an important role. For the larger part, the degree-wise power
variations are seen for the coefficient variances. There is a clean-cut
difference between the "g"-type solutions and the "l"-type solutions, the latter
yielding a factor 2–3 larger formal errors for approximately the
$l = \{12, \ldots, 30\}$ degree range. In particular for such medium-range
harmonics, the choice of the regularisation parameters appear crucial to the
formal error estimate $\hat{Q}_x^{col}$. There is also a factor two peak difference, for the
same harmonic degrees, between the JLGM–07 solutions and the JLGM–08
solutions

of Lunar Prospector-based gravity models, cf. Fig. 3.16. In other words, for models
of moderate size, like a 75×75 spherical harmonic expansion, a fairly tight data
weight is required to avoid excessive aliasing in the low-degree part of the spectrum. Expanding to 100×100 models, like LP100J and LP100K, the additional
high-degree terms enable some loosening of the weighting schemes [*Konopliv*, private communication], and, consequently, a gravity-field solution approaching a
signal-to-noise ratio of one at the very high degree end.

In a nutshell, this situation reflects an important facet of present-day lunar
gravity field modelling, already emphasised at several occasions in this book. On
the one hand, there is insufficient satellite data available to allow an unambiguous
recovery of the full selenopotential. On the other hand, there is an excellent data
set for the near-side, which absolutely requires a sizeable parameterisation, in order to allow a fully consistent solution to be derived. Another way to look at the
same problem would be to introduce other types of basis functions with a more
compact support than the fully global spherical harmonics.

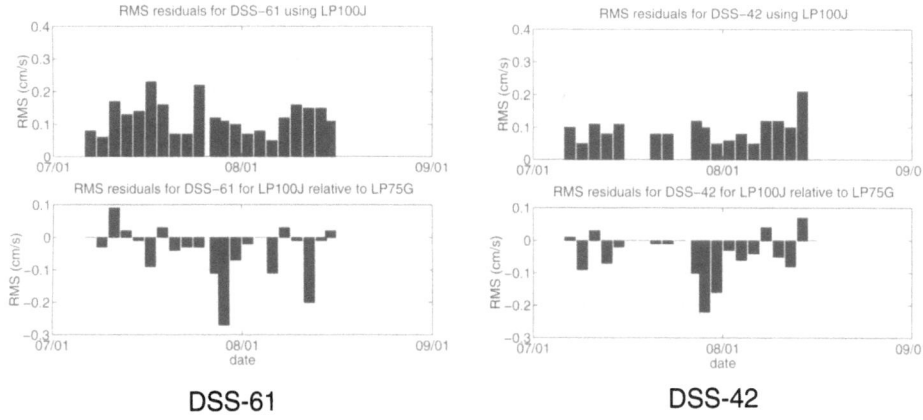

DSS-61   DSS-42

Figure 5.18   Residuals statistics for up-links by DSS-61 at Madrid and DSS-42 at Canberra, based on 2-day arcs, using LP100J

5.4.5   Discussion

In this chapter, the findings of the first European experiment in the field of low lunar orbit determination and subsequent gravity field recovery have been reported. The strengths of the Lunar Prospector orbit characteristics for near-side gravity field mapping, as well as the limitations of the data set provided by the six weeks campaign using DLR's 30 m deep space antenna at Weilheim, have been analysed. Based on orbit fit data and a variety of data weighting and constraint schemes, aimed at the exploration of the class of solutions that might possibly be derived from the observational data, a family of twelve lunar gravity field solutions have been derived.

The overall assessment techniques applied to these selenopotential solutions are all rooted in the fields of orbital mechanics, selenodesy or applied numerical analysis. That is, no attempt has been made to derive the models from a selenophysical point of view, *e.g.* through coupling with other selenophysical information, like magnetic data or seismic data, nor has it been attempted to apply regularisation schemes stemming from other selenophysical data, like *e.g.* the lunar topography and assumptions on its correlation with gravity. In other words, the assessment has been based on what is actually measured in a selenodetic experiment. Furthermore, the analysis of the Weilheim-based selenopotential solutions shows, in fact, that the JLGM–07 and JLGM–08 solutions share both the strengths and the weaknesses of other recent solutions, in particular LP75G which served as an a priori model for the present analysis. For this reason, the range of assessment tools applied to the JLGM solutions has been somewhat restricted with respect to the analysis and comparison of GLGM–2 and LP75G in Chapters 3 and 4.

Obviously, it would have been a straightforward operation to repeat the entire set of computations made in the above-mentioned chapters for the entire class of

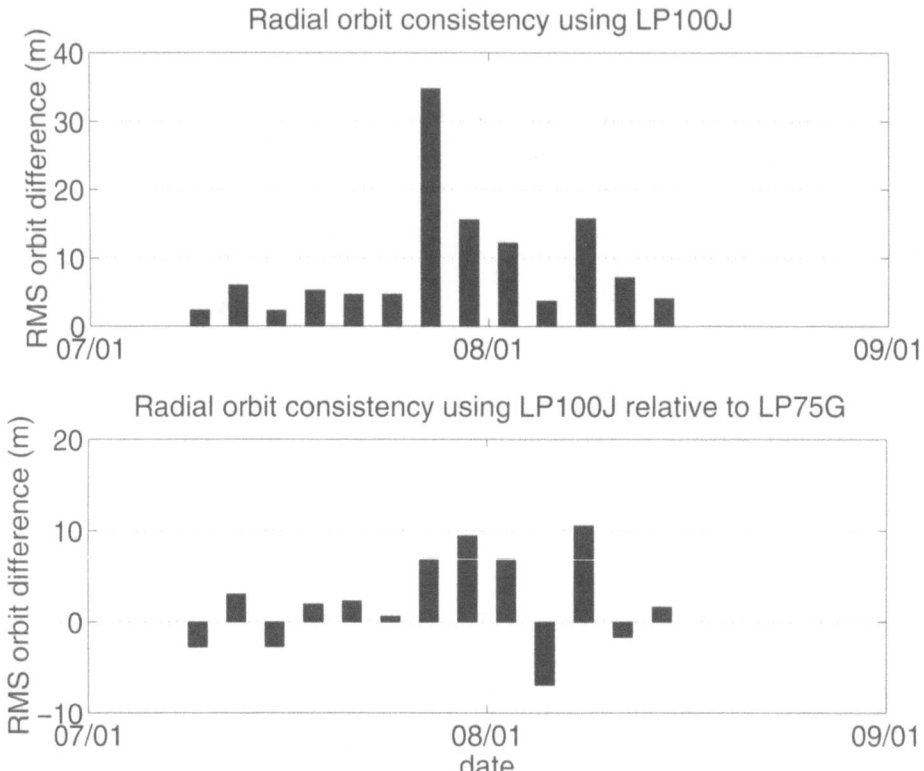

Figure 5.19   RMS radial orbit differences for 1 day overlap periods, based on 3-day
               batches for orbit computation, for LP100J. Shown are absolute values (top)
               as well as the residuals relative to those achieved with LP75G (bottom)

JLGM solutions. However, this path has not been pursued, because it would add
little to the understanding of the solutions, since most aspects of the lunar gravimetric problem and its solution are understood by means of a subset of analysis
techniques. The important aspects of signal content, and in particular the signal-
to-noise ratio, spectral leakage and correlation, as a function of data weighting
schemes and the choice of the regularisation parameter(s) are already adequately
addressed by a reduced number of assessment techniques.

   In terms of orbit computations, it has been shown that the JLGM solutions yield
improved goodness-of-fit, with respect to the *a priori* model, for both dependent
data, *i.e.* data used to derive the models, and for independent data. Consistently,
RMS values of the range rate residuals less than 1 mm/s have been achieved for 2
day data arcs and a count interval of 30 s. However, the consistency of consecutive
orbit solutions has been found to be problematic. It has been advocated that this is
a result of interruptions in the data streams, caused by the fact that the campaign

relied on a single down-link station. This conclusion is backed by the following arguments.

Using 2-way DSN Doppler data, the LP100J model, which is already introduced in Chap. 3 as a 100×100 spherical harmonic expansion and a successor of LP75G, *Carranza et al.* [1999] report 0.5 m consistencies in the radial direction. The results using the Weilheim 3-way data are depicted in Fig. 5.18 in terms of residual statistics and in Fig. 5.19 in terms of overlap consistencies. Comparing the residual data to Figs. 5.13 and 5.14 one may conclude that the residual fits of the more recent Lunar Prospector model is excellent, also for the Weilheim observations. On the other hand, the orbit consistency data do not match the 0.5 m radial accuracies reported by *Carranza et al.* [1999]. In other words, the problems of the discontinuous data stream are not limited to the JLGM solutions derived from the Weilheim data set, but most probably applies to all presently available lunar gravity models.

The Weilheim campaign has therefore brought important lessons about the lunar gravity field recovery problem. While the sample selenopotential solutions most likely cannot compete directly with solutions derived from more consistent orbit solutions, they mark an important step in European lunar gravity field analysis and demonstrate the readiness of both hardware and software components for further selenodetic investigations in the near future.

Chapter 6

Towards a global data set

'Although SST-based gravity field mapping is widely acknowledged to provide high-quality data sets in compliance with the present and near-future requirements in lunar science, within a decade or so, it is quite likely that satellite gradiometric mapping techniques will also become attractive for the lunar problem. This will happen as the cost of the gradiometer payload decreases, through heritage from terrestrial missions, and the need for high-frequency information on the selenopotential increases.'

Ibid.

6.1 Introduction

Having elaborated extensively on the nature and practical consequences of the ill-conditioning of the lunar gravimetric problem, it is only natural to devote the final chapter to the suitable remedies. Indeed, the main line of the previous chapters has been to zoom in on the practical consequences of the under-sampling of the selenopotential field, and to show that the currently available data sets do not lead to a unique and unambiguous gravity field solution. Rather, regularisation methods and the related problem of the determination of (a set of) regularisation parameters, as well as "design philosophies", *e.g.* in terms of data weighting schemes and intent of the regularisation schemes, are found to greatly influence the gravity modelling process. Likewise, the data scarcity implies that very few means are at hand to verify the solution, which only adds to the myriad of problems facing the gravity field analyst.

In terms of measurement concepts capable of measuring gravity-induced satellite orbit perturbations or some direct function of the gravitational potential over the far-side of the Moon, two remedies have already been introduced in Sect. 2.3, being SST and SGG. By the equivalence principle [*e.g., d'Inverno*, 1992], a spacecraft without contact with a ground station cannot measure the gravity field in which it orbits, unless it carries a gradiometer. The basic function of the gradiometer is to measure the components of the gravity tensor, *i.e.* the second order derivatives of the gravitational potential. Such an instrument is the key payload of the approved GOCE mission [*e.g., ESA*, 1999; *Visser*, 1999; *SID*, 2000; *ESA*, 2000], as it is essential to measure the high-frequency constituents of the terrestrial gravity field. In particular for an Earth mission, such an instrument is very demanding, in terms of complexity, hardware cost, calibration and testing, mass and operations constraints. Not only did it take the Earth sciences community more than two decades to realise such a mission, but the fact of the matter is that it would be something of an overkill for the present problems in lunar science, at least if the envisaged measurement precision levels is to be the same as for an Earth gravity field mission. On the other hand, till the present day there has been no study of a dedicated lunar gradiometry mission, making use of the design experience from Earth missions and taking into account the differences between the "lunar" and the "Earth" problems. In other words, the use of SGG measurements as a solution to the current problems are yet to be adequately analysed.

While in lunar gravimetry the fundamental problem is to acquire high-quality tracking data uniformly distributed over the Moon, the problem of downward continuation is nearly absent. Due to the lack of any significant atmosphere, it is possible to fly on a very low orbit without excessive spacecraft manoeuvring for substantial periods of time, as demonstrated by the perilune altitudes as low as 15 km during the Lunar Prospector extended mission. In other words, in terms of the eigenvalues of the Meissl scheme, cf. Fig. 2.3, there is little need to counteract the damping of the measurement signal due to the spacecraft altitude by amplifying the high-frequency information through measurements of differentials of the potential.

Another issue is that the nature of the present-day challenges in lunar science are frequently of a different scale than those met in the Earth sciences. All planets and natural satellites have their own specifics that give a characteristic flavour to the related scientific issues, and in terms of density distribution, stress distribution and overall radially-layered structure of the Moon the situation is no different. Therefore, it is not apparent that the instrumentation that currently has the best prospects in providing the high-resolution terrestrial gravity field is automatically also the preferred instrumentation of a lunar mission. In the problem of determining the global lunar gravity field it is, *e.g.*, also required that the SST (or SGG) instrumentation also provides information on the long-wavelength part of the gravity field, something that gradiometers are less capable of. Furthermore, it is maintained by some that the long-lasting history of the Earth sciences has brought the requirements for high-resolution and high-accuracy geopotential modelling to a more advanced level than has some 30 years of lunar research for the selenopotential. This is in itself not an argument against instruments capable of mapping the lunar gravity field with the highest precision. However, it is a realistic acknowledgement that the present and near-future fundamental needs of lunar science are, as will become clear in the following, addressable with SST techniques.

For these reasons, a gradiometric solution to the current problems in lunar gravimetry is unlikely. As already studied by both Europe [*e.g.*, *Flury*, 1981; *Floberghagen et al.*, 1996; *Milani et al.*, 1996] and the United States [*Konopliv*, 1991] in the past and as approved for the Japanese SELENE mission [*Kaneko et al.*, 1999; *Namiki et al.*, 1999; *Sasaki et al.*, 1999], SST-based concepts are widely acknowledged to provide the answer to the needs of the lunar science community. This chapter zooms in on SST principles and possibilities, and subsequently demonstrates, through both covariance analysis and full-scale simulations, the recovery of a high degree and order selenopotential model. The emphasis is on fully self-contained solutions that do not suffer from the need of significant regularisation, as is the case for all present-day global models. A subset of quality assessment tools applied to GLGM–2 and LP75G in Chaps. 3, 4 and 5 is used to demonstrate the expected qualitative improvement of future selenopotential solutions from SST. The full-scale simulations are therefore not intended to cover an exhaustive range of possible measurement concepts and error sources, since such studies have already been performed in the past for low–low SST [*e.g.*, *Floberghagen et al.*, 1996] and are currently under study for high–low configurations in both Japan [*Matsumoto et al.*, 2000] and in Europe [*Floberghagen*, work in progress]. Rather, the focus is on the prospective benefit to the lunar sciences, through a limited set of case studies.

6.2 SST concepts

In essence, the traditional distinction between low–low concepts and high–low concepts, which was introduced in Chap. 2, only describes the geometric relationship of the satellites in an SST configuration. In turn, the geometric constellation determines the mutual visibility of the spacecraft, and, hence, the possibility to obtain far-side data (when, where and for how long without interruption). Moreover, through the resulting orbit perturbations exerted by the forces acting on the spacecraft, the low–low and high–low concepts describe the degree of dynamical behaviour of the satellites. High-orbiting platforms are less sensitive to gravity field variations, due to the damping of the orbit perturbations with satellite altitude. Any satellite in a low orbit can therefore be said to act like a "gravity field sensor". Evidently, in order to ensure a proper sensor function, at least one of the spacecraft is required to be in a low orbit, as a low-altitude orbit is required to maximise the sensitivity of the tracking data over a vast range of orbital frequencies.

However, for the sub-satellite re-transmitting the radar signal back to the mother-craft, other considerations may apply. From a cost-effective point of view, as already mentioned in Chap. 2, a satellite performing this function function is frequently a simple, purely passive, spin-stabilised craft, which, once deployed in its orbit, may no longer be controlled. Caution is therefore taken not to jeopardise the experiment by allowing the perturbations in eccentricity grow to such a level that a hard impact on the lunar surface occurs before completion of the data collection. Such variations in the eccentricity may be quite significant in the case of the Moon, cf. Fig. 3.15, and, for non-previously flown orbits they are likely to be rather poorly determined, cf. Fig. 3.8. This is the dynamic reason why, for the sub-satellite, high orbits or at least highly eccentric orbits are sometimes preferred.

Besides the geometric aspect, SST methods are characterised by the observed signal. In a broad sense, SST may involve both radio tracking and optical tracking, as well as several link paths, leading to *e.g.* 1-way, 2-way, 3-way and 4-way observations. Although optical inter-satellite communications and tracking systems have been studied, radio tracking in various frequency bands is the only widespread tracking method for planetary orbiters. Practically, this means that SST experiment designs are based on tracking in some radio frequency band, usually S-band or X-band [*Kinman*, 1992; *Montenbruck and Gill*, 2000], but also K/Ka-band [*Milani et al.*, 1996; *Davies et al.*, 1999; *Mazanek et al.*, 2000]. While in principle both range and range rate observables are possible, range rate observations are usually preferred for low- and medium-cost missions, simply because the accurate measurement of travel times needed for range observations is difficult to carry out in a small and cost-effective on-board radar. LOS velocities, on the other hand, are derived from phase measurements of a Doppler counter. Unless the baseline between the spacecraft is too long, such measurements are usually sufficiently accurate for precise satellite geodetic instrumentation.

In terms of signal content and irrespective of the exact measurement type, there are a few additional and distinct differences between low–low and high–low constellations. First, one advantage of the low–low mode is that the required measurement accuracy has to be achieved and maintained over relatively short paths [*Rummel*, 1979]. This offers great advantages in hardware design and in the overall satellite mass budget, as systems needing to operate over vaster distances generally require more power. Second, a signal observed in a low–low mode contains a significantly greater amount of short-wavelength information, which is desirable from a high-resolution and high-accuracy gravity field recovery point of view. This fact is directly rooted in the spectral content of the measurements, as described by the Meissl scheme, Fig. 2.3.

In short, based on the above conglomerate of considerations as well as the technical difficulty in designing high-quality SST systems, the past two decades have led to several SST experiment ideas, studies and designs. The following sections detail the typical features of two such designs, one low–low, the other high–low, with emphasis on recent developments, in particular the MORO study [*Milani et al.*, 1996; *Coradini et al.*, 1996; *Floberghagen et al.*, 1996] for low–low tracking, and the SELENE mission [*Matsumoto et al.*, 1999, 2000; *Namiki et al.*, 1999; *Sasaki et al.*, 1999] for high–low concepts. A similar, but not identical, high–low concept was envisaged in the Lunarstar/GAUSS study [*Häusler*, 1998], as a proposal to NASA's Discovery programme for medium-sized science missions. The proposal included a 1-way SST with the signal path being *GAUSS sub-satellite–Lunarstar–ground sta-*

tion, and constituted an SST experiment in its simplest form. Unfortunately, this project was rather short-lived, and since the major issues in terms of the final gravity field product are close to the SELENE concept, this study shall not be discussed in any significant detail. Most of the features studied in previous mission proposals, like the Lunar Polar Orbiter [*Ananda et al.*, 1976], POLO [*Flury*, 1981] and Lunar Observer [*Konopliv*, 1991] are accommodated in these more recent analyses. A common feature of all these proposals is that *two* spacecraft are involved, *e.g.* in opposition to SST involving the GPS or TDRSS [*e.g.*, *Yunck et al.*, 1986, 1994; *Gleason*, 1991; *Bertiger et al.*, 1994; *Gold*, 1994; *Hesper*, 1994; *Melbourne et al.*, 1994; *Olson*, 1996; *Yunck and Melbourne*, 1996; *Rowlands et al.*, 1997]. The sub-satellite tracking experiment foreseen for SELENE is very likely to become a *space first* in terms of interplanetary missions.

6.2.1 Low–low configurations - the MORO study

The scientific requirement for the lunar sub-satellite (LUSS) experiment of MORO, as defined during the Phase A study (preliminary mission design), was to measure the 2-way range rate, over the entire Moon, between two relatively closely orbiting spacecraft, *i.e.* MORO *mother-craft–sub-satellite–MORO mother-craft*, with a precision of 0.1 mm/s over an integration time of 10 s. Two main drivers motivated such a precision level.

First, in terms of *local sensitivity*, the SST design was required to map 2 mGal peak gravity anomalies at the lunar surface, with a half-wavelength spatial resolution comparable to the satellite altitude of 100 km[1], or, equivalently $l_{max} \simeq 55$. As shown in Appendix D, the range-rate between two co-orbiting spacecraft, *i.e.* spacecraft that closely follow each other in the same orbit, is approximately proportional to the potential difference between the two locations along the orbit, cf. (D.18). By local sensitivity, the possibility to detect a single mass anomaly structure is intended. Using the approximate proportionality, the 0.1 mm/s precision figure was therefore determined as follows.

Assume that a gravity anomaly $\Delta g = 2$ mGal at the lunar surface is caused by a spherically shaped mascon of 10 km radius at depth $d = 50$ km. Such peak anomaly may then be the result of a density contrast with respect to the average lunar rock of $\Delta \rho \simeq 0.5$ g/cm^3. Let a local, planar coordinate frame be originated at the lunar surface, such that the horizontal distance from the mascon is denoted by x and the spacecraft altitude by $y = h =$ constant. It is therefore implicitly assumed that the two spacecraft pass directly over the mass anomaly. Let the separation between the two spacecraft be denoted by s. In this case, from (D.18), it

---

[1]A rough rule of thumb in satellite *geo*desy, based on high–low SST, says that a gravity mapping experiment based on adequate sampling is frequently capable of determining mass anomaly structures of a half-wavelength size approximately equal to the satellite altitude [*e.g.*, *Colombo*, 1989b].

holds that

$$\dot{\rho} \simeq \frac{1}{v} \left[\frac{d^2 \Delta g}{\sqrt{(h+d)^2 + x^2}} - \frac{d^2 \Delta g}{\sqrt{(h+d)^2 + (x+s)^2}} \right] \qquad (6.1)$$

For $h \simeq 100$ km, a corresponding orbital velocity $v \simeq 1.633$ km/s and a separation $s \simeq 100$ km, this leads to a peak-to-peak amplitude of the single range rate oscillation very close to 0.1 mm/s. Furthermore, the flight time needed to pass the peak signal is in the order of $\simeq 100$ s. Hence, the foreseen 10 s integration time of the Doppler measurements is sufficiently short to observe the structure of the signal. In terms of local sensitivity, the MORO experiment with a 2-way Doppler measurement precision should therefore – to the first order – be capable of mapping gravity anomalies down to approximately a 2 mGal precision level at approximately 100 km spatial resolution.

The second motivation for the precision figure was not so much a scientific reason as a technical design driver. The 0.1 mm/s precision level is comparable to the Precise Range and Range-rate Equipment (PRARE) operative on ERS–2 [*Bedrich et al.*, 1997] and it was seen as a technical challenge to develop a radar with a comparable performance, but a much smaller mass, complexity and power consumption. As such, the hardware development would also be a design challenge to push the state-of-the-art in on-board satellite radar systems.

Although the "flat Moon" assumed for the local sensitivity assessment is only legitimate as a zeroth-order approximation, since for high-accuracy applications it may no longer be assumed that x and s are small with respect to the lunar radius, it allows to relate range rate precision to detection of a mass anomaly feature. However, the gravity-induced orbit perturbation is the result of continuous density variations inside the Moon, and the overall orbit error is a cumulative effect along the orbit. In other words, the very simple static model used above is by no means sufficient to prove the feasibility of a low–low SST experiment, nor does it provide any kind of quality measure of the solution. Therefore, the local sensitivity results must be supported by some *global sensitivity* measure, as well as a corresponding quality assessment of the solutions. In this context, global sensitivity implies the recoverability of a global lunar gravity model, modelled in terms of spherical harmonic basis functions, either by means of covariance analysis, similar to the analysis of Chap 3, or by means of full-scale numerical simulations of the data reduction and gravity field model recovery process. This is the topic of Sects. 6.3 and 6.4, where the case of low–low mode tracking using integrated Doppler measurements is the primary example of the capabilities of gravity field mapping based on SST techniques.

The design of the LUSS experiment for MORO reached as far as the full Phase A study, which includes a full study of all systems, but no detailed hardware design. The block diagram of the model design implementation by *Milani et al.* [1996] is depicted in Fig. 6.1. A stable carrier at frequency $f_1 = 20$ GHz is generated on board the MORO mother spacecraft and transmitted in the forward link to LUSS. A coherent transponder aboard the latter retransmits the carrier after a suitable

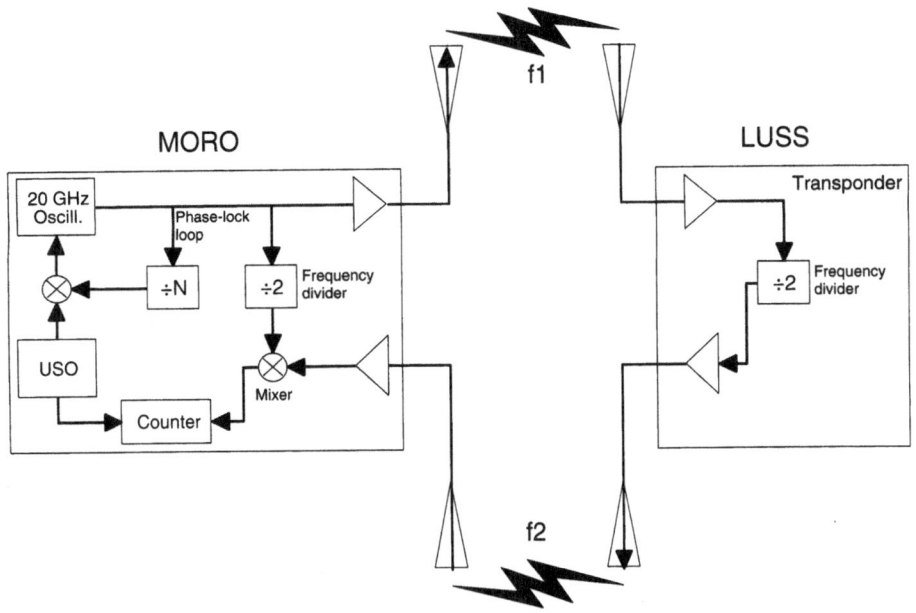

Figure 6.1   Block diagram of the MORO Doppler radar design. On the left the
transmitter/receiver unit on board the main spacecraft; on the right the
transmitter on board the sub-satellite

frequency shift to avoid crosstalk between the transmitted and received signals on
board the MORO main spacecraft. In the figure, the transponder ratio is indicated
by a "divide-by-two" frequency divider, which provides a nominal return link frequency $f_2 = 10$ GHz. The received oscillation at frequency f_2 is affected by the
Doppler shift equal to $f_D = \Delta f_1 + \Delta f_2$, where Δf_1 and Δf_2 are the shifts on the
forward and return links respectively

$$\Delta f_1 = f_1 \frac{\Delta \dot{\rho}}{2c} \quad \text{and} \quad \Delta f_2 = f_2 \frac{\Delta \dot{\rho}}{c} \tag{6.2}$$

in which $\Delta \dot{\rho}$ is the relative velocity of the two spacecraft, c the speed of light, and
the factor 2 in the first denominator takes account of the coherent transponder
turnaround ratio of $1/2$. In the receiver on board MORO, the return link is mixed
with a local reference signal at the nominal frequency f_2, which is generated by the
20 GHz oscillator using the same signal processing as the coherent transponder on
board LUSS. This allows the detection of the Doppler shift f_D. The zero-crossings
of the resulting oscillation at frequency f_D are subsequently used to drive the gate
of a counter that computes the time elapsed in a fixed number of periods of the
driving oscillation. As a result, at the end of the 10 s observation window (integration period), the counter indication is proportional to the inverse of f_D, such that
the integrated range rate observation $\Delta \dot{\rho}$ may be derived from (6.2).

The precision of the 2-way inter-satellite integrated range rate measurements envisaged for MORO depends on a number of factors. A detailed discussion of radar hardware issues is beyond the scope of the present book, but may be found in *Milani et al.* [1996]. The following is therefore limited to a very brief qualitative overview. First, there is the stability of the oscillators. The role of the Ultra-Stable Oscillator (USO) is to provide a highly accurate reference time generator to the counter as well as to provide a stable reference for the carrier frequency generation. Depending on the integration interval and on the minimum signal that must be detectable, this leads to stability requirements for the USO. Second, there is the link budget due to noise on the forward and return links, where obviously the noise level must be kept low enough as to not corrupt the measurement precision by causing erroneous counter starts and stops. Third, the frequency separation of the forward and return links, through the transponder ratio, has a significant impact on the link budget. Fourth, a critical issue is the Doppler shift induced by the motion of LUSS antennae. Since LUSS was to be a free-spinning craft, its rotational motion may not be actively controlled, which requires that the LUSS design as well as deployment procedures take due account of the need to minimise the on board antenna phase centre motion with respect to the centre of mass. *Milani et al.* [1996] actually concludes that the 0.1 mm/s precision level may be achieved throughout the experiment lifetime, provided hardware specifications for the USO and LUSS transponder ratio are met and the LUSS antennae mounts are well calibrated to align the LUSS phase centre with the centre of mass.

6.2.2   High–low configurations - the SELENE experiment

The SELENE mission foresees in a large spacecraft, hereafter called Orbiter, in a 100×100 km, 90° inclination near-circular orbit, as well as two smaller satellites, named RSAT and VSAT, in eccentric orbits of larger semi-major axis, cf. Table 6.1. The orbital elements are chosen on the basis of lifetime predictions of the smaller sub-satellites as well as orbit maintenance and instrument operation considerations for the main spacecraft. Altogether, the mission provides a selenodetic instrument package consisting of

- 2-way Earth-based Doppler tracking of Orbiter, RSAT & VSAT: *station-Orbiter/RSAT/VSAT-station* during the entire 1 year experiment duration using domestic tracking stations operated by the Japanese space entities ISAS and NASDA

- 4-way Doppler tracking using RSAT to obtain far-side data: *station-RSAT-Orbiter-RSAT-station*

- differential VLBI involving the Orbiter and a radio source on board VSAT as well as, in a more conventional mode, involving a quasar and VSAT

- a highly accurate laser altimeter which will be used for the profiling of the Moon. The crossover mode is unfortunately not available as a data source for

	a	e	i	Ω	ω
Orbiter	1838.0 km	0.0	90°	tbd	133.0°
RSAT	3000.0 km	0.9345	90°	tbd	133.0°
VSAT	2200.0 km	0.9130	90°	tbd	133.0°

Table 6.1   Tentative baseline epoch orbital elements for the sub-satellites of the SELENE mission. The right ascension of the ascending node of both satellites will be determined by the launch date. Satellite lifetime issues are involved with the choice of the argument of perilune, and this angle may also be adjusted in the near future

orbit and gravity field estimation since – by nature of the 90° orbits – the tracks are collinear rather than crossing

The key instrument for the global sampling of the gravity field is obviously the 4-way Doppler radar which utilises the transponder function of RSAT in eccentric orbit to track the Orbiter while over the far-side. The signal type is much like that of the TDRSS satellites [*Marshall et al.*, 1996; *Visser and Ambrosius*, 1996; *Luthke et al.*, 1997; *Rowlands et al.*, 1997] utilised for the communication with and tracking of spacecraft in low Earth orbit.

Although specific requirements in terms of accuracy and spatial resolution have not been publicly announced, a key element of the overall SELENE concept is to be able to provide the first and global data set that puts researchers in the position to resolve the major remaining "mysteries" in lunar science. In other words, the focus is on the overall package as a whole, and not a specific stand-alone instrument, and the quality of the final gravity field experiment will eventually depend largely on the ability to combine the data of the three satellites in an optimal fashion. It is sought to optimise the sensitivity to the orbit perturbations over an as vast as possible range of orbital frequencies, by combining long-arc RSAT and VSAT 2-way Doppler data for an optimal sensitivity to the long-wavelength constituents, and relatively short-arc 2-way and 4-way Doppler data of the Orbiter for the higher frequencies. One of the major scientific challenges of the gravimetric experiment is therefore how to combine these data sources, by choosing the optimal arc length strategy as well as data processing strategy in order to extract the frequency-wise information that the various data sets contain.

The differential VLBI data using two near-perpendicular baselines have the prospective benefit of giving near-instantaneous position determination of the satellites, and therefore aid in the short-arc orbit determination of the Orbiter. During the nominal 1-year mission, attitude correction manoeuvres at the frequency of up to three manoeuvres per day are required. Irrespective of the telemetry information on post-thrusting attitude, these manoeuvres are likely to cause parasitic

satellite	data type	precision	integration time
Orbiter	2-way Doppler	2.0 mm/s	10 s
RSAT	2-way Doppler	0.2 mm/s	60 s
VSAT	2-way Doppler	0.2 mm/s	60 s
Orbiter/RSAT	4-way Doppler	1.0 mm/s	10 s
Orbiter/VSAT	ΔVLBI[a]	1.0 mm	120 s
Quasar/VSAT	ΔVLBI[a]	1.0 mm	120 s

[a]Doubly differenced 1-way range involving three stations for a near North-South baseline and a near East-West baseline

Table 6.2   Tracking specifications for the SELENE selenodetic instrument package

residual orbit effects, for which the differential VLBI observations may be help-ful. After the nominal mission, there is the opportunity to leave the spacecraft unthrusted in any way for considerable periods of time, in which the role of the ΔVLBI observations in the gravity field solutions might become less important.

The specifications of the radar tracking of the satellites are given in Table 6.2. Because the precision level of the 4-way Doppler radar is inferior to the design specifications for MORO, and, moreover, the tracking is of the high–low type, it is expected that the maximum spatial resolution of a self-contained gravity field solution is somewhat inferior to that of a MORO-type experiment. cf. the Meissl scheme in Fig. 2.3. However, such qualitative measures directly depend on mis-sion duration as well as possible near-future design upgrades in both the ground segment and the space segment. Currently on-going simulation studies are aiming at the accurate prediction of the resolution limit from the overall SELENE selen-odetic instrument package [*Matsumoto et al.*, 2000].

6.3 Covariance analysis of low–low selenopotential recovery

Least squares error analysis, or covariance analysis, provides a convenient means for investigating the accuracy with which a specified set of unknowns may be recovered from a given set of observations. The fundamental strength of all such approaches is that, since principally only the left-hand side of the normal equation (or sequential filter) is required, the error analysis may be performed without an actual experiment (simulated or real) being carried out. The analysis tools may therefore be applied to pre-mission performance studies, and as such play a role in the optimisation of the experiment. For example, given some orbit and instrument characteristics, based on technical or practical mission considerations, one is in the position to derive the resulting accuracy and resolution of the estimated spherical harmonic coefficients, as well as derived quantities such as selenoid heights or gravity anomalies.

A basic requirement for such an analysis is a linear model relating the observations to the unknowns. Evidently, there are several ways in which to derive such a linear model. So-called fully dynamic "brute-force" numerical approaches, such as those used to derive GLGM–2, LP75G and the Weilheim-based selenopotential solutions of Chap. 5, constitute one fundamental option. In this case, linear equations are formed for the first-order corrections to an *a priori* selenopotential model. The normal matrix (or square root information filter) relates the unknown corrections to the measurements. Error propagation, discussed in Chap. 4, in this case yields the variance-covariance matrix of the selenopotential coefficients.

However, several considerations favour (semi-)analytical approaches, based on a justifiable set of assumptions and approximations. For one, the estimation of an extended number of spherical harmonic coefficients from a rather formidable number of observations is a numerically challenging task. So far, lunar gravity models are restricted to a maximum degree and order of about 100, cf. Chap. 2. Such a model contains about 10,000 coefficients and the normal matrices on which they are based are therefore of rank $\sim 10,000$ (provided there are no numerical singularities). Although such a matrix is manageable and invertible on present-day supercomputers, it is a tedious task to apply the full matrix in any sort of least squares error analysis. End-to-end simulations of the entire experiment, for a conglomerate of mission scenarios and instrument system parameters, therefore easily become a non-feasible option. Particularly attractive are therefore approximate methods that are characterised by a significant reduction in the numerical effort, compared to the straightforward full-scale approach.

This is exactly the strength of (semi-)analytical spherical harmonic analysis of SST data. Based on certain assumptions for the underlying linear(ised) model, they facilitate rapid error estimation from low–low SST tracking. Similarly, equivalent formulae may be established for other linear functions of the spherical harmonic coefficients defined from the same set of assumptions, and this has led to a widespread use of the method for related measurement concepts, in particular for SGG. The core idea is to use three assumptions (or approximations) in combination with LPT to decouple the normal matrix according to order m and parity $l - m$, which causes the normal matrix to attain a block-diagonal structure. In turn, such a block-diagonal matrix may be inverted efficiently, *i.e.* in a matter of seconds or minutes rather than hours or even days, which may be required for a full-scale numerical approach. Provided the underlying linear model is adequate and the assumptions are justifiable for the problem at hand, such tools therefore greatly facilitate the error propagation, so as to yield the *a posteriori* block-diagonal variance-covariance matrix of the selenopotential coefficients. Another strength of the analytical approach is obviously that it retains all the advantages of the LPT in providing insight in the time-wise or spatial distribution of the error, cf. Chap. 3.

After an introduction of the method and a sketch of the SST low–low observation equations, the following presents a least squares error analysis of gravity field recovery from low–low SST range rate observations (integrated Doppler). The immediate goal is to explore the class of solutions and locate the solution that minimises the overall error, and therefore maximises the experiment performance. In view of the discussion in the previous chapters, it is clear that the "quality" depends on the application. In the present context, one is investigating self-contained, unregularised SST-based solutions, and quality is therefore measured in terms of the resulting covariance. The test case is a MORO-type mission with two co-orbiting spacecraft in a 100 km polar orbit. It is assumed that the SST constellation is characterised by two degrees of freedom, being the *in-plane* separation in the radial and along-track directions and the *out-of-plane* in the cross-track direction.

Obviously, such least square error analysis relies fully on the validity of the linear model. Although the theory has proved useful and adequate for pre-mission studies, a verification of the thus-derived error estimates by means of full-scale simulations is mandatory. To this end, a complementary full-scale simulation of the experiment is presented in Sect. 6.4.

6.3.1   Time-wise approach in the time-domain - the Colombo method

The time-wise approach for spherical harmonic analysis of SST data was originally proposed by *Colombo* [1984], based on a heritage of earlier work [*e.g., Colombo*, 1981; *Kaula*, 1983; *Wagner*, 1983]. Rightfully, the approach is therefore sometimes referred to as the *Colombo method*. The method was conceived during a period when SST techniques were considered, by a majority of the satellite geodetic community, to be the primary means for mapping the finer resolution gravity field. Due to the successful application of the method in error studies and its prospects as a possible technique for rapid reduction of actual data, refinements and variants [*Colombo*, 1986; *Schrama*, 1986; *Wagner*, 1987b; *Wiejak et al.*, 1991] as well as new applications of the method for other observables, in particular the field of SGG [*Rummel and Colombo*, 1985; *Colombo*, 1989a; *Schrama*, 1991], quickly appeared. In the subsequent years, the method was extensively used – and still is – in the study of SGG-based gravity field mapping [*e.g., Visser*, 1992; *Visser et al.*, 1995b; *Visser*, 1999; *Koop*, 1993; *Rummel et al.*, 1993; *SID*, 2000; *ESA*, 2000; *Klees et al.*, 2000] and in the study of gravity field recovery from GPS observations [*Visser et al.*, 2000]. In a sense, the application of the Colombo method to SGG observables is perhaps more straightforward than its use with low–low SST observations, since in the former case the orbit problem and the high-resolution gravity field problem may often be treated separately. However, the validity of the methods, also for SST data types, has been proven for terrestrial applications in the past [*e.g., Visser*, 1992, 1999; *SID*, 2000]. As will be shown in the following, their application in lunar gravimetry is also warranted. Due to the widespread use of the Colombo method, as well as its documentation in the aforementioned works, in the following only the main line of thought is outlined.

The name-tag "time-wise" refers to the fact that in this approach to spherical harmonic analysis of SST data the observable (whether it is range measurement or range rate measurement) is regarded as a function of time, rather than a function of selenographical location. The latter would lead to the so-called "space-wise" approach, which is a complementary technique. Although the difference between the two methods is merely a matter of representation, in the case of the Moon, the time-wise view is the only method which facilitates rapid global gravity field recovery. This is primarily due to the significant role of orbit errors and correlations between orbit parameter and gravity field parameter estimates. Space-wise methods are not discussed in any detail in this book. However, the concept is to make use of the spatial correlation of the measurements to assemble spatially contiguous observations, which may be used for cross-checking of observations and for the reduction of the data noise through block- or grid-averaging [*ESA*, 2000]. The space-wise method offers advantages in the treatment of measurements which are distant in time, but close in space, and is therefore related to least squares collocation [*e.g., Moritz*, 1989], see also *Tscherning* [2001] for a recent application of the method. However, the solution of the satellite orbit problem and the correlations between the orbit parameters and the gravity field are usually neglected (although not always, cf. *Rummel and Colombo* [1985] for a counter-example), which makes the method unattractive for the lunar problem at the present stage. The idea of the time-wise approach, on the other hand, is to consider the entire set of SST measurements, obtained throughout the mapping campaign, as a time series, *i.e.* a mere sequence of data points as a function of time. The mathematical principle behind the method is rotational symmetry of the measurements (and therefore the orbits of the spacecraft) distributed over the lunar sphere. In the following, the fundamental SST observation is the inter-satellite range ρ. Integrated Doppler measurements (range rates), as envisaged for MORO, may be straightforwardly computed from first-order differences of two consecutive ranges, divided by the sampling interval Δt.

The time-wise error analysis is based on three assumptions and approximations, the rational and necessity of which will be discussed in the following. They are:

- the orbits of the satellites are circular;
- the data are distributed regularly along the orbit;
- the data collection period is equal to an integer multiple of repeat periods;

The need for a circular orbit is, conform the theory outlined in Chap. 3, a fundamental prerequisite for the use of the Hill equations. Hence, in the time-wise view the potential $U(t)$ along the circular reference orbit is rewritten, from (3.7), as

[*Schrama*, 1991; *Sneeuw*, 1991; *Koop*, 1993]

$$U(t) = \sum_{k=-l_{max}}^{l_{max}} \sum_{m=0}^{l_{max}} [A_{km} \cos \psi_{km}(t) + B_{km} \sin \psi_{km}(t)] \tag{6.3}$$

where

$$A_{km} = \sum_{l=l_{min}[2]}^{l_{max}} \frac{GM}{a} \left(\frac{a_e}{a}\right)^l \overline{F}_{lm}^k \begin{bmatrix} \overline{C}_{lm} \\ -\overline{S}_{lm} \end{bmatrix}_{\substack{l-m \text{ even} \\ l-m \text{ odd}}} \tag{6.4}$$

$$B_{km} = \sum_{l=l_{min}[2]}^{l_{max}} \frac{GM}{a} \left(\frac{a_e}{a}\right)^l \overline{F}_{lm}^k \begin{bmatrix} \overline{S}_{lm} \\ \overline{C}_{lm} \end{bmatrix}_{\substack{l-m \text{ even} \\ l-m \text{ odd}}}$$

are the so-called *lumped coefficients* and

$$\psi_{km}(t) = \psi_{km}^0 + \dot{\psi}_{km}t = \psi_{km}^0 + k(\dot{\omega} + \dot{M}) + m(\dot{\Omega} - \dot{\theta}) \tag{6.5}$$

In the above summations, the [2] indicates that the summation occurs in steps of two over degrees of equal parity. The new index $k = l - 2p$ and the normalised inclination functions $\overline{F}_{lm}^k$ therefore relate to the inclination functions used in Chap. 3 as

$$\overline{F}_{lm}^k = \overline{F}_{lm(l-k)/2} \tag{6.6}$$

Similarly, for a repeat orbit of zero eccentricity ($q = 0$), it holds from (3.11) and (3.12) that

$$\dot{\psi}_{km}(t) = f_{km}(\dot{\omega} + \dot{M}), \quad f_{km} = k - m\frac{N_m}{N_r} \tag{6.7}$$

where N_r and N_m are relative primes. Furthermore,

$$l_{min} = \max(|k|, m) + \delta$$

with

$$\delta = \begin{bmatrix} 0 \\ 1 \end{bmatrix}_{\substack{k-\max(|k|,m) \text{ even} \\ k-\max(|k|,m) \text{ odd}}}$$

If the data sampling is regular, in other words if there are no gaps in the data collection and the sampling occurs at a constant sampling rate Δt, (6.3) may be regarded as a Fourier series, in which the lumped coefficients correspond to the Fourier coefficients. This is the prerequisite for the second assumption.

The lumped coefficient formalism may be straightforwardly extended to any linear function $y(t)$ of the selenopotential coefficients which has the general form (6.3). For the lumped coefficients it holds for all such functions [*Schrama*, 1991]

$$\begin{bmatrix} A_{km} \\ B_{km} \end{bmatrix} = \sum_{l_{min}[2]}^{l_{max}} \begin{bmatrix} \alpha_{lm} & \beta_{lm} \\ \beta_{lm} & -\alpha_{lm} \end{bmatrix} \begin{bmatrix} H_{lmk}^y \\ G_{lmk}^y \end{bmatrix} \tag{6.8}$$

where

$$\alpha_{lm} = \begin{bmatrix} \overline{C}_{lm} \\ -\overline{S}_{lm} \end{bmatrix}_{\substack{l-m \text{ even} \\ l-m \text{ odd}}} \quad \text{and} \quad \beta_{lm} = \begin{bmatrix} \overline{S}_{lm} \\ \overline{C}_{lm} \end{bmatrix}_{\substack{l-m \text{ even} \\ l-m \text{ odd}}} \tag{6.9}$$

and H_{lmk}^y and G_{lmk}^y are the so-called *transfer* or *sensitivity* coefficients. The notation G_{lmk}^y should not be confused with the eccentricity functions introduced in Chap. 3. For the sake of completeness, it deserves mentioning that there exists an alternative formulation for (6.8) which makes use of so-called cross-track inclination functions [*Balmino et al.*, 1996]. These are, however, not discussed here. Since the problem at hand is developed around circular reference orbits, the eccentricity functions play no role. For the potential $U(t)$, one has for example

$$H_{lmk}^U = \frac{GM}{a} \left(\frac{a_e}{a} \right)^l \overline{F}_{lm}^k \tag{6.10}$$
$$G_{lmk}^U = 0$$

By definition, the lumped coefficients for various linear functions y differ only in the sensitivity coefficients. A key to the application of the Colombo formalism for SST observations is therefore to derive the necessary transfer coefficients for a linear observation equation, which in turn uniquely determine the lumped coefficients.

In contrast to SGG observables, which are the second-order derivatives of U and are given by analytical formulae, the low–low SST observables are derived in terms of the periodical orbit perturbations induced by the gravity field. Hence, the idea is to develop an SST observation equation in terms of $\{\Delta r^{(0)}, \Delta \tau^{(0)}, \Delta c^{(0)}\}$. The analytical, linear SST model is then found by inserting the transfer coefficients for the orbit perturbations. These coefficients are straightforwardly found by inspection of (3.13) and (3.14) and yield the following lumped coefficients, in the radial and the along-track directions,

$$\begin{bmatrix} A_{km} \\ B_{km} \end{bmatrix}_{\Delta r^{(0)}} = \sum_{l_{min}[2]}^{l_{max}} a \left(\frac{a_e}{a} \right)^l \overline{F}_{lm}^k \frac{(l+1)f_{km} - 2k}{f_{km}(f_{km}^2 - 1)} \begin{bmatrix} \alpha_{lm} \\ \beta_{lm} \end{bmatrix} \tag{6.11}$$

$$\begin{bmatrix} A_{km} \\ B_{km} \end{bmatrix}_{\Delta \tau^{(0)}} = \sum_{l_{min}[2]}^{l_{max}} a \left(\frac{a_e}{a} \right)^l \overline{F}_{lm}^k \frac{2(l+1)f_{km} - (f_{km}^2 + 3)k}{f_{km}^2 (f_{km}^2 - 1)} \begin{bmatrix} \beta_{lm} \\ -\alpha_{lm} \end{bmatrix} \tag{6.12}$$

and in the cross-track direction,

$$\begin{bmatrix} A_{km} \\ B_{km} \end{bmatrix}_{\Delta c^{(0)}} = \sum_{l_{min}[2]}^{l_{max}} a \left(\frac{a_e}{a} \right)^l \frac{1}{2(1 - f_{km}^2)} \left\{ (f_{km} + 1) \left[\frac{\overline{F}_{lm}^{k-1}}{\sin i} [(k-1)\cos i - m] - \overline{F}_{lm}^{\prime k-1} \right] \right.$$
$$\left. - (f_{km} - 1) \left[\frac{\overline{F}_{lm}^{k+1}}{\sin i} [(k+1)\cos i - m] + \overline{F}_{lm}^{\prime k+1} \right] \right\} \begin{bmatrix} \beta_{lm} \\ -\alpha_{lm} \end{bmatrix} \tag{6.13}$$

See also *Wagner* [1985], *Wagner* [1987a] and *Schrama* [1991] for similar formulae, albeit in a slightly different notation.

The topic at hand is least squares error analysis. However, it deserves mentioning that a fast Fourier transform (FFT) applied to a time series of measurements would principally yield the lumped coefficients from which the spherical harmonic coefficients may be estimated. Such an approach is called the "time-wise approach in the frequency domain". In this case, the evaluation of the cosine/sine functions is avoided, and the lumped coefficients of the function $y(t)$ may be used as "pseudo-observations" in the actual least squares estimation. Such a computation yields the optimal harmonic coefficients if there is a unique mapping between the lumped coefficients and the Fourier coefficients. This requirement is fulfilled if the signal is truly periodic, in other words if the time series is obtained along a repeat orbit, and the number of orbital revolutions in the repeat cycle N_r is greater than two times the maximal degree of the selenopotential solution l_{max}. This explains the need for the third assumption.

The error analysis in the framework of the time-wise approach in the time-domain, on the other hand, requires the formation of a normal matrix based on the linear observation equation $y(t)$. From (6.3) and equivalent formulae for other linear functions $y(t)$ it is obvious that such a normal matrix $\mathbf{A}^T\mathbf{A}$ is generated from sums of products of trigonometric (sine/cosine) functions (and the lumped coefficients). With T_{rep} being the repeat period and Δt the sampling interval, it holds, based on the assumption of regular sampling,

$$N_{obs} = \frac{T_{rep}}{\Delta t}$$

where N_{obs} is the total number of observations made in the repeat period. From the definition of the repeat orbit (3.17), it also holds that

$$T_{rep} = N_r \frac{2\pi}{\dot{\omega} + \dot{M}} = N_m \frac{2\pi}{\dot{\Omega} - \dot{\theta}}$$

Hence, it follows from (6.7) and neglecting the constant phase shift (or initial condition) ψ_{km}^0 that

$$\psi_{km} = f_{km} N_r \frac{2\pi}{T_{rep}} j\Delta t = 2\pi f_{km} N_r \frac{j}{N_{obs}}, \quad j = 0, \ldots, N_{obs} - 1 \qquad (6.14)$$

The components of the normal matrix $\mathbf{A}^T\mathbf{A}$ are formed by inner products over the N_{obs} measurement points. Since the signal is periodic and N_r and N_m are relative primes, it holds [*Colombo*, 1984]

$$\sum_{j=0}^{N_{obs}-1} \cos 2\pi f_{k_1 m_1} N_r \frac{j}{N_{obs}} \cos 2\pi f_{k_2 m_2} N_r \frac{j}{N_{obs}} = \begin{cases} 0 & m_1 \neq m_2 \text{ or } k_1 \neq k_2 \\ \frac{N_{obs}}{2} & m_1 = m_2, k_1 = k_2 \end{cases} \qquad (6.15)$$

Identical results apply to inner products of sine functions, while all sine/cosine cross-terms sums are zero. Since the lumped coefficients of a certain order m_1 do not contain selenopotential coefficients common with lumped coefficients of order

$m_2 \neq m_1$ this allows a separate treatment of each order m. In other words, if the normal matrix is organised per order, this matrix has a block-diagonal structure, depicted in Fig. 6.2, since the correlation with other blocks is zero. In addition, given the requirement that $k_1 = k_2$ to have a non-zero inner product, a second consequence is that even and odd degrees l are uncorrelated. The maximum dimension of one such m-block is therefore $l_{max}/2 + 1$. Consequently, the inverse of the normal matrix, *i.e.* the variance-covariance matrix may be rapidly obtained from the computation of the inverse of each moderately-sized m-block.

If the lumped coefficients are also stochastically uncorrelated, the inclusion of a diagonal $\mathbf{Q}_y^{-1}$ to form a weighted normal matrix $\mathbf{A}^T\mathbf{Q}_y^{-1}\mathbf{A}$ preserves this block-diagonal structure. Frequency-dependent errors are also possible through a power spectral density function. This is, *e.g.*, an issue in SGG, where the gradiometer performance is frequency-dependent. However, in SST applications, frequency-dependent diagonal elements of $\mathbf{Q}_y$ are generally no issue, due to the nature of the radar measurements. Having computed the inverse of an m-block, the square root of the diagonal elements straightforwardly yields the formal error for a spherical harmonic coefficient.

6.3.2   The linear low–low SST observation equation

The derivation of a low–low SST range and range-rate observation (or signal) equation has received significant attention over the past two decades, and in particular during the 1980's. The following outlines the geometric relationships needed in order to derive a linear observation equation for a low–low SST signal in terms of the radial, along-track and cross-track orbit perturbations of the two spacecraft with respect to their circular reference orbits. Substitution of the lumped coefficients (6.11) through (6.13) subsequently yields the sought observation equation. The block-diagonal error analysis technique described in Sect. 6.3.1 may then be applied to obtain the formal spherical harmonic coefficient error estimates. It deserves mentioning that, very recently, also non-linear observation equations have been derived [*Cui and Lelgemann*, 2000]. In this case, however, the m-blocks are not purely rectangular. Instead, each m-block consists of one rectangular sub-block and two trapezoidal sub-blocks. These very recent results have yet to be implemented and tested; moreover, the topic at hand is restricted to least squares error analysis based on the linear error propagation model.

The overall geometry of the problem is depicted in Fig. 6.3. An orbit-connected Cartesian coordinate frame is introduced, in which the orbital plane of satellite S_1 defines the $\xi\zeta$-plane, and the η direction is chosen as to complete a right-hand triad, in the order $\{\xi, \eta, \zeta\}$. In general, these axes do not coincide with the seleno-centric pseudo-inertial reference frame, since the orbit inclination may be arbitrarily chosen. The two satellites are assumed to be separated by an in-plane angle ΔM and by a smaller out-of-plane angle $\Delta\Omega$. All other orbit parameters are assumed to be identical for the two satellites. Originally, for low–low SST only the in-plane angle was considered [*e.g.*, *Wolff*, 1969; *Kaula*, 1983; *Wagner*, 1983; *Colombo*, 1984;

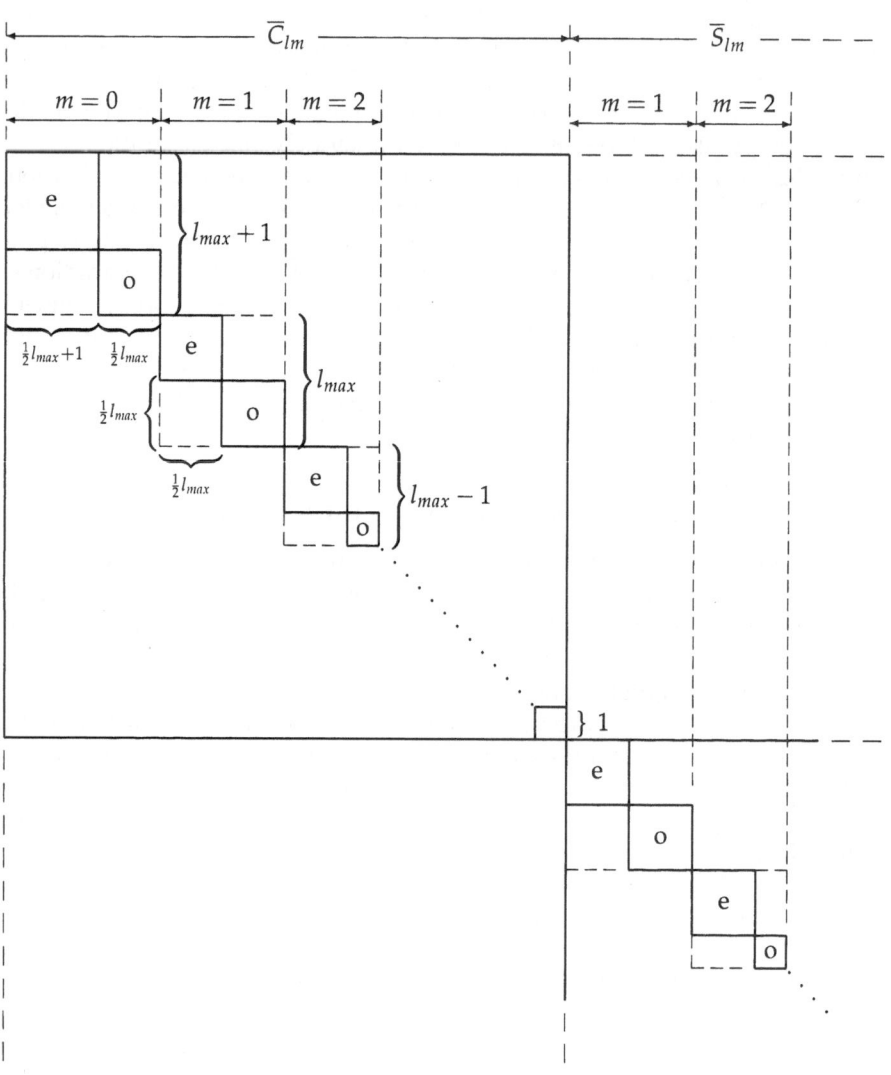

Figure 6.2   The block-diagonal structure of the normal matrix for any linear function $y(t)$
of the spherical harmonic coefficients, based on a circular reference orbit and
uniform sampling over an entire repeat period. The symbol "e" denotes even
degrees l; *vice versa*, "o" denotes odd degrees. Adopted from *Koop* [1993]

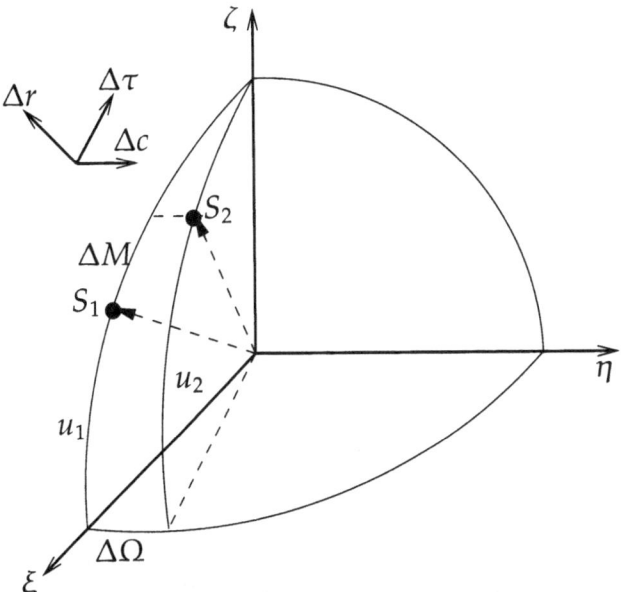

Figure 6.3   Orbit geometry of a two-satellite low–low SST experiment

Schrama, 1986]. Such a constellation is relatively easy to realise, also for simple non-propulsive relay craft deployed from a mother spacecraft, *e.g.* using a spring mechanism. However, it has since been recognised that a small additive out-of-plane separation of the two satellites adds information in the cross-track direction [*e.g., Wagner, 1987b; Floberghagen, 1995; Floberghagen et al., 1996*].

The nominal orbits of both spacecraft are circular with semi-major axis a and inclination i. Due to the circularity of the orbit, the in-plane position of the space-craft is given by the argument of latitude $u = \omega + f$, where f is the true anomaly. Since $e = 0$, one may write

$$u_1 = u = \omega + M \quad \text{and} \quad u_2 = \omega + M + \Delta M = u + \Delta M$$

Using the notation of Chap. 3, the selenopotential-induced orbit perturbations yield the Cartesian satellite positions

$$\xi_1 = (a + \Delta r_1)\cos u - \Delta\tau_1 \sin u$$
$$\eta_1 = \Delta c_1 \tag{6.16}$$
$$\zeta_1 = (a + \Delta r_1)\sin u + \Delta\tau_1 \cos u$$

and

$$\xi_2 = (a + \Delta r_2)\cos(u + \Delta M)\cos\Delta\Omega - \Delta\tau_2\sin(u + \Delta M)\cos\Delta\Omega - \Delta c_2\sin\Delta\Omega$$
$$\eta_2 = (a + \Delta r_2)\cos(u + \Delta M)\sin\Delta\Omega - \Delta\tau_2\sin(u + \Delta M)\sin\Delta\Omega + \Delta c_2\cos\Delta\Omega$$
$$\zeta_2 = (a + \Delta r_2)\sin(u + \Delta M) + \Delta\tau_2\cos(u + \Delta M) \tag{6.17}$$

The square of the inter-satellite range ρ^2 is formed by computing the 2-norm of the difference between the above two Cartesian position vectors

$$\rho^2 = \| \{ \xi_2 - \xi_1, \eta_2 - \eta_1, \zeta_2 - \zeta_1 \} \|_2^2 \tag{6.18}$$

In view of the LPT it is assumed that the perturbations are small with respect to ρ, such that linearisation techniques are valid. Cross-terms and second-order terms of orbit perturbations may therefore be neglected, which yields

$$
\begin{aligned}
\rho^2 \simeq\; & 2a^2 \left[1 - \cos(u + \Delta M) \cos u \cos \Delta\Omega - \sin(u + \Delta M) \sin u \right] \\
& + 2a(\Delta r_1 + \Delta r_2) \left[1 - \cos(u + \Delta M) \cos u \cos \Delta\Omega - \sin(u + \Delta M) \sin u \right] \\
& + 2a\Delta\tau_1 \left[\cos(u + \Delta M) \sin u \cos \Delta\Omega - \sin(u + \Delta M) \cos u \right] \\
& + 2a\Delta\tau_2 \left[\sin(u + \Delta M) \cos u \cos \Delta\Omega - \cos(u + \Delta M) \sin u \right] \\
& + 2a\Delta c_1 \left[- \cos(u + \Delta M) \sin \Delta\Omega \right] \\
& + 2a\Delta c_2 \cos u \sin \Delta\Omega
\end{aligned}
\tag{6.19}
$$

Using standard properties of trigonometric functions, it can be shown that the dominating term of the right-hand side of (6.19) is given by $2a^2(1 - \cos \Delta M)$. Dividing both sides by this dominating term generates an expression of the form

$$\frac{\rho^2}{2a^2(1 - \cos \Delta M)} \simeq 1 + \epsilon, \quad \epsilon \ll 1$$

the square root of which may be approximated by the first-order Taylor expansion

$$\frac{\rho}{\sqrt{2}a\sqrt{1 - \cos \Delta M}} \simeq 1 + \frac{1}{2}\epsilon$$

Since one is exclusively interested in periodic range variations due to the periodic orbit perturbations, and not in range signals merely due to the orbit geometry, this leads to the linearised function

$$
\begin{aligned}
\Delta\rho\sqrt{2}\sqrt{1 - \cos \Delta M} =\; & \\
& (\Delta r_1 + \Delta r_2) \left[1 - \cos \Delta M + (1 - \cos \Delta\Omega) \cos u \cos(u + \Delta M) \right] \\
& + \Delta\tau_1 \left[-(1 - \cos \Delta\Omega) \sin u \cos(u + \Delta M) - \sin \Delta M \right] \\
& + \Delta\tau_2 \left[-(1 - \cos \Delta\Omega) \cos u \sin(u + \Delta M) + \sin \Delta M \right] \\
& + \Delta c_1 \left[- \cos(u + \Delta M) \sin \Delta\Omega \right] \\
& + \Delta c_2 \cos u \sin \Delta\Omega
\end{aligned}
\tag{6.20}
$$

Equation (6.20) gives the linear low–low range signal in terms of the orbit perturbations of S_1 and S_2. Substitution of the lumped coefficients in the three directions $\{r, \tau, c\}$ from (6.11) and (6.13) yields the low–low observation equation. It can be shown that the linearisation error in (6.20) remains smaller than approximately 2%

as long as $\Delta\Omega$ is smaller than $\Delta M/3$. Moreover, in the frequent case of a *co-planar* configuration (6.20) reduces to

$$\Delta\rho = (\Delta r_1 + \Delta r_2) \sin\frac{\Delta M}{2} + (\tau_2 - \tau_1) \cos\frac{\Delta M}{2} \tag{6.21}$$

which is well-known from the literature on low–low SST.

As already mentioned, linearised range rate (Doppler) observations may be formed from

$$\Delta\dot{\rho} = \frac{\Delta\rho(t_i) - \Delta\rho(t_{i-1})}{\Delta t} \tag{6.22}$$

where i is the measurement index and Δt is the integration time of the Doppler measurement or, equivalently, the sampling interval of the time series. In other words, there is a straightforward extension from the range observable to integrated Doppler measurements.

6.3.3   Formal errors based on low–low range rate measurements

The low–low range rate observation equation described in Sect. 6.3.2, using the lumped coefficient block-diagonal approach described in Sect. 6.3.1, has been implemented for a variety of MORO-class 100 km polar co-orbiting SST constellations. More precisely, the performance of the selenopotential recovery from a single repeat cycle of one sidereal month (or 27.32166 days) in which the spacecraft at 99.431 km altitude perform 334 orbital revolutions at $i = 90°$. Hence, the repeat cycle is characterised by $N_m = 1$ and $N_r = 334$, which yields principally two global sweeps of the lunar surface (considering ascending and descending orbits). These are the same types of orbits as used for the orbit analysis of present-day gravity field models in Chap. 3 and typical for the low-risk phase of any lunar mapping mission.

The approach is first to determine the optimal in-plane separation angle ΔM, based on co-planar ($\Delta\Omega = 0$) geometries. In a second instance, for the optimal in-plane angle, the influence of various out-of-plane separations is investigated. *Wagner* [1987b] introduced the term *en echelon* for low–low SST configurations in which $\Delta\Omega \neq 0$. A perhaps better term is *butterfly configuration*, since, seen from any of the two spacecraft, the orbit of the other spacecraft traverses its orbital plane twice per orbital revolution. In the following, the solutions shall therefore be referred to as *co-planar* or *butterfly*, depending on whether or not the SST observation contains cross-track information.

Co-planar constellations

Logically, a prime aspect of the use of analytical formulae for gravity field recovery error prediction from any type of measurement is the verification of the method. The use of the Colombo method for least squares error analysis based on low–low

SST measurements has been verified by a number of authors for terrestrial applications in the past [*e.g., Colombo*, 1986; *Wagner*, 1987b; *Visser*, 1999]. More abundant are the studies of the validity of the method for gradiometry-based gravity field recovery [*e.g., Rummel and Colombo*, 1985; *Schrama*, 1991; *Visser*, 1992; *Visser et al.*, 1995b; *Visser*, 1999; *SID*, 2000; *ESA*, 2000]. For the present purpose, the formal error estimates of a 70×70 selenopotential solution derived from the time-wise Colombo method are compared with formal errors derived from a fully numerical integration and parameter estimation approach, which is described in Sect. 6.4. The comparison, for a $3°$ in-plane separation and both spacecraft nominally on 100 km, $90°$ polar orbits, is shown in Fig. 6.4. This figure depicts the per-degree RMS amplitude of the formal errors as well as the difference of the converged iterated solution with respect to the assumed to be "truth" model in the numerical case. For the full-scale analysis, uncorrelated noise according to the measurement precision of 0.1 mm/s has been added to the simulated measurements.

Overall, there is an excellent agreement between the formal errors derived from the Colombo approach and those from the full-scale numerical scheme. In other words, it may be concluded that the Colombo method is an accurate and fast tool for least squares error prediction. This conclusion is quasi-perfectly matched by the excellent fit of the formal errors with the difference between the converged selenopotential solution and the *truth* model, which shows that the stochastic and the deterministic analysis tools yield consistent error measures. Although, as already mentioned, this has been proven for the mapping of the terrestrial gravity field by means of SST or SGG in the past, the extension of the methods to the lunar problem (or mapping of any other celestial body) is not straightforward, because the approximation errors are different, *e.g.* due to the larger variations in orbital eccentricity with respect to the circular reference orbits as well as the deviations of the actually flown trajectories from the assumed repeat orbits. The regularity of the sampling was not interrupted in the present test, *i.e.* the effect of data gaps on the formal error prediction has not been analysed.

However, there is a certain discrepancy for the low-degree terms. This feature is frequently observed in terrestrial gravity field analysis, and is related to the lack of state vector estimation in the analytical approach. As outlined in Chap. 3, the LPT is equivalent to the estimation of a single state vector for the entire repeat cycle. Obviously, in the fully dynamic, numerical approach, satellite state vectors are estimated for each arc, of which there are several in a 27-day repeat period. These state vector estimates in turn correlate with the harmonic coefficients, in particular the lower degree and order terms. As was also shown in Chap. 3 the orbit errors of lunar orbiters are clustered mainly around the lower frequencies, and are for a large part due to the low-degree harmonics. Therefore, the formal error prediction is sensitive to the orbit determination strategy to adjust for the low-frequency part of the orbit error, which explains the difference seen in Fig. 6.4.

The main question one wants to address by means of the least squares error analysis is the maximum resolution and ultimate accuracy of the low–low SST-

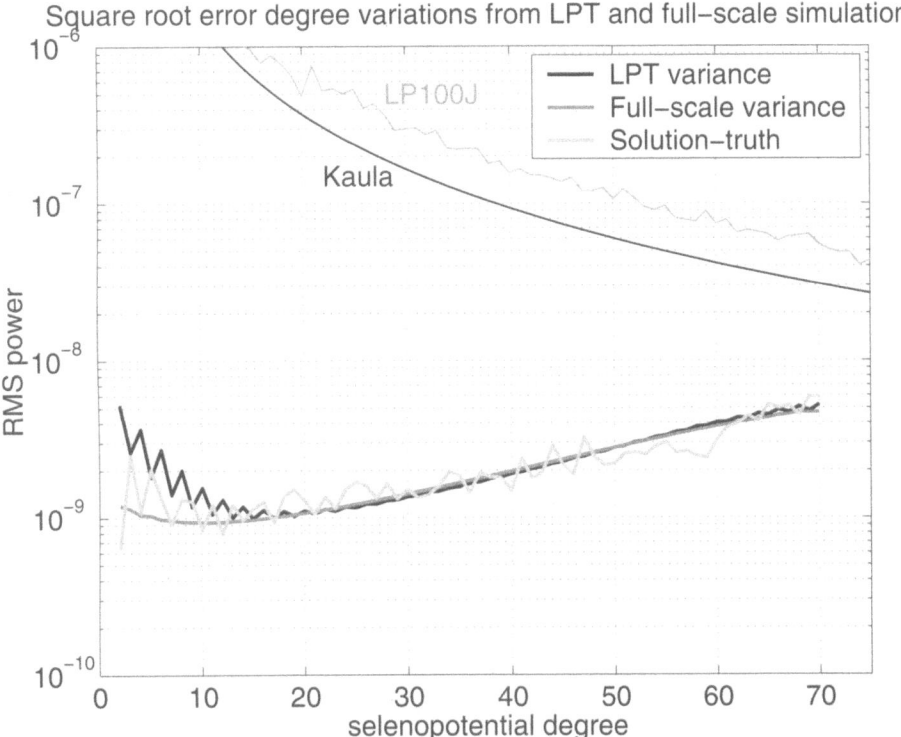

Square root error degree variations from LPT and full–scale simulations

Figure 6.4 — Comparison of degree-wise RMS formal error spectra from $3°\,\Delta M$ co-planar low–low SST at 90° inclination and 100 km altitude derived from the time-wise Colombo method based on LPT and a fully dynamic numerical approach, including integration of the satellite orbits and the variational equations. Also shown is the difference between the converged numerical selenopotential solution based on low–low SST and a truth model on which the gravity field recovery simulation is based

based gravity field estimation. In addition, one is interested in particular features of the solution, which may provide additional insight in the accuracy of individual coefficients or groups of coefficients. These are, in general, functions of the measurement type, measurement accuracy, orbit geometry and solution strategy. Of particular interest is naturally the improvement SST measurements can provide, with respect to the present-day situation. In a second instance, one wants to assess the error in derived selenophysical products, such as gravity anomalies and selenoid heights.

The degree-wise error spectra of the co-planar SST-based selenopotential coefficient errors are depicted in Fig. 6.5, for ΔM values between 3° and 10°. Shown is also the degree signal variation of LP100J, as well as the Kaula-rule that has been used to constrain the high-degree part of the gravity field since 1993, cf. Chap. 4.

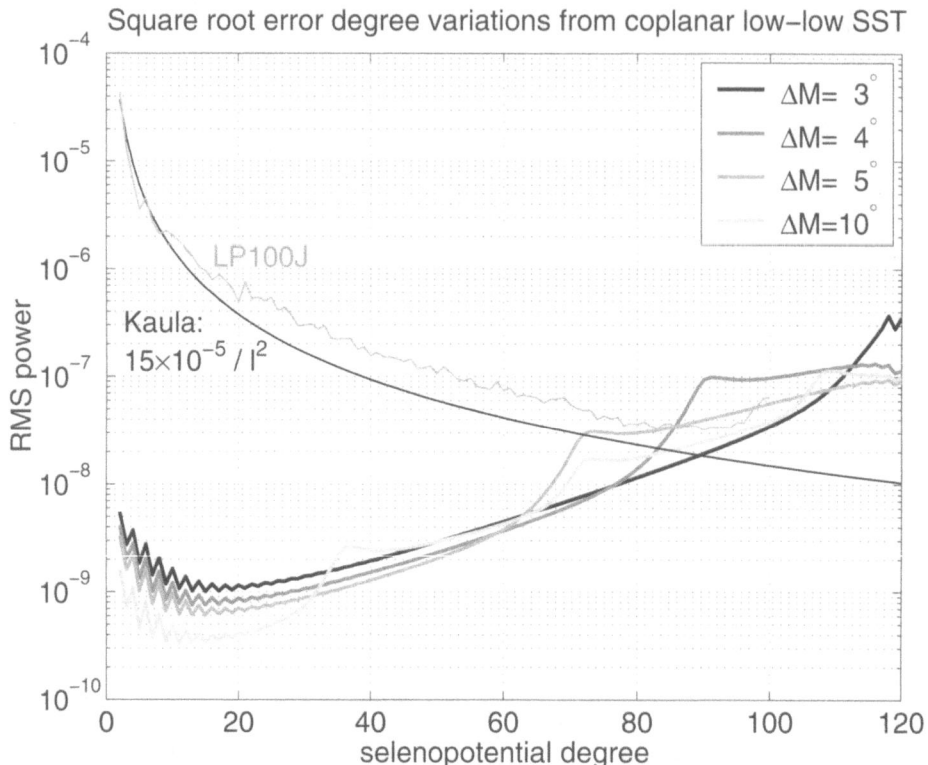

Figure 6.5   Degree-wise RMS error spectra of selective co-planar low–low SST based
                solutions up to degree and order 120 based on the time-wise Colombo
                method

Several features of the error are apparent. First, there is an exponential increase of
error at high degrees. This is a direct consequence of the attenuation effect due to
the satellite altitude above the lunar surface. In turn, the error magnitudes result
from a limited precision of the Doppler radar, combined with the fact that SST data
are unidirectional, and provide orbit perturbation data along the LOS only. A second typical feature of any low–low SST experiment are furthermore the significant
sampling gaps near degrees l being integer wave numbers

$$l \simeq j \times \frac{360°}{\Delta M}, \quad j = 1, 2, \ldots \tag{6.23}$$

for which the SST signal is nearly cancelled. The inevitably occurrence of such conditions in turn manifests itself in terms of an increased coefficient error level. Since,
for such separation angles, the two satellites move along the orbit *in phase* with the
particular wave number of the gravity field, this condition is sometimes referred

to as a *resonance*[2]. Clearly, resonant conditions should be avoided, or otherwise be removed by filtering of the measurements and the normal matrix in order to suppress the resonant terms.

By (6.23), a cancellation of the low–low range rate occurs at lower degrees for larger in-plane angles. On the other hand, larger values of ΔM provide greater sensitivity to the long-wavelength gravity field. This sensitivity is therefore a function of the inter-satellite baseline. For very small values of ΔM, the two spacecraft are experiencing largely the same orbit perturbations due to the long-wavelength part of the selenopotential, and, consequently, measurements between closely orbiting platforms may not yield improvements to the same part of the gravity field. This is exactly the reason why a gradiometer is chosen to obtain the ultimate precision for the highest frequencies of the gravity field, since in this case the baseline is virtually zero (order of 1 m). A uniaxial gradiometer is therefore sometimes viewed as an extreme case of low–low SST with a very short baseline. The consequence of this phenomenon is that it leads to an interesting trade-off. To achieve the smallest overall commission error, aiming for a maximum spatial resolution model, relatively small along-track separation angles must be chosen, which are contradictory to the demands for the long-wavelength potential. These are very real problems in lunar gravimetry, since no single spherical harmonic function has been fully sampled. Ideally, the mission planning of a lunar low–low experiment should therefore foresee in a variation of the in-plane separation in the course of the mission, or, possibly, in the deployment of more than one relay satellite, each of which orbiting at difference angles with respect to the mother spacecraft. Nonetheless, it is an outstanding fact that, irrespective of the choice of in-plane separation, the SST-based solutions are self-contained, *i.e.* with no need for numerical regularisation, with a several orders of magnitude improvement in the formal errors, compared to the present-day situation. Comparing Fig. 6.5 with Figs. 3.1, 3.16 and 5.17 there is, first of all, no dependency on regularisation to bound the formal error. Second, in a direct comparison, there is a minimum one order of magnitude improvement for all harmonic degrees up to and including 60, as well as an approximate factor ~ 5 improvement between degrees 60 and 100.

If the Kaula-rule is interpreted as a measure of the true amplitude of the signal, the maximum resolvable harmonic degree is found to be in the order of 70 to 90, depending on the in-plane separation angle, and corresponding to half-wavelength spatial resolutions of between 78 km and 61 km on the lunar surface. Based on the frequency content of the Lunar Prospector Doppler data, cf. Chap. 5, which clearly suggests that 100 km altitude data contain valuable information up to and exceeding degree 100, an in-plane separation angle close to 3° appears a good compromise, since it minimises the overall commission error. The intrinsic error level at lower degrees is acceptable (orders of magnitude improvement with respect to the

---

[2]The term *resonance* is in the author's opinion unfortunate. Resonant effects are normally associated with strong effects that are easily observable, and a geometry-driven near-cancellation of the SST signal due to a particular phasing of the two spacecraft in a low–low SST configuration is in fact rather the opposite. The term is nevertheless maintained for reasons of compatibility with other works.

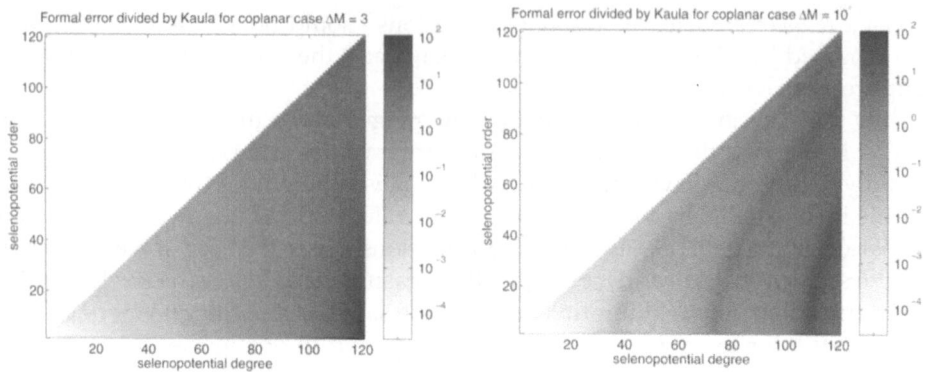

Figure 6.6   Per-coefficient formal error divided by Kaula's rule $15 \times 10^{-5}/l^2$ for $\Delta M = 3°$ (left) and $\Delta M = 10°$ (right) co-planar low–low SST solutions up to degree and order 120

current situation), and, at the same time, this separation angle enables resonance-free solutions up to $l \simeq 105$ to 110. For $\Delta M = 3°$ the maximum resolvable degree is $l_{max} \simeq 90$.

Fig. 6.6 shows the coefficient-wise error for co-planar configurations $\Delta M = 3°$ and $\Delta M = 10°$. The error is normalised with respect to the Kaula rule $15 \times 10^{-5}/l^2$ used for the recent gravity field models, cf. Table 4.1, so as to relate the formal error to a noise-to-signal ratio measure. The exponential increase of error with increasing degree and order is confirmed, and also on a per-coefficient basis it is readily seen that a 3° spacing leads to good observability of all harmonic coefficients up to degree and order 90. At the same time, if it is desirable to include orbit errors beyond 90 cpr, the same figure shows that regularisation is necessary, for which a careful study using methods like those discussed in Chap. 4 is necessary. Most SST solutions presented in this book are all self-contained in the sense that it is assured that signal-to-noise ratio of all solved-for parameters is better than one, and that the spherical harmonic basis functions are properly sampled from a polar orbit.

In the case of the 10° spacing, the immediately striking feature is the "propagation" of the resonance condition through the harmonic range. The characteristic pattern of the peak errors is given by the roots of the fully-normalised associated Legendre functions with an inter-distance equal to the satellite spacing of 10°. These functions therefore directly dictate which harmonic coefficients are unobservable with the given ΔM.

A key to the assessment of the SST experiment is obviously to test these accuracy figures against the mission requirements. In order to relate the achievable spatial resolutions for given inter-satellite separation angles to the accuracy requirements for derived selenophysical products, the global RMS gravity anomaly and selenoid height commission errors, respectively, are given in Tables 6.3 and 6.4 for the full range of separation angles between 1° and 10°. The gravity anomaly

ΔM	$l_{max} = 60$	$l_{max} = 70$	$l_{max} = 80$	$l_{max} = 90$
1°	3.700	7.345	14.267	27.178
2°	1.885	3.765	7.364	14.148
3°	1.305	2.643	5.280	10.516
4°	1.046	2.222	5.167	29.001
5°	0.974	3.934	14.381	24.722
6°	2.241	5.652	9.880	17.243
7°	2.323	4.188	7.511	13.753
8°	1.825	3.337	6.264	17.926
9°	1.498	2.846	8.109	18.282
10°	1.289	2.974	8.423	14.939

Table 6.3 Cumulative gravity anomaly commission errors as a function of SST geometry and maximum degree l_{max} for co-planar low–low SST-based gravity field solutions at 90° inclination and 100 km altitude. All figures are in mGal

and selenoid height errors are also shown in Fig. 6.7, for the same subset of solutions as for the degree-wise error plots. In all computations, the errors in all coefficients from degree two and order zero upward are taken into account.

For satellite spacings not corrupted by resonant conditions, the relationship between overall commission error and l_{max} is approximately exponential. At the maximum resolution level $l_{max} = 90$ or a half-wavelength of 61 km the gravity anomaly error amounts to 10.516 mGal and the selenoid height error to 1.392 m, in both cases for the 3° spacing. However, at the required spatial resolution for the MORO mission, the same SST configuration delivers a gravity anomaly error of 1.305 mGal and a selenoid height error of 0.279 m, which is well within the envelope of a "few mGal" which is considered a prerequisite for advanced local and regional studies of the selenopotential. These tables and figures furthermore illustrate the previously discussed trade-off between spacial resolution requirements, optimal use of signal content and the in-plane separation angle. Obviously, next to the commission error, the overall total error contains an "omission error", which in the case of the Moon is quite difficult to estimate, cf. Sect. 6.3.4. Some might argue that approximate power rules, such as a Kaula rule, are available for this purpose, but in view of the very distinct degree variation differences at the high end of the degree spectrum outlined in previous chapters, the use of a Kaula rule for omission error prediction is considered unreliable, even for order of magnitude computations. Nonetheless, the optimal use of low–low SST data requires the careful selection of the geometric constellation in relation to the mission focus and mission goals, as it is quite apparent that different parts of the gravity field motivate the use of different geometries. From the pragmatic point of view of optimal use of

ΔM	$l_{max} = 60$	$l_{max} = 70$	$l_{max} = 80$	$l_{max} = 90$
1°	0.798	1.310	2.176	3.626
2°	0.405	0.670	1.121	1.884
3°	0.279	0.468	0.800	1.392
4°	0.222	0.390	0.765	3.577
5°	0.203	0.640	2.081	3.275
6°	0.424	0.957	1.500	2.323
7°	0.463	0.736	1.152	1.848
8°	0.378	0.594	0.959	2.280
9°	0.316	0.507	1.172	2.377
10°	0.273	0.512	1.237	1.983

Table 6.4   Cumulative selenoid height commission errors as a function of SST geometry and maximum degree l_{max} for co-planar low–low SST-based gravity field solutions at 90° inclination and 100 km altitude. All figures are in metres

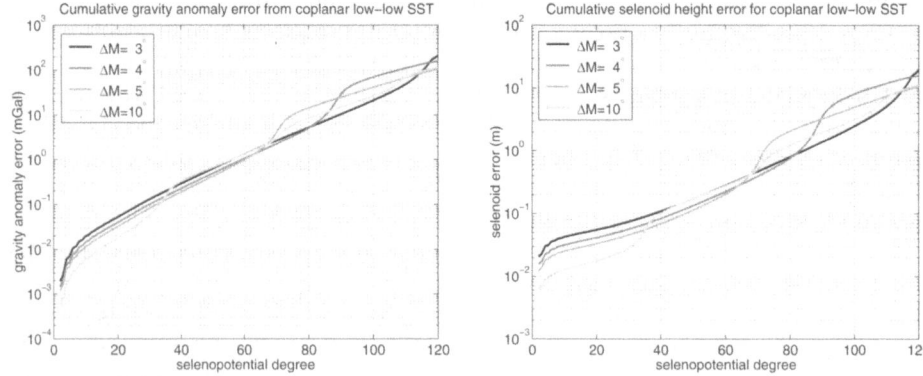

Figure 6.7   Cumulative RMS gravity anomaly commission (left) and selenoid height (right) errors based on low–low co-planar SST at 90° inclination and 100 km altitude

all frequencies in the tracking data, all the results indicate that a 3° in-plane angle and a maximum degree and order close to 90 minimise the overall error. In relation to the pre-mission spatial resolution requirements defined for MORO, these results suggest that the "rule of thumb" prediction that the maximum achievable half-wavelength resolution from satellite-based gravimetric mapping is approximately equal to the satellite altitude might be modified, as there are strong indicia that a 40% improvement is possible.

Butterfly constellations

The butterfly SST constellations offer the possibility to improve the observability of lumped harmonics through the introduction of cross-track information in the observations. In terms of in-plane separations, Sect. 6.3.3 provides evidence for an optimal spacing of 3° in the orbital plane. The following provides results for the effect of varying out-of-plane separation angles in terms of the commission selenopotential recovery error.

Fig. 6.8 gives the degree-wise error variations of the butterfly selenopotential solutions, and also relates them to the purely co-planar case. It is readily seen that there is a significant improvement, for the higher-degree coefficients, where the impact is growing with the amount of cross-track information present in the data. For $\Delta\Omega = 1.0°$, or 33% of the in-plane angle, the improvement may be as big as 40%. In other words, a significantly improved signal-to-noise ratio may be achieved simply by adjusting the right ascension of the ascending node Ω of one of the spacecraft by a moderate amount of $\sim 1°$. The results indicate that even greater improvements may be achieved for larger values of $\Delta\Omega$. Obviously, there exists a large class of possible solutions, and the mission-specific optimisation is largely dependent on the possibility to alter or freely choose the orbital parameters. However, practically, in a moderate-cost science mission, it is unlikely that there will be the freedom to independently choose the orbits of the two spacecraft. Therefore, results are not shown for orbits which have, *e.g.* perpendicular orbital planes, or orbits that pass directly underneath each other in order to maximise the radial sensitivity.

Likewise, Fig. 6.9 illustrates that the per-coefficient estimation error improvement due to the out-of-plane information is not limited to any particular type or range of harmonic coefficients, but is consistent throughout the entire observable harmonic range, compare Fig. 6.6. However, the most spectacular improvement is achieved for the sectorial and near-sectorial terms, which in a purely co-planar configuration are obviously ill-sampled. Hence, the overall impact of the butterfly type of geometry is i) to enhance the signal-to-noise ratio and therefore allow the unambiguous determination of more spherical harmonic coefficients and, hence, a better recovery of the selenopotential; and ii) to better distribute the lumped frequency orbit perturbation data over the harmonic coefficients for a given l_{max}.

This trend is also evident in the selenophysical products of gravity anomalies and selenoid heights, shown in Tables 6.5 and 6.6, as well as the corresponding Fig.

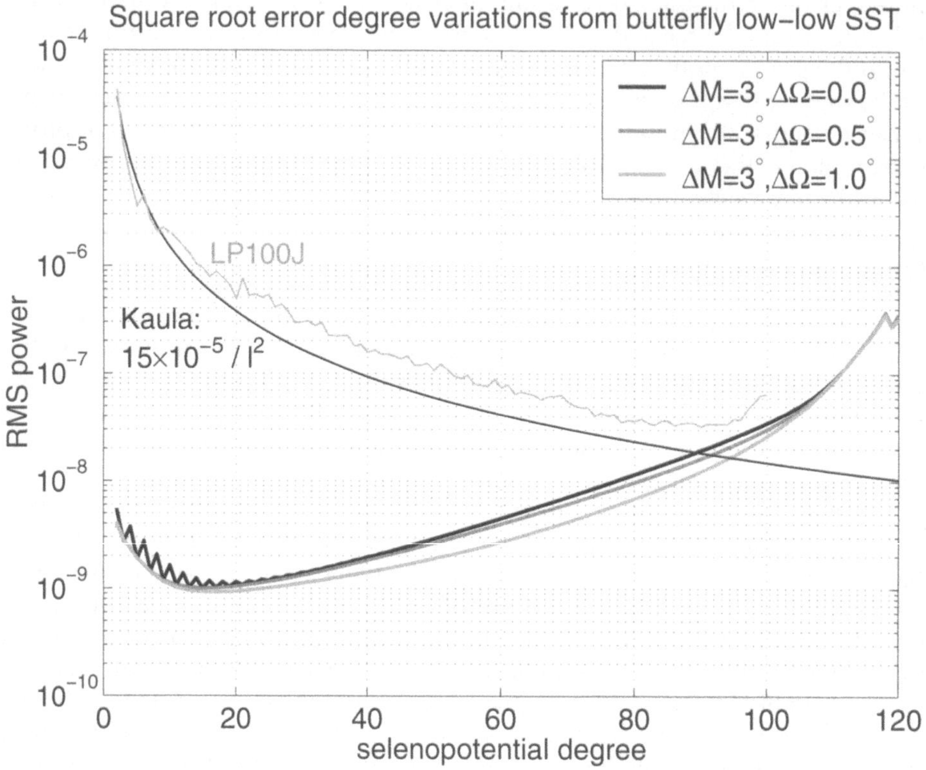

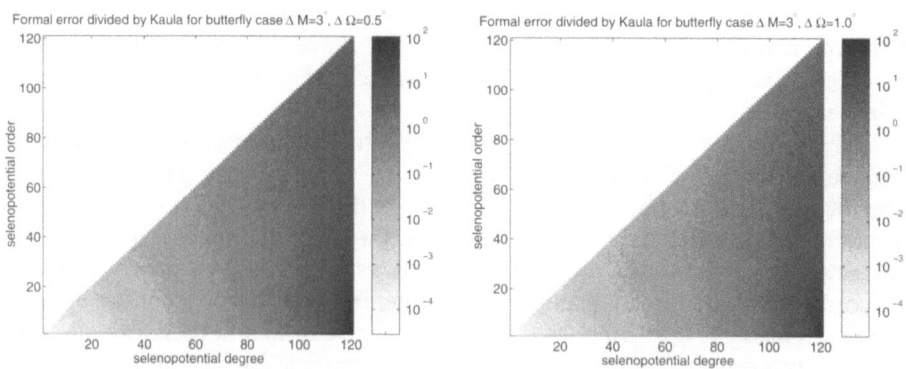

Figure 6.8   Degree-wise RMS error spectra of selective butterfly low–low SST based
solutions up to degree and order 120

Figure 6.9   Per-coefficient formal error divided by Kaula's rule $15 \times 10^{-5}/l^2$ for butterfly
low–low SST solutions up to degree and order 120. The in-plane separation
ΔM is 3° and the out-of-plane separation is respectively 0.5° (left) and 1.0°
(right)

$\Delta\Omega$	$l_{max} = 60$	$l_{max} = 70$	$l_{max} = 80$	$l_{max} = 90$
0.0°	1.305	2.643	5.280	10.516
0.5°	1.175	2.315	4.505	8.825
1.0°	0.828	1.586	3.119	6.390

Table 6.5   Cumulative gravity anomaly commission errors as a function of the out-of-plane separation $\Delta\Omega$ and maximum degree l_{max} for butterfly low–low SST-based gravity field solutions with an in-plane separation ΔM of 3°. As before, both spacecraft orbit at 90° inclination and 100 km altitude. All figures are in mGal

$\Delta\Omega$	$l_{max} = 60$	$l_{max} = 70$	$l_{max} = 80$	$l_{max} = 90$
0.0°	0.279	0.468	0.800	1.392
0.5°	0.252	0.412	0.686	1.172
1.0°	0.183	0.287	0.477	0.845

Table 6.6   Cumulative selenoid height commission errors as a function of the out-of-plane separation $\Delta\Omega$ and maximum degree l_{max} for butterfly low–low SST-based gravity field solutions with an in-plane separation ΔM of 3°. As before, both spacecraft orbit at 90° inclination and 100 km altitude. All figures are in m

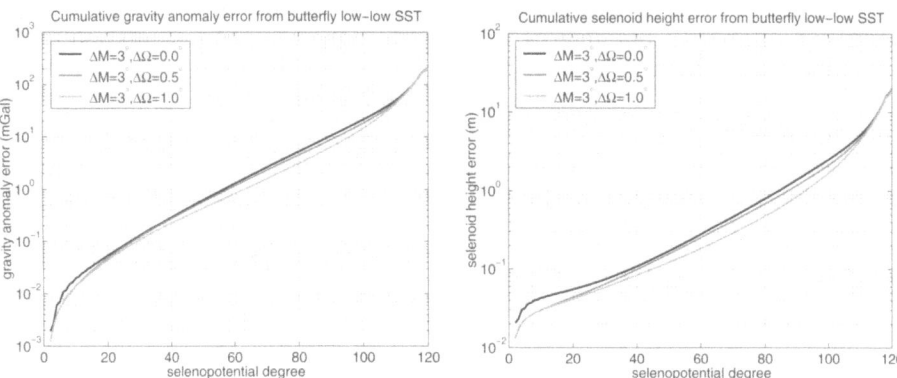

Figure 6.10   Cumulative RMS gravity anomaly (left) and selenoid height (right) commission error based on low–low butterfly SST at 90° inclination and 100 km altitude. The co-planar case is included for reference

6.10. In terms of gravity anomaly errors it is seen that the overall commission error is reduced to 6.390 mGal for $\Delta\Omega = 1°$, compared to the 10.516 mGal for co-planar SST and $l_{max} = 90$. Similarly, there is a corresponding 40% improvement in the selenoid height error, which is down to 0.845 m from 1.392 m.

6.3.4   A note on omission errors

With any selenopotential solution there are principally two kinds of errors involved, and the trade-off between these fundamentally determines the optimum truncation limit l_{max}. First, there are the commission errors of the limited set of estimation parameters, assuming that no other parameters are relevant. These are the error sources discussed in the previous section, and also to be discussed in the full-scale simulation of Sect. 6.4. However, in particular near the natural truncation limit determined by the sensitivity of the overall tracking experiment, errors due to the *omitted* parameters are of considerable importance.

In geopotential modelling, these are in the first place the infinite set of selenopotential coefficients beyond l_{max}. In view of experience from terrestrial gravity field modelling using *in situ* gravimetric surface measurements, and also considering the slow decay of the Kaula rule at high degrees, these omission parameters exhibit a slow decline of influence with high values of $l > l_{max}$. *Vice versa*, commission errors increase rapidly with l_{max} since the sensitivity at satellite altitude falls exponentially with l_{max}. The total error is the sum of commission error, with zero omitted terms, and omission error, assuming perfectly known estimation parameters. The trade-off between these respectively estimated and – in the lunar case – largely unknown errors, directly leads to the problem of determining the optimum truncation limit l_{max}, by minimising the total error.

In an actual mission situation the problem is even more difficult to assess since the omitted parameters not only affect the projected omission error directly, but also the estimated parameters through cross-coupling and aliasing. More precisely, the estimation parameters will absorb higher-degree information beyond the truncation limit, which reduces the capacity of the same higher degrees to cause omission error. The problem is therefore not only to assess the largely unknown omission effect, assuming no correlation with the estimation parameters, but also to assess the cross-coupling or aliasing that is an inevitable component of the gravity field estimation problem. Finally, due to the aliasing of non-estimated parameters in the estimated parameters, the omission error will change as well, which in fact turns the above-mentioned two-member class of errors into a class of four. Once more, lunar gravimetry takes a problem to an extreme, since there is hardly any surface gravity data available. In terrestrial applications, these are usually the primary data source for the assessment of the direct omission effect. Moreover, there is no wealth of satellite data and models of different maximum degree and order developed by means of different modelling approaches, which would at least allow for a semi-independent evaluation of gravity field models, and especially high-frequency information content.

Such a status quo obviously makes the qualitative prediction of the total error rather cumbersome. In particular the fact that present-day gravity field models yield widely different degree-wise amplitude spectra at the high-degree end of the harmonic range discourages their use in the prediction of omission errors. While not underestimating their importance near the truncation limit of self-contained selenopotential solutions, as presented in Sect 6.3.3, omission errors are therefore disregarded in the present analysis.

6.4   Full-scale simulation of low–low selenopotential estimation

The goal of a full-scale simulation of an experiment of any kind is to test and verify its performance under realistic circumstances. Within the context of low–low SST-based selenopotential recovery, full-scale simulations are carried out by means of iterative loops through the entire processing chain of the SST measurements, including aspects of orbit determination, in order to assess the overall performance of the gravity field recovery process. Essentially, the process is therefore identical to the one applied to the integrated Weilheim-down-link Doppler measurements in Chap. 5, albeit using simulated measurements of SST type. The natural focal points for such simulations are the quality assessment, using tools similar to those of Chaps. 3 and 4, as well as purely numerical aspects related to the convergence of the solution. Realistic force models and measurement models as well as stochastic error models are therefore essential ingredients of the simulation schemes.

A second aspect of such simulations is the capacity to include error sources of various nature into the simulation campaign, and therefore, in contrast to least squares error analysis, include systematic error components in the overall data reduction scheme. The following details a study of gravity field recovery based on low–low SST integrated Doppler observations. Next to serving as a benchmark for the semi-analytical tools applied in Sects. 6.3.3 and presented in Fig. 6.4, these simulations are intended to establish and verify reliable accuracy figures for the overall potential field estimation. Obviously, since there is a near one-to-one match with the error prediction from the semi-analytical Colombo scheme, the aspect of validation using state-of-the art numerical models is of the utmost importance since error prediction in itself may be computed efficiently by simpler means. For similar tests involving the recovery of the terrestrial gravity field from SGG measurements, the reader is referred to *SID* [2000].

A realistic estimation and error prediction scheme obviously requires an accurate knowledge of the error sources, which is an extremely complex issue. Frequently, in precise satellite geodetic applications, models of particular physical phenomena related to a particular spacecraft or a particular radar are developed as soon as it is realised that the available models are *de facto* limiting factors for the experiment performance. Next to inherent radar hardware issues, in lunar gravity field modelling such error sources frequently stem from the surface forces acting on the spacecraft. Due to the extremely thin lunar exosphere, in contrast to terrestrial gravity field modelling experiments, such effects are largely limited to effects of solar radiation pressure, the equilibrium radiation of the Moon itself and to a possible asymmetric thermal radiation of the spacecraft themselves. In other words, for a maximum quality of the gravity field product, improvements and dedicated modelling efforts for a range of radiative forces, including the effects of spacecraft geometry and ray paths, are required. It is therefore a matter of effort to derive ultimate accuracy figures from a given SST configuration. Such dedicated efforts have already started for lunar mapping missions. The past few years have seen the first geometric and material property models of a relay satellite in lunar orbit [*Kubo-oka*, 1999; *Kubo-oka and Sengoku*, 1999], as well as the effect of lunar albedo and lunar thermal radiation on the precise orbit determination and subsequent gravity field estimation [*Floberghagen et al.*, 1998, 1999]. Other improvements are expected as the pre-mission analysis of the selenodetic experiments foreseen for SELENE continue to proceed.

It would go beyond the scope of the present analysis to detail all the aspects of a mission-specific high-accuracy gravity field recovery experiment. The target at hand is, much like for the least squares error analysis, to illustrate the striking benefit a well-defined global data set can mean to the lunar sciences. Moreover, as already mentioned, such case studies have been published in the literature on low–low SST [*Floberghagen et al.*, 1996], as well as for high–low SST [*e.g.*, *Konopliv*, 1991; *Matsumoto et al.*, 1999]. Frequency-dependent measurement errors as well as systematic error sources are therefore not discussed in any significant detail. In the following, the main characteristics of the deterministic data reduction of low–low SST range rate measurements are therefore given for a limited set of two test cases only:

- a case with noise-only data corruption; and

- a case with a significant uncompensated systematic error

The former is identical to the solution presented in Fig. 6.4. However, since the full inverse is available, error propagation includes the correlations between all estimation parameters. In the latter case, the systematic error is chosen to be a fixed, non-adjusted 10% error in the direct solar radiation pressure acting on both spacecraft. This is a severe error, which one will normally not encounter in a real mission situation, since, for any precise geodetic/selenodetic/planetodetic mission, a considerable effort is necessarily devoted to the reduction of non-gravity field related phenomena, which would otherwise alias into the gravity field product. Evidently, the crucially important factor is the frequency characteristics of the individual physical phenomena to be modelled, since matching frequency spectrum envelopes makes decorrelation of the solar radiation scaling parameters and the corresponding constituents of the gravity field coefficients impossible. It should also be clear from the previous that the actual error is normally somewhere between the two extremes outlined above.

A more elaborate parameter identification analysis may be found in *Flobergha-*

	a	e	i	Ω	ω	M
S_1	1838000.0 m	0.0	90°	0.0°	0.0°	0.0°
S_2	1838000.0 m	0.0	90°	0.0°	0.0°	3.0°

Table 6.7   Baseline epoch two-satellite osculating elements for the low–low SST gravity
field recovery experiment. Other test cases include a range of inclinations
between 80° and 100°

gen et al. [1996], who detail a study of the effect of orbit and measurement related
parameters on the overall experiment performance. A crucial aspect of such parameter identification studies is the apparent need for some moderate amount of
Earth-based tracking to provide a stable datum for the satellite orbits. Since the
short-baseline low–low range rate measurements contain information on the relative motion of a system of *two* satellites, it is required that one degree of freedom is
fixed by other, more appropriate, tracking data. This is the role of the Earth-fix in
the tracking segment. Other parameters include the optimal choice of arc length,
which is a trade-off between observability of orbit perturbation frequencies and
possible second-order effects which increase with time, and the required *a priori*
knowledge of the satellite orbits. The latter is an issue of practical importance if
the contact with the relay satellite is lost for a considerable amount of time, since
proper orbit solutions are required to make optimal use of subsequent SST data.

6.4.1   The simulation setup

As already outlined, the simulation covers a tracking period of 28 days, which is
divided in four arcs of 7 days. The initial osculating elements are shown in Table
6.7, for a 1 January, 2003 at 00:00:00 UTC epoch. For the entire 28 days period,
simulated measurements are generated using the GEODYN II software. The force
model is identical to the one used for the orbit modelling of the Lunar Prospector
spacecraft in Chap. 5, cf. Table 5.1, with the exception that the reference gravity
field is given by the GLGM–2 model. Integrated 2-way Doppler measurements are
generated between the S_1 mother spacecraft and the S_2 target spacecraft performing the relay function at 10 s integration intervals, to which uncorrelated noise of
0.1 mm/s is added. Non-interrupted tracking is assumed, which leads to a regular
sampling of the force field. The spacing between adjacent tracks due to the lunar
rotation is 1.054°.

In addition to the SST measurements, some moderate amount of conventional
Earth-based Doppler tracking is required to unambiguously determine the satellite
orbits. To this end, a two-station ground segment was chosen, making use of deep
space antennas located in Perth, Australia, and in Madrid, Spain. These stations

Low-low 70x70 normal matrix

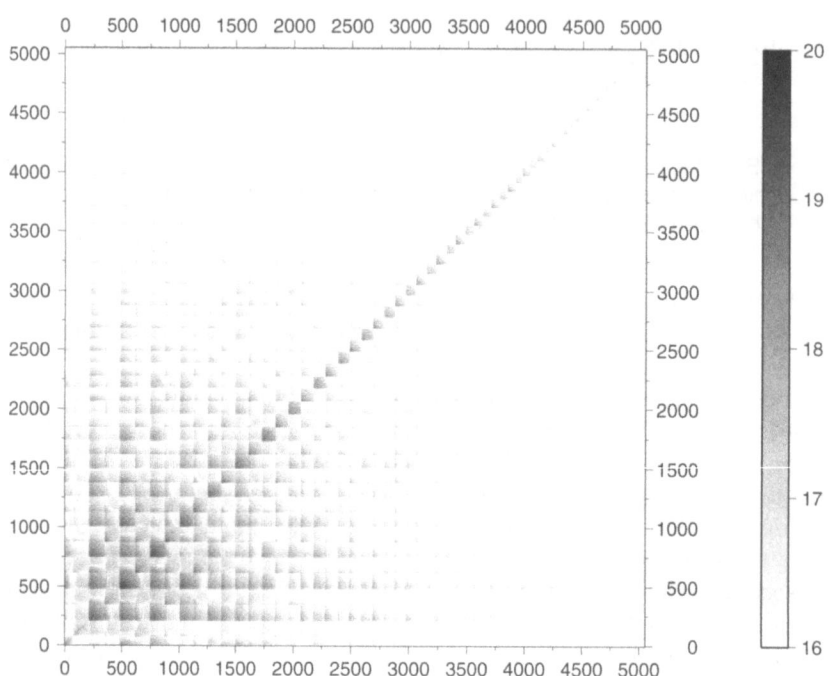

Figure 6.11 Log_{10} combined normal matrix from 28 days of low–low SST tracking at 90°
inclination and 100 km altitude. The matrix is organised per order m,
beginning in the lower left corner with $m = 0$ coefficients, and progressing
with higher order $\{\overline{C}_{lm}, \overline{S}_{lm}\}$ coefficient pairs towards the upper right corner

track the S_1 mother spacecraft exclusively, since it is assumed that a high-power
transponder system is too demanding - in terms of mass and cost - for a small, free-
flying companion spacecraft S_2 performing the relay function. In other words, the
simulations are carried out with minimisation of the cost and complexity of the
relay function in mind. In this concept, S_1 is tracked by the deep space antennae
by means of 2-way Doppler measurements of the LOS velocity component at the
sampling rate of one measurement per 60 s and a precision level of 0.3 mm/s.
Such precision figures are well within reach by current radar systems, cf. Fig. 5.7.
Conventional 2-way range measurements are not considered for the present study.

   For the data reduction, again using the GEODYN II software, the processing
of each 7-days arc is characterised by the estimation of a state-vector and a solar
radiation pressure scaling coefficient. After convergence of these arc-dependent

70x70 solution correlation coefficients

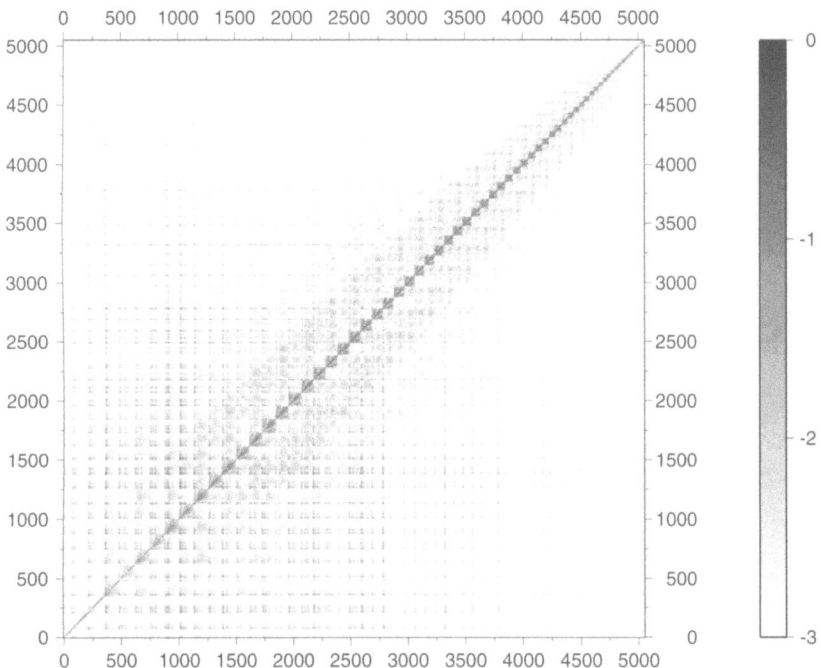

Figure 6.12   Log$_{10}$ absolute correlation coefficients from 28 days of low–low SST tracking
at 90° inclination and 100 km altitude

parameters, a normal equation is generated for the gravity field parameters. All
four such matrix systems are then combined to form a *combined normal equation* for
the gravity field parameters, making use of the partitioning procedure described
in Chap. 5 to eliminate the effect of the arc-dependent parameters. After solution
of the common parameters, the arc-dependent parameters may be solved for, from
which a second iteration over the whole data set using the updated selenopotential solution may be performed. This process is executed until given convergence
criteria for the first-order linear corrections to the *a priori* selenopotential model are
achieved. In the present case, the selected convergence criterion is that the change
in RMS value of the residuals must change by no more than 2% between successive iterations, which for the lunar low–low SST case generally is satisfied after 3
global iterations.

The normal matrix entries of the third and final iteration for a 70 × 70 solution,
are shown in Fig. 6.11, and the corresponding correlation coefficients are depicted

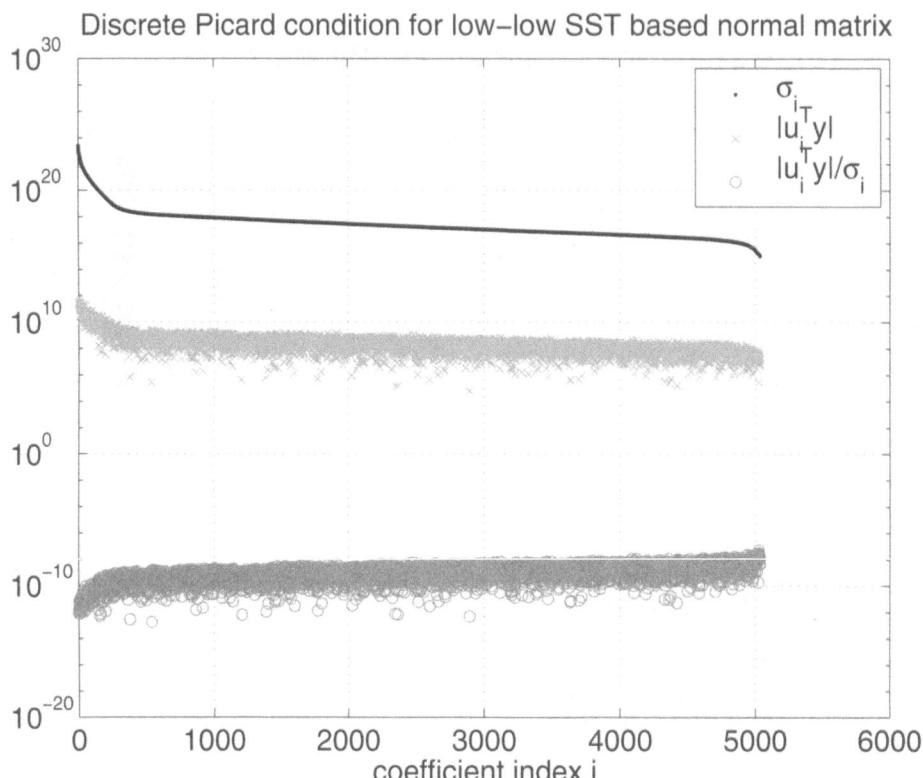

Figure 6.13   Picard plot for the unregularised normal equation based on low–low
              co-planar SST at 90° inclination and 100 km altitude

in Fig. 6.12. As was the case for the correlation coefficients of present-day models shown in Fig 3.7, the normal matrix is organised per order m, with lower order m-blocks at the lower left end and the higher order m-blocks at the upper right end of the plot. From the 28 days of tracking on a nominal $\sim$ 100 km orbit, it is apparent that the block-diagonal structure is dominant. However, for even orders $m = \{2, 4, 6, 8\}$ and slowly decaying for higher, even values of l, relatively high off-diagonal values are present, which indicate a non-negligible level of correlation between harmonic orders m, as shown in Fig. 6.12. The presence of such high-value off-diagonal elements is related to the differences between the numerical case and the Colombo method. Numerically, the orbit is non-circular, with altitude variations along the track in the order of 20 km. Second, a tracking campaign of 28 days is assumed, which is different from the 27.32166 days repeat cycle assumed in the analysis using LPT. And finally, in the numerical case, non-gravitational forces are included which destroy the symmetry of the orbit, making it strictly non-repetitive, although the deviations are relatively small. In other words, the

SST data stream in an actual mission situation is not a truly periodic signal.

Since only a lumped frequency data set of moderate length is available, the number of unique frequencies is naturally limited. Correlation between m-blocks is a direct evidence of inability to uniquely distribute the tracking data information over the orbital frequencies. In other words, an improved frequency-wise distribution of the satellite tracking data over the harmonic coefficients requires the sampling of different orbital frequencies, rather than more data of exactly the same type. Practically, this may be managed by adjusting the length of the repeat cycle, by adjusting the spacecraft altitude. Note that cross-track manoeuvres to alter the right ascension of the ascending node in order to sample the selenopotential, in a second repeat cycle, between the tracks of the first repeat cycle does not reduce the correlation level between harmonic coefficients. In this case, the normal matrix entries become twice as large, which simply reduces the formal errors by a factor of $\sqrt{2}$.

In terms of the discrete Picard plot and the conditioning of the combined normal matrix shown in Fig. 6.13, it is readily seen that the low–low SST-based normal matrix outperforms the GLGM–2 and LP75G models of Fig. 4.1 in several ways. First, the condition number of the unregularised 70×70 matrix is 2.282×10^8, compared to 2.098×10^{17} for GLGM–2 and 7.652×10^{15} for LP75G, which shows that the SST data set vastly outperforms the current state-of-the-art based on conventional tracking by deep space stations. Second, inspecting the significantly slower decay of the singular values as well as the Fourier coefficients, which lead to a better compliance with the Picard criterion as expressed by the $\left| \mathbf{u}_i^T \mathbf{y} / \sigma_i \right|$ coefficients, there appears to be no numerical problems hampering the ability to compute well-defined unregularised solutions up to degree and order 70, other than the problem of spectral leakage discussed above. Such a fundamental improvement of the numerical properties of the normal equations derived from SST compared to conventional Doppler tracking, due to the regularity and high quality of the sampling of the potential field, is exactly the reason why SST provides the answer to present-day needs in lunar gravimetry. Based on the one-to-one relationship with the predictions from least squares error analysis, it may also be concluded that the numerical solutions support the results based on the LPT that the actual resolution limit is close to $l_{max} = 90$, based on an SST integrated Doppler precision level of 0.1 mm/s and a purely polar orbit.

6.4.2   Gravity field recovery results

The spherical harmonic coefficient formal errors are depicted in Fig. 6.14 for two inclination cases of 90° and 100°, respectively, using the same colour legends as Figs. 6.6 and 6.9 for purposes of direct comparison. Similar to the degree-wise plots of Figs. 6.4, there is, also for individual coefficients, a near-perfect match between the analytical prediction and the deterministic experiment. In the case of a 100° inclination, there is an under-sampling of the polar areas, which is also called the *polar gap problem*, since 10° spherical caps around both poles remain uncovered

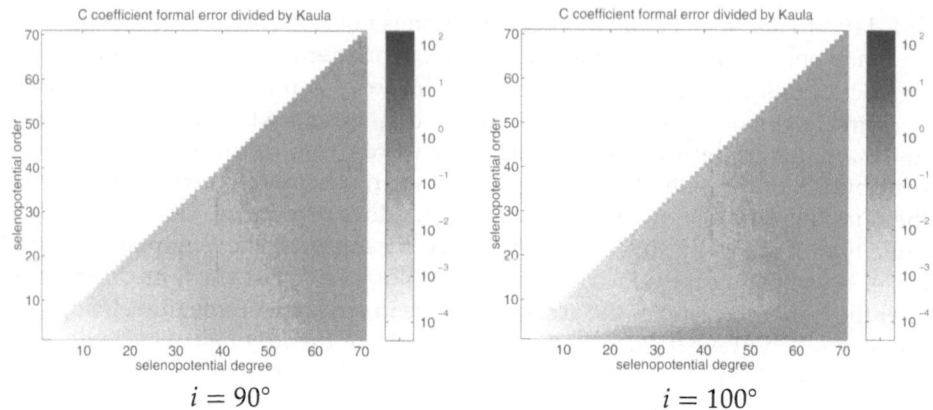

$$i = 90°$$ $$i = 100°$$

Figure 6.14 $\overline{C}_{lm}$ formal errors divided by Kaula's rule $15 \times 10^{-5}/l^2$ from the converged full-scale gravity field recovery simulation of a 70×70 selenopotential solution, based on a $\Delta M = 3°$ co-planar low–low SST configuration. Results are shown for 90° inclination and 100° inclination, where the latter case illustrates the effect of the incomplete sampling of the spherical harmonic basis functions. The scale is identical to the least squares error analysis of Fig. 6.6

inclination	min	max	RMS
80°	0.01	16.51	2.66
85°	0.01	0.91	0.48
90°	0.01	0.93	0.46
95°	0.01	0.92	0.47
100°	0.01	15.16	2.38

Table 6.8   Selenoid height error characteristics from selenopotential recovery at inclinations between 80° and 100°. All numbers in units of metres and the grid size is 1° by 1°

with measurements. Such an under-sampling of the potential field is manifested in an increased error level at lower orders, compared to the case of a global sampling, and is a feature frequently discussed in the context of terrestrial gravity field missions, where mission constraints, or the desire to fly on sun-synchronous trajectories, which are non-polar by definition, ultimately lead to a slight under-sampling of the gravity field. As such, these plots are an argument against the utilisation of non-polar orbits for gravity mapping campaigns, even if, *e.g.*, such trajectories have been flown by previous spacecraft tracked from Earth-based stations, such as Lunar Prospector.

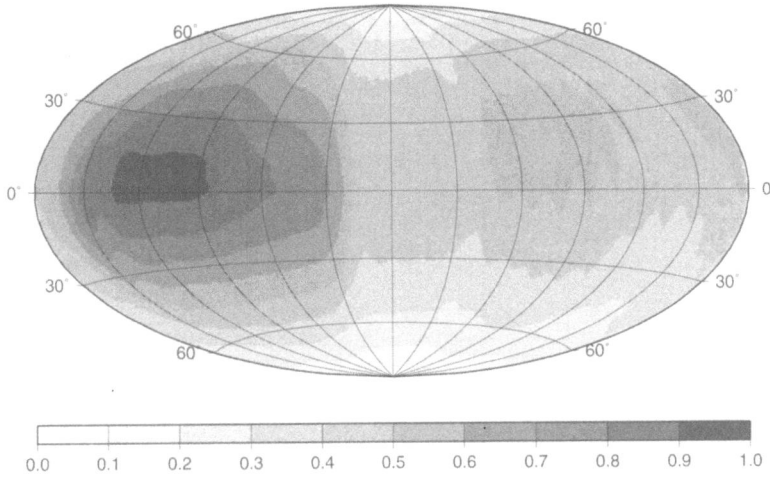

Figure 6.15   Selenographical distribution of the formal selenoid height errors in metres
from the variance-covariance matrix of low–low SST-based 70×70
selenopotential solution

Similar trends are seen in terms of derived selenophysical products. Table 6.8
illustrates the behaviour of the 70×70 commission selenoid height error as a function of the two-satellite inclination value. In the case of $i = 100°$, the maximum
error rises significantly, around the unsampled polar areas, causing significantly
larger maximum error values. For the sampled regions of the Moon, the error remains consistent with those from globally sampled basis functions, which keeps
the overall RMS error value at an acceptably low level.

Along the same line, the selenographical distribution of the errors, depicted in
Fig. 6.15 and to be compared with Figs. 4.7 and 4.8, illustrate the impact of the SST
measurements for such products, not only in a global RMS figure, but in terms of
their spatial distribution. Comparing the selenoid height error results with those
of GLGM–2 and LP75G, which moreover are bounded due to the regularisation
scheme rather than actual data, there is roughly a factor 25 and 18 improvement,
respectively. The fact that the error remains slightly larger over the far-side is due
to the limited amount of ground tracking, which provides a slightly better data
set for the near-side gravity field recovery. The fact that satellite track effects may
still be seen in the near-side solution may provide additional insight into the correlation problem. Based on the earlier observation that four weeks of low–low
SST from a single nominal altitude of 100 km are insufficient to decorrelate all
coefficient-wise orbit perturbation contributions over the entire harmonic range

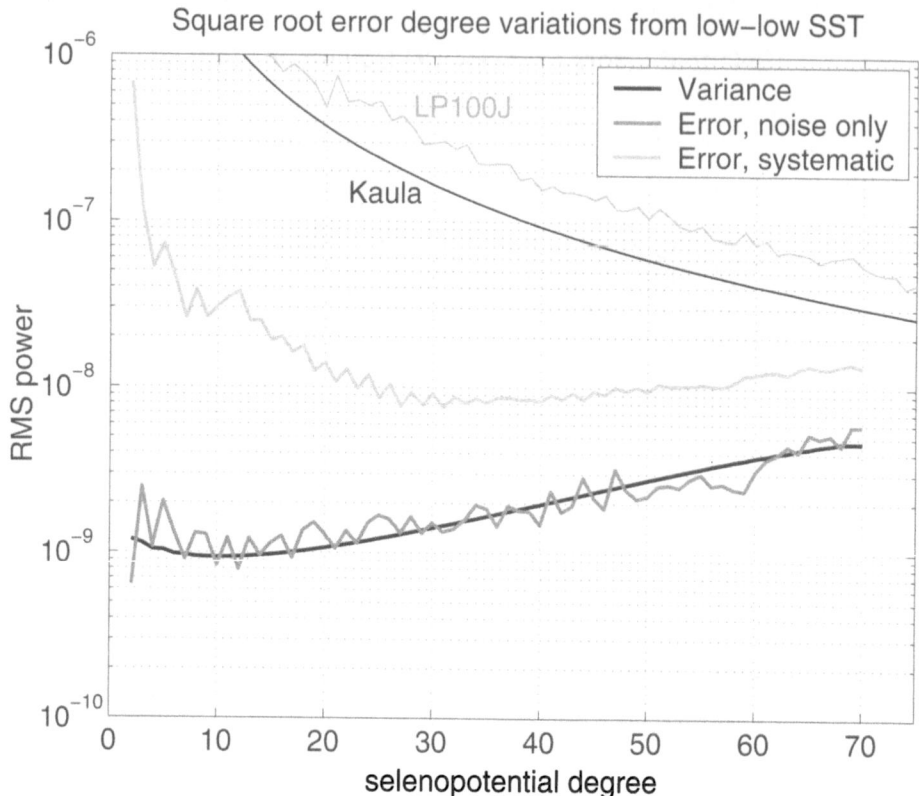

Figure 6.16   Degree-wise RMS formal error spectra from $3°\Delta M$ co-planar low–low SST at
90° inclination using a fully dynamic numerical approach, including
integration of the satellite orbits and the variational equations, for the case of
noise-only data corruption and the case of a 10% fixed error in the solar
radiation pressure coefficient

of frequencies, for a given l_{max}, both a denser sampling (leading to better spatial decorrelation) and a different set of orbital frequencies in subsequent repeat cycles (through a different choice of orbits) will improve the correlation. It also deserves mentioning that the very reason why GLGM–2 and LP75G prove to suffer less from correlation is that the stringent regularisation in itself imposes a strong diagonal normal matrix structure. Another way to improve the situation is to apply the *butterfly* type of constellation, which includes cross-track orbit perturbation information, and therefore helps separate the lumped frequency data over the spherical harmonic coefficients in a better way.

Finally, the sensitivity of the solution quality with respect to non-gravitational force model deficits is investigated by means of the test case of a 10% non-adjusted error in the scaling parameter for the direct solar radiation pressure. Fig. 6.16 depicts the degree-wise error variations in direct comparison with the uncorrelated noise-only case. Comparison of the two curves reveals that the larger part of the differences is found at the lower degrees. In fact, it may be shown that the amplitude frequency spectra of radiation pressure induced orbit errors peak in the 0 – 2 cpr range [*Floberghagen et al.*, 1998], which in fact indicates that the radiation pressure effects, if not properly modelled alias in the low-frequency gravity field constituents. In other words, the modelling of radiation pressure effects is of the utmost importance for the quality of the low-frequency gravity field estimation. A serious problem is that a large part of the gravity-induced orbit error is also seen in this frequency range, cf. Fig. 3.2 and 5.11. If the radiation pressure error is severe, the spectral correlation may become a limiting factor of the experiment performance.

The non-adjustment of these four parameters (one for each 1 week arc), on the other hand, increases the RMS commission error in selenoid height from 0.46 m in the case of uncorrelated noise to 1.84 m. Likewise, the gravity anomaly commission error increases from 2.64 mGal to 6.89 mGal for a 70×70 solution. Although the systematic error case at hand is conservatively pessimistic with respect to a likely actual mission performance, it is illustrative of the interlink between the precision of the data on the one hand, which actually make a high-accuracy and high-resolution gravity field recovery possible by virtue of the signal content of the sampled data, and, on the other hand, force model defects of non-gravitational nature. Only through effort, material property measurements and ray path modelling will it be possible to approach the theoretical precision level of the uncorrelated noise-only case. In fact, this is nothing short of an open invitation to SST experiment planners to dedicate serious efforts to the non-gravitational aspects of the precise orbit determination, such as to minimise the aliasing with the selenophysical end product. The real performance of a well-defined low–low SST experiment will therefore be somewhere between these two extremes. This is a promising prospect, since even the most pessimistic figures are compatible with the experiment requirements. However, it is clear that achieving the ultimate quality demands an effort from both gravity field analysts and orbit modelers.

Chapter 7
Epilogue

Science, just like life, is probably best understood backwards. Research is also never complete, since, by its nature, it not only provides answers, but at the same time triggers more questions. Naturally, this epilogue therefore contains the most important conclusions and recommendations that follow from the analysis presented in the previous chapters.

7.1   Conclusions

The problem

In retrospect and summary, the global lunar gravimetric problem, *i.e.* the problem of determining, analysing and interpreting the gravitational field of the Moon, has a wonderfully ambiguous nature, which is also reflected in the title of this work. *The Far Side* does not merely refer to the far-side of the Moon, which is the hemisphere inaccessible to standard spacecraft tracking techniques using Earth-based deep space antennae. It also hints to the problem itself.

A numerical analyst or a mathematical geodesist would, by inspection of the unregularised information matrices that contain all the information on the coefficients of a spherical harmonics model provided by the tracking data of spacecraft in low lunar orbit, easily conclude that only a very limited number of spectral components (expressed in terms of singular values and vectors) is observable. In other words, from a purely numerical or mathematical point of view, only a restricted, relatively low degree and order self-contained spherical harmonic model is directly retrievable from the available data sets spanning more than thirty years of lunar exploration.

Nonetheless, two factors support the conclusion that, while such a view is strictly correct, it does not necessarily provide the optimal model based on the same data. For one, from a physical point of view, the gravity field does not exhibit infinitely large variations over the unsampled regions, as also demonstrated by the observed orbit behaviour of lunar satellites. In this case, in spite of a severe under-sampling of the selenopotential, through its induced orbit perturbations on Moon-orbiting satellites, the global orbit measures actually demonstrate encouraging metre-level orbit consistencies. Such consistencies are defined as the orbit differences between overlapping portions of successive arcs. Second, the very high-accuracy and high-resolution sampling of the lunar near-side that has been achieved over the past few years, in particular the Doppler measurements from the Lunar Prospector mission, constitutes a unique data set which undeniably contains highly valuable information on the entire near-side. In order to exhaust all of this information, the parameterisation of the gravity field must span both low-frequency and high-frequency components. Evidently, in the framework of global spherical harmonic basis functions this implies an extended series expansion. By treating regularisation as a compensation for the under-sampling of the problem, using "tuned" or "dedicated" regularisation schemes, present-day models are therefore found to yield "reasonable" or "possible" values for mass anomaly structures over the entire lunar sphere. On the other hand, it is only logical to conclude that such a problem asks for a different, non-global set of basis functions as a replacement for the spherical harmonics framework.

Exactly here lies the ambiguous nature of the problem. A formalistic view, where the lunar gravimetric problem is treated as a purely numerical (analysis) problem, does not pay due respect to the physics involved. At the same time, orbit analysis, which is based on the solution of a dynamic problem, may be misleading to the end user, in the sense that good orbit determination figures clearly do not mean that also individual spherical harmonic coefficient estimates are good or even reasonable. Similar arguments straightforwardly follow for physical products based on the same coefficients, such as selenoid heights of gravity anomalies. It is therefore not without reason that the title of this book is inherited from Gary Larson's wonderful comic *The Far Side*, which stimulates the imagination of the reader by projecting people, animals and aliens in stretched situations. Just as in the comic, the perspective on the present-day situation in global lunar gravimetry depends on the point of view.

It is clear that, as the global lunar gravimetric problem is posed today, an optimal solution probably does not exist. Objective criteria required to be met by the solution are not (and cannot be) unambiguously defined. Measurement fit or orbit consistency are reasonable measures for the overall force model quality in some "lumped" sense, however, by their connection to orbital arc length they do not provide much information on the spatial variations in gravity field model quality. Similarly, by tuning the regularisation parameters, the models can turn out quite different. Consequently, it is possible to focus modelling efforts directly on a particular type of application or subsequent selenophysical usage. Accordingly, the problem is ill-posed in more than a strict mathematical sense. In a mathematical framework, regularisation is used to counteract effects of inaccurate or incomplete sampling. However, in the present problem, it must – for a global solution – also account for the missing physics of the problem.

This ambiguity of the solution space generally leads to varying conclusions and subsequent interpretations, that are not only a consequence of the coming of some additional new data (which was the case *e.g.* in 1994 with Clementine, as well as in 1998/99 with Lunar Prospector), but also the overall view on the gravity field data reduction process. This should be taken as a clear warning to everyone involved in subsequent selenophysical interpretation of the gravity field models, such as end users who use the models in modelling *e.g.* the crustal structure, Moho topography or correlations between topography and gravity.

Indeed, the problems in lunar gravimetry summarised above are not new. In fact, they were formulated by Kaula as far back as in 1969, shortly after the discovery of the lunar mascons. Nonetheless, the ambiguity they cause, and the fact that new data tend to be seen as a new truth even though the tracking concepts have remained basically the same, make their validity still stand. Therefore, and ever since, the lunar science community has lobbied for a dedicated gravity field experiment that would overcome all of these problems, and actually unambiguously determine all the coefficients of a model spanning both the low-frequency domain and the high-frequency domain with high accuracy. The primary role of this book has been to illustrate most, if not all, of the above mentioned problems, by applying available quality assessment tools to existing gravity field models. In a subsequent stage, it has aimed to outline and study in detail one possible path towards self-contained solutions based on global data of uniform quality. To this end, the use of low–low SST for global gravity field estimation has been studied.

Scientific rationale and history

The rationale and history of the lunar gravimetric problem were detailed in Chap. 2. Primarily, the focus was on gravity-related science *of* the Moon, and the use of gravity field models in engineering applications, like spacecraft navigation. It was argued that the full benefit of the gravitational data is only found in the combination with other data sets, such as seismographic and topographic profiling, as well as lunar laser ranging. Such a combined approach will, provided the sampling is uniform and accurate, allow the tackling of the major selenophysical challenges, starting from the determination of a possible iron-rich core, ending at the very high-frequency signals from localised mass anomaly features. Flight dynamics related rationales are found in the pure navigation area, such as lander missions, as well as in the need for accurate orbits for observations that are directly related to orbit quality, *e.g.* topographic profiling using a laser altimeter.

The history of lunar gravimetry goes back to 1966, when the flattening of the selenopotential was first determined. The available data sets include the Lunar Orbiter missions, Apollo 15 & 16 sub-satellites, Clementine and Lunar Prospector, as well as the Soviet Luna missions. Recent solutions include GLGM–2, developed by NASA/GSFC based on Clementine and historical data, and a series of Lunar Prospector models, denoted by LPXXXn, developed by JPL. The LPXXXn models also include historical data from the Apollo and Lunar Orbiter era. For the latter,

"XXX" denotes the maximum degree and order and "n" denotes version number within the XXX "series" of models. Notable solutions are LP75G and LP100J.

Orbit analysis of current models

By application of linear perturbation theory (LPT) to the problem of orbit error analysis in Chap. 3, it has been possible to characterise the quality of currently available models, exemplified by GLGM–2 and LP75G. These two models were deliberately chosen. For one, their difference illustrates the difference between the pre-Lunar Prospector and the post-Lunar Prospector view on the gravity field of the Moon, as it has been presented to the scientific community. Second, and perhaps more importantly, these two models have been developed by different research groups, using different tools and solution strategies. In other words, the differences are not merely due to additional data going into LP75G, but also reflect the entire approach to the global gravity field estimation problem. The fact that the models are found to be largely different, in particular for the high end of the degree scale where there is roughly a one order of magnitude difference in signal amplitude, is therefore a clear indication that data alone do not lead to a full understanding of the gravity field of the Moon.

The analysis was conveniently divided into a time-wise part and a space-wise part. The time-wise analysis evidently characterises the orbit error in the time domain (or as a function of orbital frequency), whereas the space-wise analysis projects the orbits error onto the lunar sphere. In both modes, the published variance-covariance matrices of GLGM–2 and LP75G as well as the difference between the two models were used as complementary measures for the intrinsic selenopotential error. Overall, both approaches confirm that GLGM–2 is derived with primarily non-orbit related applications in mind. Rather, it is aimed to provide a meaningful gravity field solution for global or coarse-scale selenophysical applications, by imposing a deliberate damping of the high-degree power. LP75G, on the other hand, tends to be more suitable for orbit determination purposes (it obviously contains more high-frequency data), and, by similar arguments also focuses on regional to local-scale physics. Signs of aliasing at the degrees above $l = 100$ furthermore hint to the existence of significant gravity signatures beyond the maximum spatial resolution of present-day models. This might be a confirmation that the high-frequency gravity field constituents may be a target area for future selenodetic experiments.

The same two models were also subject to long-term orbit prediction studies, which proved that mission analysis applications are still significantly hampered by inconsistent orbits produced with recent selenopotential models. Reliable orbit predictions are therefore still problematic, in particular for previously not flown orbits. On the other hand, for the low polar orbit flown by Lunar Prospector, based on both predictions and error studies, there is good reason to believe that the behaviour is well understood at the present time.

Ill-conditioning of the lunar gravimetric inverse problem

In Chap. 4, a detailed analysis was presented of the numerical issues of the lunar gravimetric problem. It was focused on the ill-conditioning of the problem, as presently defined based on Doppler observations using Earth-based antennae, by means of singular value decompositions and the test of the compliance with the so-called discrete Picard condition. The multiple role of regularisation, and how it is taken to an extreme in the case of global lunar gravimetry from basically a single hemisphere data coverage, was furthermore described and analysed. A crucial aspect of regularisation concerns its relation to the underlying estimation scheme. It was argued that the use of Kaula-rule regularisation schemes in the framework of least squares collocation or Bayesian estimation might not be legitimate, since there is a clear danger that the *a priori* information is wrong. In that case, a regularisation error, or bias, is introduced in the solution, which should be duly accounted for in quality assessment studies. In other words, a straightforward error propagation to yield a variance-covariance matrix may no longer be a valid representation of the solution quality.

It was demonstrated that the inclusion of the bias in the error propagation, in the framework of *biased estimation*, the *mean square error matrix* quality measure may deviate from the simple error variance-covariance matrix. This fact not only illustrates the role of the regularisation in determining the solution (and hence the magnitude of the bias), but also that the entire view of the estimation process influences the current understanding of the selenopotential.

Although the assessment of the bias itself is a major problem, since it depends on the exact unbiased solution of the linearised estimation problem, and therefore cannot be computed, it may be estimated. As such, it is possible to investigate whether the solution error assessment is sensitive to the regularisation error. For GLGM–2, the difference between the signal-to-noise ratios using formal errors and the square root of the diagonal elements of the mean square error matrix amounts to up to nearly one order of magnitude. This is particularly valid for the degree range above 40. For LP75G, the differences are smaller due to the higher signal power and the looser constraints.

Similarly, for related selenophysical products, such as *selenoid heights*, *i.e.* the height of the reference equipotential surface above the best-fitting ellipsoidal figure of the Moon, it is seen that the root-mean-square error may vary by as much as 50%, depending on the size of the bias. In other words, estimating reliable figures for the quality of a state-of-the-art selenopotential solution is a far from trivial problem.

The second part of the chapter dealt with the search for the optimal regularisation parameter, *i.e.* the scaling parameter of the Kaula rule, that would lead to the optimal selenopotential solution based on measurements alone. In this case, lunar physics are disregarded since it is exactly the physical structure of the Moon that one tries to derive from the measurements. Using premature power laws, based on internal stress, orbit determination measurement fits and possible experience from other, similar problems increases the risk for circular reasoning, and is in fact a major cause for the ambiguity of the global lunar gravimetric problem, as discussed previously.

Several such methods exist, and are continuing to be developed by numerical analysts. Two particular methods, being the *L-curve* criterion and the *quasi-optimality* condition have been applied to the normal matrices of GLGM–2 and LP75G. Both methods seek to minimise the overall error by balancing the measurement-driven error and the regularisation error. Generally, these criteria do not compensate for the extreme under-sampling of the potential field, and therefore yield solutions rather close to the unregularised solution. A consequence for solutions based on regularisation parameters derived using the L-curve and quasi-optimality criteria is therefore that they generally yield nearly self-contained solutions for the well-sampled near-side, while on the other hand the far-side solutions are characterised by large-normed and "useless" signal and error amplitude variations. In other words, the physics of the Moon, as best illustrated by the continuous orbit behaviour, is not reflected in such solutions. The label "useless" does, however, not mean that the results are non-illustrative. In fact, they are very clear indications of the severe ill-conditioning of the lunar gravimetric problem, and, therefore, the primary beacons signalling that caution should be taken when interpreting present-day selenopotential models.

European lunar gravity field modelling efforts

Chapter 5 dealt with European efforts to model the gravity field. Using passive 3-way Doppler measurements of Lunar Prospector collected over a period of six weeks in the summer of 1998, precise orbit solutions as well as a range of adjustments to the LP75G model were computed. The up-link was provided by the stations of the NASA Deep Space Network and the down-link was received by the 30 m antenna at Weilheim/Lichtenau, Germany, operated by the German Space Operations Center of the German Aerospace Establishment. Next to providing first-hand experience with the orbit and gravity field estimation problems using real data, the goal of the experiment was to verify the entire processing chain from raw data to gravity field model coefficients and selenoid heights. In other words, all aspects of hardware and software performance were tested and verified in order to prove readiness for near-future dedicated lunar gravity field mapping campaigns, such as SELENE.

The fundamental result from the campaign was that in the presence of relatively large uncertainties in the dynamical models, such as the gravity field itself, continuous tracking is of paramount important. With intermittent tracking using a single down-link station the residual fit of the measurements may be encouraging and give proof of adequate performance in both software and hardware components. However, without truly consistent orbits, promising data fits alone are no guarantee for stable normal equations. Put otherwise, a future gravity field mapping experiment should in the first place ensure that the orbits on which the normal equations are based (and irrespective of the type of observable) are near-continuous.

Along the lines of previous thought, the Weilheim observations were used to explore the solution space, by tuning the regularisation parameters, the data weights and the contribution of the new data relative to the *a priori* observation. As such, these experiments are another illustration of the fact that it is relatively easy to derive a selenopotential solution that is not driven by data alone.

Low–low SST-based gravity field modelling

The final analysis chapter of this book dealt with SST techniques and their use in obtaining a global sampling of the selenopotential with uniform accuracy. Obviously, the overall goal of studying the gravity field recovery on the basis of SST measurements was to show that SST measurements provide a very efficient and cost-effective means to overcome the present problems in global lunar gravimetry. It should nevertheless be noted that SST does not provide the *only* solution to this problem. Measurements of the gradients of the gravity field, the so-called gravity tensor, accomplished by means of a *gradiometer* constitute another fundamental option. However, satellite gradiometry is not studied in any detail in this work, and its application in lunar gravimetry is left for future study.

Obviously, the performance of any SST mission depends on a range of parameters, such as, among others, instrument specifications, orbital height, inclination and spacing between the spacecraft. In this book a low–low configuration of two spacecraft was assumed, in which both spacecraft follow a nominal polar orbit at 100 km altitude. Furthermore, a Doppler radar was assumed to deliver 2-way range rate measurements at the precision level of 0.1 mm/s at a sampling rate of 10 s. These specifications coincide with the preliminary design of the gravimetric instrumentation foreseen for MORO in 1996.

It was demonstrated that by proper definition of a so-called *low–low* mode SST experiment, the selenopotential may be determined unambiguously up to approximately degree and order 90 or a spatial resolution at the lunar surface of about 60 km, compared to, roughly speaking, degree and order 10 (546 km) in the present situation. Such a solution would, on a global scale, allow the first ever self-contained gravity mapping of the far-side, *i.e.* without the use of constraints (or regularisation) in any form. At the same time, the data would contain very high-frequency information on orbital perturbations which would allow the study of local features, such as transition zones, small craters, etc. at unprecedented spatial resolutions.

The optimal spacing between the two spacecraft in a low–low configuration, both in-plane and out-of-plane, depends on the desired spatial resolution of the solution. Large angular separations between the spacecraft imply a greater sensitivity to the long-wavelength components of the gravity field. In other words, the high-frequency components which define the finer scales of the gravity field are poorly determined with such a configuration. *Vice versa*, small angular spacings improve the short-wavelength sensitivity at the cost of loss of accuracy at longer wavelengths. In other words, there is a trade-off between resolution and accuracy over the entire harmonic spectrum.

For a maximum spatial resolution, an approximately 3° in-plane separation should be chosen. An additional out-of-plane separation includes cross-track information in the data, and therefore the robustness of the estimation problem. A small difference in right ascension of the ascending node of about 1° is found to be adequate for this purpose.

In terms of selenophysical products, the expected root-mean-square selenoid height error for a maximum degree and order 90 amounts to 0.85 m, whereas the corresponding error in gravity anomalies is found to be 6.39 mGal. Compared to LP75G, which is a degree and order 75 solution, relies heavily on regularisation and still yields 7.15 m in terms of selenoid height errors, this is a remarkable improvement. For a maximum degree and order 70, the SST-based solution yields 0.29 m and 1.59 mGal, respectively. These results are confirmed by both the so-called *Colombo method*, which is based on linear perturbation theories and by full-scale numerical simulations of the experiment.

Systematic error sources are evidently going to play a crucial role in the gravity field mapping based on SST, just as they do for any other application based on a dynamic solution to the satellite orbit problem. In the vicinity of the Moon, solar radiation pressure is the primary perturbation factor, next to the incomplete knowledge of the gravity field. In order to test the sensitivity of the low–low SST-based gravity field solution to errors in this non-conservative force model, the recovery process was also simulated with a deliberate 10% error in the direct radiation pressure model. Despite remaining well below the actual signal strength, in this case, the error increases by one to two orders of magnitude at the low-degree harmonic range. In all fairness, this is a "worst-case" scenario, since no attempt was made to recover the correct parameter from the available measurements prior to solving for the gravity field. Nonetheless, the sensitivity of low-degree selenopotential, *i.e.* the low-frequency components, to errors in this non-conservative force is a strong incentive for planners of future gravity mapping campaigns using SST to invest in force models.

7.2   Recommendations for further research

Related to the problem of unambiguously recovering the global gravity field of the Moon, which has been thoroughly discussed throughout this book, a number of recommendations for future work stand out.

For one, it is obviously important that a self-contained data set soon becomes available. If possible, this should be the result of a measurement campaign that provides near-continuous tracking, since consistent orbit solutions are a prerequisite for stable combined normal equations. This is particularly valid as long as the force model quality, in particular that of the gravity field, is modest. The importance of the selenodetic experiments foreseen for SELENE (or similar future missions) can therefore not be overestimated. It is emphasised once again that any effort in that direction should be encouraged.

Second, *high–low* mode SST should be investigated as a direct counterpart to the low–low modem studied in this book. This is in fact the preferred instrumentation for SELENE, since there are some intrinsic advantages in terms of relay satellite (and therefore experiment) lifetime duration. However, the high–low mode does not provide continuous tracking, with due consequences for the orbit determination. Hence, it would be interesting to have reliable figures that directly compare the two variations of SST.

Third, the use of satellite gradiometry as an alternative to SST for the lunar case deserves a careful study. While it remains a fact that gradiometers generally outperform SST techniques at the very fine scales of the gravity field, their performance in a lunar mission, through a dedicated instrument design could lead to interesting conclusions. For example, already at present there are indicia of a sizeable power of the selenopotential at degrees beyond 100. Therefore, as lunar science continues to advance, one might soon be in a situation where the determination of the high-frequency end of the selenopotential becomes a specific mission target.

Fourth, although the simulated SST-based gravity field solutions indicate that the near-future solutions are fully self-contained, in the sense that they require no numerical regularisation, as long as the truncation degree and order corresponds to a signal-to-noise ratio of one, the role of regularisation deserves further in-depth analysis. Current lunar models are based on *Kaula's rule of thumb*, scaled to the Moon, and frequently scaled in order to optimise either the measurement fit, or the amplitude spectrum in a certain harmonic range. This book has shown that these empirical scaling parameters greatly affect the final result. It might therefore be illustrative to approach the regularisation problem with a non-Kaula based scheme.

Fifth, once a self-contained selenopotential solution has been derived on the basis of some set of SST (or SGG) measurements, it is likely that the orbit solutions for Lunar Prospector and possibly other low-orbiting spacecraft may be improved with respect to the current situation. This obviously requires a re-computation of all satellite orbit arcs. In particular, the very low tracking data obtained during the extended mission phase of Lunar Prospector, where the periapsis altitude was as low as 15 km, have proven to contain a significant amount of high-frequency information. In this regard, it is likely or at least quite possible that these data may yield more information of the high-frequency gravity field than known today, once a decent global data-based SST solution has been derived.

Sixth, and finally, the available tracking data provide an excellent sampling of the near-side, and only a marginal sampling of the far-side near the limbs and poles. Nevertheless, for physical reasons, and in particular due to the link with the satellite orbit, the problem is posed as a global problem. A complementary approach would be to derive a regional or local solution, based on a satellite orbit solution from a global model. This could be done by means of boundary element techniques, where the regional gravity field is solved for as a boundary value problem. In that case, it would be interesting to zoom in the differences between the regional solution and the same portion of global, fully dynamic satellite-based solution. In other words, the question to be answered is whether the global modelling approach is truly optimal also for the near-side, or whether the smoothing effect of regularisation removes interesting peak mass anomalies.

Fundamentals of selenography

This appendix details some quint-essentials of selenography, and serves as a basic reference for lunar surface features and a first-order map of the lunar geology (selenology). Figures A.1 and A.2 schematically show, in a simple cylindrical projection, the characteristic features of the Moon and relate the names to selenographical location. A more detailed description of the quintessential maria, basins and craters, including their location, approximate size and geophysical characteristics is given in near-stenographical form in the extended tables. Most of the tabulated information is extracted and/or derived from *Spudis* [1996].

Lunar maria

Name and Location	Description
Mare Crisium (10-25° N, 50-70° E)	Mascon mare near the east limb; low to very low titanium basalts, extruded around 3.4 Gyr ago
Mare Fecunditatis (5° N - 20° S, 40-60° E)	Complex, shallow mare made up of low-, moderate-, and high-titanium basalts, extruded about 3.4 Gyr ago
Mare Humorum (18-30° S, 31-48° W)	Mascon mare on the southwestern near-side of the Moon, filled with moderately high-titanium basalts, 3.2–3.5 Gyr old
Mare Imbrium (15-5O° N, 40° W-5° E)	Mascon mare on the near-side, deeply filled with low- and high-titanium basalts; age: from 3.3 to less than 2 Gyr old

Mare Nectaris (10-20° S, 30-40° E)	Mascon mare on the central near-side; low-titanium basalts covering very high titanium basalts; age: 3.8–3.5 Gyr
Mare Nubium (10-30° S, 5-25° W)	Complex, shallow mare; low- and high-titanium lava flows; age: 3.3–3.0 Gyr
Mare Serenitatis (15-40° N, 5-20° E)	Mascon mare; very high titanium lavas around the margins and centre of very low titanium lava; age: 3.8–3.3 Gyr
Mare Smythii (5° N-5° S, 80-95° E)	Mascon mare, very shallow; moderate-titanium lava; possibly extremely young (1–1.5 Gyr)
Mare Tranquillitatis (0-20° N, 15-45° E)	Complex, shallow, irregular mare; site of the first lunar landing; old (3.8 Gyr), with very high titanium lavas
Oceanus Procellarum (10° S-60° N, 10-80° W)	Complex, shallow, irregular mare; largest on the Moon; many compositions, with ages including the youngest lavas on the Moon (less than 1 Gyr old)
Sinus Medii (3° S-5° N, 5° W-5° E)	Small patch of mare near the exact centre of the lunar near-side; site of the Surveyor 6 landing (1967)

Lunar craters

Name and Location	Description
Albategnius (11.2° S, 4.1° E; 136 km)	Large crater in the central highlands, sketched by Galileo in 1610
Alphonsus (13.4° S, 2.8° W; 119 km)	Old crater with three dark, volcanic cinder cones on its floor
Archimedes (29.7° N, 4.0° W; 83 km)	Crater flooded by mare basalt, demonstrating time span between Imbrium basin and its mare fill
Aristarchus (23.7° N, 47.4° W; 40 km)	Very fresh crater excavating highland debris from beneath mare basalt cover
Compton (56.0° N, 105.0° E; 160 km)	Small central-peak-plus-ring basin near lunar north pole
Cone (3.5° S, 17.5° W; 370 m)	Small, fresh crater excavating Fra Mauro breccias that were sampled on the Apollo 14 mission, 1971
Copernicus (9.7° N, 20.0° W; 93 km)	Relatively young impact crater south of Mare Imbrium; defines Copernican stratigraphic system

Descartes
(11.7° S, 15.7° E; 48 km)

Old crater in central highlands, near the landing site of Apollo 16, 1972

Eratosthenes
(14.5° N, 11.3° W; 58 km)

Unrayed crater near Mare Imbrium; defines Eratosthenian stratigraphic system

Flamsteed P
(3.0° S, 44.0° W; 112 km)

Old crater flooded by some of the youngest (1 Gyr) lavas on the Moon; site of the Surveyor 1 landing in 1966

Fra Mauro
(6.0° S, 17.0° W; 95 km)

Old crater covered by ejecta from the Imbrium impact basin; near the landing site of Apollo 14, 1971

Herigonius
(13.3° S, 34.0° W; 15 km)

Small crater north of Mare Humorum, near some of the most spectacular sinuous rilles in the maria

Hortensius
(6.50° N, 28.0° W; 15 km)

Small crater, near which occur many small lunar shield volcanoes

Kopff
(17.4° S, 89.6° W; 42 km)

Unusual crater, long thought to be volcanic, in Orientale basin; may have been created by an impact into a semi-molten melt sheet

Lamont
(5.0° N, 23.2° E; 175 km)

Ridge ring system in Mare Tranquillitatis, formed over a two-ring basin

Letronne
(10.6° S, 42.4° W; 120 km)

Crater largely flooded by mare basalt in Oceanus Procellarum

Lichtenberg
(31.8° N, 67.7° W; 20 km)

Rayed crater that is partly covered by a very young mare lava flow, possibly less than 1 Gyr old

Linné
(27.7° N, 11.8° E; 2 km)

Very fresh, bright crater in Mare Serenitatis, reported before the space age to appear and disappear

Ritter
(2.0° N, 19.2° E; 29 km)

Forms together with Sabine an unusual twin impact crater in Mare Tranquillitatis, similar in morphology to Kopff

Sabine
(1.4° N, 20.1° E; 30 km)

See Ritter

Shorty
(20.0° N, 31.0° E; 110 m)

Small impact crater at the Apollo 17 landing site, 1972; excavated dark mantle ash from beneath a layer of highland debris

Sulpicius Gallus
(19.6° N; 11.6° E; 12 km)

Crater near a large exposure of dark mantle deposits.

Theophilus
(11.4° S, 26.4° E; 100 km)

Large crater on the edge of Mare Nectaris.

Tsiolkovsky
(20.4° S, 129.1° E; 180 km)

Spectacular, mare-filled crater on the lunar farside.

Tycho (43.3° S, 11.2° W; 85 km)	Fresh, prominent rayed crater on the near-side of the Moon; rays extend across entire hemisphere; central peak exposes deep-seated rocks
Van de Graaff (27.0° S, 172.0° E; 234km)	Double crater on the far-side; site of a major geo-chemical anomaly caused by its location just inside the rim of South Pole Aitken basin

Lunar basins

Name and Location	Description
Crisium basin (17.5° N, 58.5° E; 740 km)	Nectarian-age multiring basin; ejecta possibly sampled by the Luna 20 mission
Humorum basin (24° S, 39.5° W, 820 km)	Nectarian-age basin south of Procellarum
Imbrium basin (33° N, 17° W, 1,150 km)	Major large basin on the Moon; defines base of Imbrian System; formed 3.84 Gyr ago; its ejecta was the sampling objective of the Apollo 14 and 15 missions
Nectaris basin (16° S, 34° E; 860 km)	Defines base of Nectarian System; possibly sampled on the Apollo 16 mission in 1972; age: 3.92 Gyr
Orientale basin (20° S, 95° W; 930km)	Youngest large, multiring basin on the Moon, formed sometime after 3.84 billion years ago; its interior and exterior deposits were used as a guide to interpret older, degraded basins
Procellarum basin (26° N, 15° W; 3,200 km)	Alleged impact basin, supposedly the largest on the Moon; Clementine laser altimetry data do not support its existence
Schrödinger (75.6° S, 133.7° E; 320 km)	Type example of a two-ring basin, near the south pole of the Moon; formed after Imbrium basin but before Orientale basin
Serenitatis basin (27° N, 19° E; 900 km)	Nectarian-age multiring basin, sampled and explored by Apollo 17 mission in 1972; age: 3.87 Gyr
South Pole Aitken basin (56° S, 180° E; 2,500 km)	Largest, deepest (over 12 km) impact crater known in the solar system; oldest basin on the Moon; absolute age unknown (4.3 Gyr??)

Other surface features

Name and Location	Description
Apennine Bench (25-28° N, 0-10° W)	Refers to a relatively elevated region near Archimedes and just inside the rim of Imbrium basin; includes light-toned
Apennine Mountains (15-30° N, 10° W-50° E)	Large mountain chain making up the southeastern rim of the Imbrium basin
Cayley plains	Light-toned, smooth highland plains, first defined in the central near-side but having moonwide distribution; probably a form of impact ejecta from the youngest major basins; may cover ancient mare lavas in some areas
Cordillera Mountains (10-35° S, 80-90° W)	Arcuate mountain chain that makes up the rim of the Orientale basin
Hadley-Apennines (26° N, 4° W)	Informal name given to the region of the Apollo 15 mission exploration; includes mare, Hadley Rille, and Apennine highlands
Hadley Rille	See Rima Hadley
Marginis swirls (15° N, 90° E)	Light-toned swirls north of Mare Marginis; origin unknown
Marius Hills (10-15° N, 50-60° W)	Complex area of small domes, cones, and sinuous rilles in Oceanus Procellarum; the dome-like swell may indicate that this region is a large lunar shield volcano
Reiner Gamma (7° N, 59° W)	Bright, swirl-like deposit in Oceanus Procellarum; origin unknown
Rima Bode II (13° N, 4° W)	Cleftlike vent and linear trench outline vent system for a large, regional blanket of dark volcanic ash
Rima Hadley (25° N, 3° W)	Long sinuous rille starting in the highlands and emptying into the maria; probably a lava channel and/or tube
Rümker Hills (41° N, 58° W)	Complex of cones and domes in Oceanus Procellarum, similar to Marius Hills but much smaller
Taurus-Littrow (20° N, 31° E)	Informal name given to the region of the Apollo 17 mission exploration; includes mare, dark mantle, and the highlands of the Serenitatis basin
Tranquillity Base (1° N, 23° E)	Site, in Mare Tranquillitatis, of man's first landing on the Moon, Apollo 11, July 20, 1969

Figure A.1   The near-side of the Moon. In red the landing spots of Soviet Luna missions,
and in blue the corresponding landing areas of the U.S. Surveyor (S), Ranger
(R) and the Apollo Lunar Module (A) spacecraft. The map also provides a
first-order indication of the ages of the lunar crust, with brownish colours
indicating ancient, primary crust of old age, and lighter purple-tinted colours
indicating younger, secondary crust. Intermediate ages are indicated in tints
of yellow. Very light, gray colours are used for the maria. Thin-lined
near-circular perimeters in various colours indicate crater locations and
approximate shape. Source: [*Verger*, 1992]

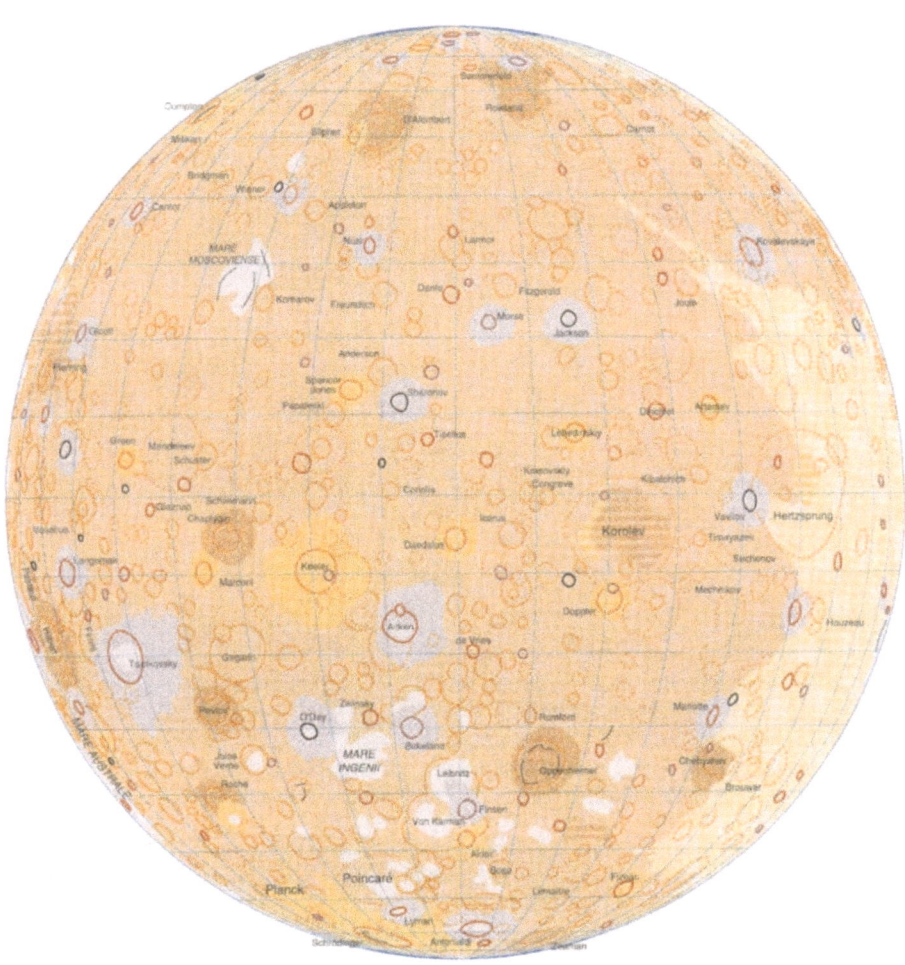

Figure A.2   The far-side of the Moon. The colour indications are identical to those used
for the near-side in Fig. A.1. Notice that no spacecraft has performed a
landing at the far-side. The site of the hard impact of Lunar Prospector near
the south pole is obviously not visible in the present projection. Source:
[*Verger*, 1992]

The generalised singular value decomposition (GSVD)

The GSVD of the matrix pair $(\mathbf{A}, \mathbf{L})$ is a generalisation of the SVD of $\mathbf{A}$ in the sense that the generalised singular values of $(\mathbf{A}, \mathbf{L})$ are essentially the square roots of the generalised eigenvalues of the matrix pair $(\mathbf{A}^T\mathbf{A}, \mathbf{L}^T\mathbf{L})$ [*Hansen*, 1998a].

In order to keep the exposition simple, assume that $\mathbf{A} \in \mathbb{R}^{m \times n}$ and $\mathbf{L} \in \mathbb{R}^{p \times n}$ with $m \geq n \geq p$ and that $\mathcal{N}(\mathbf{A}) \cap \mathcal{N}(\mathbf{L}) = 0$ and that both $\mathbf{A}$ and $\mathbf{L}$ have full row rank. Then the GSVD is a decomposition of $\mathbf{A}$ and $\mathbf{L}$ in the form

$$\mathbf{A} = \mathbf{U} \begin{pmatrix} \Sigma & \mathbf{0}_{p \times n-p} \\ \mathbf{0}_{n-p \times p} & \mathbf{I}_{n-p} \end{pmatrix} \mathbf{X}^{-1}, \quad \mathbf{L} = \mathbf{V} \left(\mathbf{M}, \mathbf{0}_{p \times n-p} \right) \mathbf{X}^{-1} \tag{B.1}$$

where the columns of $\mathbf{U} \in \mathbb{R}^{m \times n}$ and $\mathbf{V} \in \mathbb{R}^{p \times p}$ are orthonormal, *i.e.* $\mathbf{U}^T\mathbf{U} = \mathbf{I}_n$, and $\mathbf{V}^T\mathbf{V} = \mathbf{I}_p$; $\mathbf{X} \in \mathbb{R}^{n \times n}$ is nonsingular, and Σ and $\mathbf{M}$ are $p \times p$ diagonal matrices with elements

$$0 \leqslant \sigma_1 \leqslant \cdots \leqslant \sigma_p \leqslant 1, \quad 1 \geqslant \mu_1 \geqslant \cdots \geqslant \mu_p > 0$$

normalised such that

$$\Sigma^T\Sigma + \mathbf{M}^T\mathbf{M} = \mathbf{I}_p \quad \text{or likewise} \quad \sigma_i^2 + \mu_i^2 = 1$$

Then the *generalised singular values* γ_i of $(\mathbf{A}, \mathbf{L})$ are defined as

$$\gamma_i = \sigma_i/\mu_i, \quad i = 1, \ldots, p \tag{B.2}$$

For historical reasons [*Hansen*, 1998a], and opposite to the standard singular values, they appear in non-decreasing order. Since

$$\mathbf{X}^T\mathbf{A}^T\mathbf{A}\mathbf{X} = \begin{pmatrix} \Sigma & 0 \\ 0 & \mathbf{I}_{n-p} \end{pmatrix} \quad \text{and} \quad \mathbf{X}^T\mathbf{L}^T\mathbf{L}\mathbf{X} = \begin{pmatrix} \mathbf{M}^2 & 0 \\ 0 & 0 \end{pmatrix} \tag{B.3}$$

the pairs $\left(\gamma_i^2, \mathbf{x}_i\right)$ are the generalised eigensolutions of the pair $\left(\mathbf{A}^T\mathbf{A}, \mathbf{L}^T\mathbf{L}\right)$ associated with the p finite generalised eigenvalues. Likewise, the first p columns of $\mathbf{X} = \left(\mathbf{x}_1, \ldots, \mathbf{x}_n\right)$ satisfy

$$\mu_i^2 \mathbf{A}^T\mathbf{A}\mathbf{x}_i = \sigma_i^2 \mathbf{L}^T\mathbf{L}\mathbf{x}_i, \quad i = 1, \ldots, p$$

and, hence, $\mathbf{A}^T\mathbf{A}\mathbf{x}_i = \gamma_i^2 \mathbf{L}^T\mathbf{L}\mathbf{x}_i$. In other words, the vectors $\mathbf{x}_i$ are *generalised singular vectors* of the pair $(\mathbf{A}, \mathbf{L})$. For $p < n$ the matrix $\mathbf{L} \in \mathbb{R}^{p \times n}$ always has a nontrivial null-space $\mathcal{N}(\mathbf{L})$ [*Hansen, 1998a*]. The last $n - p$ columns $\mathbf{x}_i$ of $\mathbf{X}$ satisfy

$$\mathbf{L}\mathbf{x}_i = 0, \quad i = p + 1, \ldots, n$$

and are therefore basis vectors for the null-space $\mathcal{N}(\mathbf{L})$.

Similar to the SVD, the GSVD of $(\mathbf{A}, \mathbf{L})$ provides three new sets of linearly independent basis vectors - being the columns of $\mathbf{U}$, $\mathbf{V}$ and $\mathbf{X}$ - such that the two matrices $\mathbf{A}$ and $\mathbf{L}$ become diagonal when transformed into these new bases. The two sets of basis vectors for the columns of $\mathbf{U}$ and $\mathbf{V}$ are orthonormal. Unfortunately, there is no simple and straightforward relationship between the generalised singular values and vectors and their ordinary counterparts. However, in the case that $\mathbf{L}$ is well-conditioned (has a "small" condition number) it may be shown that the matrix $\mathbf{X}$ is also well-conditioned [*Hansen, 1998a*]. Consequently, the diagonal matrix $\mathbf{\Sigma}$ directly contains information on the ill-conditioning of $\mathbf{A}$.

Some useful coordinate transformations

A typical problem in the handling of the orbital equations of motion is that of coordinate transformations, predominantly that of transforming a given set of co-ordinates from an inertial frame to a rotating (in inertial space), body-fixed frame and *vice versa*. For example, the equations of motion are usually solved in some pseudo-inertial reference frame while the gravitational attraction due to the lu-nar mass distribution is given as a function of selenographical coordinates. The evaluation of the force exerted on the spacecraft therefore requires transformation formulae to be readily available. Furthermore, several types of both inertial and body-fixed systems are in use. The purpose of this appendix is to briefly outline some of the coordinate systems and transformation algorithms applicable to the analysis of lunar satellite orbits and to the lunar gravimetric problem. Since the theory and practise of coordinate reference systems is a mature scientific disci-pline in its own, the reader is advised to consult the specialised literature for a more comprehensive discussion. A selection of important references is provided in the subsequent text.

Irrespective of the exact choice of inertial and body-fixed reference frames, the transformation between any such pair may be described by the general matrix-vector equation

$$\mathbf{r}_{BF} = \mathbf{E}\, \mathbf{r}_I \tag{C.1}$$

where $\mathbf{r}_{BF}$ is the position in the rotating, body-fixed (in the present context: Moon-fixed), rotating frame, $\mathbf{r}_I$ is the position vector in the inertial frame and $\mathbf{E}$ is a 3×3 rotation matrix with time-dependent coefficients that describe the rotation.

Until recently, the perhaps most widely used inertial frame for satellite orbit integration was the pseudo-inertial frame described by the Earth mean equator and equinox of J2000, in short *EME2000* [*Seidelmann*, 1992]. The J2000 epoch is a short-hand notation for *Julian Date* 2451545.0 (or *Modified Julian Date* 51544.5 or 1.5 January, 2000 *Barycentric Dynamic Time*). Access to the EME2000 system is provided by the so-called FK5 star catalogue, which provides precise positions and proper motions of some 1 500 stars for the J200o epoch and in the given reference frame. More recently, based upon certain dynamical difficulties in the definition of the ecliptic plane and the equinox, it has been decided to replace EME2000 with a new *International Celestial Reference System*, or in short ICRS. The practical realisation of the ICRS is designated the *International Celestial Reference Frame* (ICRF), and IAU has adopted this frame for use from 1998 onwards [*Feissel and Mignard*, 1998].

For Earth satellite orbits, correcting the EME2000 or ICRF coordinates for the precession parameters of a given epoch transforms from EME2000/ICRF into the mean equator and equinox of the epoch Julian Date. A subsequent correction for nutation of the Earth rotational axis provides the transformation from mean equator and equinox of date to the true equator and equinox of date, or in short the "true-of-date" system. Adopted IAU parameter values for the precession and nutation matrices are given in *Seidelmann* [1992]. A brief discussion of the geometric transformations and the transformation parameters is also given in *Seeber* [1993] and in *Montenbruck and Gill* [2000]. Alternatively, the nutation parameters are also provided by the JPL Development Ephemeris files [*Standish et al.*, 1995; *Standish*, 1998].

In the case of lunar and planetary orbiters the situation is, although physically the same, somewhat different. For one, the IAU provides recommended values for the rotational elements for the Sun, the planets and their natural satellites, hence also the Moon [*Davies et al.*, 1992, 1996]. These expressions are derived from an analytical libration theory, using the current state-of-the-art gravity field parameters. Tabulated parameters include the right ascension and declination of the north pole (α_0, δ_0) as well as the orientation of the prime meridian. These parameters define the direction of the ascending node of the body's equatorial plane on the standard equatorial plane of the Earth, the standard Earth equator being the mean equator of J2000. This intersecting line of nodes located at $(\alpha_0 + \pi/2, 0)$ defines the so-called *IAU vector*, which in turn serves as a reference for the computation, at any epoch, of the prime meridian W. Note that this reference is independent from the particular choice of the prime meridian of a celestial body; it merely defines a reference point in pseudo-inertial space with respect to which W may be computed. Further details on this procedure may be found in *Green* [1985], *Seidelmann* [1992] and *Davies et al.* [1996].

The transformation from EME2000 or ICRF into the IAU system, which is defined by the celestial body's equatorial plane and the IAU vector, is given by [*Seidelmann*, 1992]

$$\mathbf{R}_{\mathrm{IAU}} = \mathbf{R}_x(\pi/2 - \delta_0)\mathbf{R}_z(\alpha_0 + \pi/2) \qquad (C.2)$$

where $\mathbf{R}_x$ and $\mathbf{R}_z$ are the elementary rotation matrices about the x and z axes, respectively. The IAU system is a pseudo-inertial system aligned with the body's equator, and is therefore not a body-"fixed" system in the rigourous sense of the word. It is also not strictly a true-of-date system, since the IAU parameters do not include nutation for all bodies. However, in the case of the Moon, the tabulated parameters include the secular effect of precession on the equatorial planes of the celestial body, as well as the physical librations of the Moon. Finally, the transformation from the IAU reference frame to the body-fixed (rotating) frame described by the body's equator and prime meridian is given by

$$\mathbf{R}_W = \mathbf{R}_z(W) \tag{C.3}$$

In terms of the IAU parameters, the complete transformation from EME2000 or ICRF to the body-fixed frame is therefore described by

$$\mathbf{r}_{BF} = \mathbf{R}_W \mathbf{R}_{IAU}\, \mathbf{r}_{EME2000/ICRF} \tag{C.4}$$

An alternative to the IAU system for the transformation from EME2000 or ICRF to the body-fixed, rotating *selenocentric* reference frame is provided by recent JPL Development Ephemeris. DE-403 [*Standish et al.*, 1995] and DE-405 [*Standish*, 1998] include lunar libration angles. Basically, these are based on a highly precise numerical integration of the Moon's rotational motion, and therefore allow the computation of

$$\mathbf{r}_{Moon\text{-}fixed} = \mathbf{L}\, \mathbf{r}_{EME2000/ICRF} \tag{C.5}$$

through

$$\mathbf{L} = \mathbf{R}_z(\psi)\, \mathbf{R}_x(\theta)\, \mathbf{R}_z(\phi) \tag{C.6}$$

where $\{\phi, \theta, \psi\}$ are the libration angles of the Moon. Their detailed description may be found in *Seidelmann* [1992] and *Newhall and Williams* [1997].

Since all transformations discussed here are described by an orthonormal rotation matrix, it goes without saying that their inverse is given by the transposed matrix.

The Euler-Lagrange equation
and the range rate SST signal

This appendix details the use of the Lagrangian formalism of classical mechanics in order to derive an approximate relationship between the inertial velocity and the body-fixed gravitational potential. This enables the derivation of a signal equation for the line-of-sight range rate signal between two spacecraft in a co-orbiting low-low SST configuration in terms of the selenopotential and other forces acting on the spacecraft. The spacecraft are considered to be a point-mass or "cannonball"-type satellites, which implies spacecraft geometry plays no role in computing the orbit. A fundamental result is that the low-low range rate is approximately proportional to the potential difference along the orbit.

D.1   The Euler-Lagrange formalism

The Euler-Lagrange equations for the motion of a particle in an inertial system read

$$\frac{d}{dt}\left(\frac{\partial L}{\partial \dot{q}_k}\right) - \frac{\partial L}{\partial q_k} = 0 \qquad (D.1)$$

where $\{q_k\}$ are generalised position coordinates, the "dot" denotes derivation with respect to time t and L is the *Lagrangian* of the problem [*e.g.*, *d'Inverno*, 1992; *Arnold*, 1980; *Arnold et al.*, 1997]. The Euler-Lagrange equations, which are derived from d'Alembert's principle of virtual work and from the variational principle of Hamilton, constitute a system of three second-order differential equations, corresponding to the three degrees of freedom in Euclidean space, *i.e.* $k \in \{1, 2, 3\}$. Their main advantage is that they are not tied to one particular choice of coordinates.

In geodesy and geophysics, opposite to common practise in physics, the gravitational force is traditionally written as the *positive* gradient of a scalar *forcing function* $V(q_k; t)$. The main example in this book is obviously how the gravitational attraction due to the mass distribution of the Moon governs the motion of a satellite in lunar orbit through $\ddot{\mathbf{r}} = \nabla U$. Similar reasoning generally holds for all other masses in the universe, as they contribute to the conservative force field in which the spacecraft moves. The forcing function V may therefore principally contain other conservative effects, *e.g.* third-body perturbations due to the planets and the Sun as well as the dynamical effects of tides. The sign convention is furthermore that the potential is negative at the lunar surface and gradually increasing towards zero at infinity. In this sign convention, the *potential energy* is given by the negative of V. For conservative problems, that is problems in which the force exerted on the mass particle is only a function of the generalised position q_k, it then holds that $L = T + V$, where T is the kinetic energy.

Similarly, if the problem at hand includes non-conservative forces, *i.e.* forcing effects that depend of the generalised velocity coordinates $\{\dot{q}_k\}$, the Langrangian may be written as $L = T + V + G$, and the components of the generalised non-conservative force $\mathbf{g}$ are given by

$$g_k = g_k(\dot{q}_k, q_k, t) = \frac{\partial G}{\partial q_k} - \frac{d}{dt}\left(\frac{\partial G}{\partial \dot{q}_k}\right)$$

In the framework of orbital mechanics, the vector $\mathbf{g}$ may obviously contain the effects of direct radiation pressure, albedo, infra-red planetary radiation, orbit manoeuvring, parasitic effects of attitude correction manoeuvres as well as a range of minor forces, such as thermal effects or meteoritic impacts, all depending on the desired level of precision. Atmospheric drag is not included in this list since it is of negligible concern for lunar orbiters.

For each coordinate q_k this leads to

$$\frac{d}{dt}\left(\frac{\partial(T+V)}{\partial \dot{q}_k}\right) - \frac{\partial(T+V)}{\partial q_k} = \sum_j \mathbf{g}_j^T \frac{\partial \mathbf{x}_j}{\partial q_k} \qquad (\text{D.2})$$

where $\mathbf{g}_j$ is the jth non-conservative force acting on the particle, and therefore $\sum_j \mathbf{g}_j = \mathbf{g}$. The notation using the inner product on the right-hand side suggests the use of Cartesian coordinates as the choice of q_k. In this case,

$$\{q_k\} = \mathbf{x} = \mathbf{x}(q_k; t), \qquad k = 1, 2, 3$$

and

$$T = \frac{1}{2}\|\dot{\mathbf{x}}\|_2^2 \quad \text{and} \quad V = V(\mathbf{x}; t)$$

where the norm $\| \cdot \|_2$ is the standard 2-norm. Note that in this notation $x_1 = x; x_2 = y; x_3 = z$ in the notation of Chap. 2 and Appendix C. Also notice that the potential energy must be allowed to be time-dependent. This is due to the fact that the problem is described in an inertial coordinate system. In other words, the potential energy terms must account for the lunar rotation, as well as all other explicit time-varying effects in the inertial frame. For example, treating the particle motion as a two-body problem, neglecting all conservative effects except the static selenopotential, it holds that $V = U$ plus a term including the rotation of the selenopotential in inertial space. Notice also that by carrying out the explicit computation of the Euler-Lagrange equations for the given coordinate choice, one arrives back at Newton's second law of motion

$$\frac{d\dot{\mathbf{x}}}{dt} = \ddot{\mathbf{x}} = \nabla V + \mathbf{g}$$

which is usually the equation from which the orbit problem is studied as an initial-value problem.

D.2 Satellite velocity and the conservative forcing function in inertial space

In order to derive the range rate signal equation for a co-orbiting low-low SST configuration as a function of the selenopotential parameters, it is illustrative to introduce the scalar

$$H = H(\mathbf{x}, \dot{\mathbf{x}}; t) = T - V$$

This will allow to relate the spacecraft velocity $\dot{\mathbf{x}}$ to the forcing function V in inertial space. As an aside, with the above definition of potential energy, H is the so-called *Hamiltonian* of the particle motion problem, if and only if $\mathbf{g} = 0$. In that case, H represents the total energy. Taking the time-derivative of H, one has

$$\frac{dH}{dt} = \sum_{k=1}^{3} \frac{\partial H}{\partial x_k} \frac{dx_k}{dt} + \sum_{k=1}^{3} \frac{\partial H}{\partial \dot{x}_k} \frac{d\dot{x}_k}{dt} + \frac{\partial H}{\partial t} \tag{D.3}$$

$$= -\sum_{k=1}^{3} \frac{\partial V}{\partial x_k} \dot{x}_k + \sum_{k=1}^{3} \frac{\partial T}{\partial \dot{x}_k} \ddot{x}_k - \frac{\partial V}{\partial t}$$

$$= -\sum_{k=1}^{3} (\ddot{x}_k - g_k) \dot{x}_k + \sum_{k=1}^{3} \dot{x}_k \ddot{x}_k - \frac{\partial V}{\partial t}$$

$$= \sum_{k=1}^{3} g_k \dot{x}_k - \frac{\partial V}{\partial t}$$

which after time integration yields

$$T - V = \sum_{k=1}^{3} \int_{t_0}^{t_1} g_k \dot{x}_k dt - \int_{t_0}^{t_1} \frac{\partial V}{\partial t} dt + C_0 \tag{D.4}$$

where C_0 is the constant of integration. For V containing only a static potential in inertial space (no lunar rotation and no time-dependency of the selenopotential over the integration interval; the latter is of course not expected on physical grounds since the integration intervals are short) and in the absence of non-conservative forces, (D.4) would constitute the law of energy conservation. The satellite velocity furthermore enters the equation via the *kinetic energy*. Hence, it holds that the potential in inertial space results from the integral equation

$$V = \frac{1}{2}\|\dot{x}\|_2^2 - \sum_{k=1}^{3} \int_{t_0}^{t_1} g_k \dot{x}_k \, dt + \int_{t_0}^{t_1} \frac{\partial V}{\partial t} dt - C_0 \tag{D.5}$$

D.3   The forcing function in body-fixed rotating coordinates

Next, the goal is to relate the satellite velocity to the Moon-fixed, rotating (in inertial space) selenopotential, given by (2.2). In order to simplify the analysis, it is assumed that the forcing function V contains the lunar gravitational potential only, *i.e.* $V = U$. The analysis is therefore of a qualitative rather than highly precise nature, as both third-body effects and tidal forces are neglected.

Since V is in an inertial frame, this requires a transformation from the fixed lunar to the fixed celestial frame. *Jekeli* [1999] describes this transformation in terms of longitude and co-latitude angles $\{\lambda, \pi/2 - \phi\}$ for a point-mass in orbit around the Earth. In this case, there is a direct mapping between the geocentric body-fixed frame using latitude and longitude $\{\lambda_E, \phi_E\}$ and the geocentric celestial frame $\{\alpha_E, \delta_E\}$, since the two systems make use of the same reference plane and pole, being the true-of-date Earth equator and north pole. If the Earth rotation rate is assumed to be constant, for Earth orbiters it therefore holds that

$$\lambda_E = \alpha_E + \Delta\lambda_P + \Delta\lambda_N - \omega_E t \tag{D.6}$$
$$\phi_E = \delta_E + \Delta\phi_P + \Delta\phi_N \tag{D.7}$$

where α_E is the EME2000 right ascension of a satellite in Earth orbit, δ_E is the corresponding EME2000 declination and ω_E the rotation rate, cf. Appendix C. The differential corrections are correction terms for precession (P) and nutation (N) in order to facilitate the transformation at any given epoch.

In the case of a lunar orbiter[1], the geometry of the problem is different, since the lunar equator is inclined to the Earth equatorial plane. Several pseudo-inertial systems are in use, most of which are described in Appendix C. If the inertial frame of choice is the IAU system, the rotation between the Moon-fixed selenocentric frame and the IAU system is described by a rotation about the rotational axis of the Moon $\mathbf{R}_x(W)$. The angle $W(t)$ is given by the IAU/IAG/COSPAR working group reports on cartographic coordinates and rotational elements of the planets and satellites [*Davies et al.*, 1992, 1996]. This rotation is, just like the case of the Earth orbiting satellite, dominated by a constant component caused by the lunar rotation amounting to 13.17635815° per day. In fact, all other parameters affecting the computation of $W(t)$, like the lunar physical librations, the precession of the

---

[1]similar reasoning obviously applies to spacecraft in orbit around the planets, as well as any other natural satellite, asteroid or comet whose rotational elements are known

lunar pole or the deceleration of the lunar rotation rate as it moves further away from Earth, are either several orders of magnitude smaller or periodic over time periods several orders of magnitude longer than that of the lunar rotation. In other words, to a sufficiently accurate (for qualitative analysis) level of accuracy, one may write

$$\lambda \approx \hat{\alpha}_{IAU} - \dot{W}t \tag{D.8}$$

$$\phi \approx \hat{\delta}_{IAU} \tag{D.9}$$

where $\{\hat{\alpha}_{IAU}, \hat{\delta}_{IAU}\}$ are true-of-date inertial coordinates in the IAU system and $\dot{W} = dW/dt$ is constant. The " ˆ " and the subscript "IAU" indicate that these angles are not the usual right ascension and declination, but a set of coordinates on the body-fixed sphere introduced to describe a similar concept based on the equatorial plane of any body for which there exists an IAU system. The label "true-of-date" obviously indicates that precession of the lunar pole as well as the lunar physical librations are included.

The explicit partial derivative of V with respect to time now reads

$$\frac{\partial V}{\partial t} = \frac{\partial V}{\partial \lambda}\frac{\partial \lambda}{\partial t} + \frac{\partial V}{\partial \phi}\frac{\partial \phi}{\partial t} = \frac{\partial V}{\partial \hat{\alpha}_{IAU}}\frac{\partial \lambda}{\partial t} + 0 \tag{D.10}$$

$$= - \dot{W}\frac{\partial V}{\partial \hat{\alpha}_{IAU}}$$

Getting back to Cartesian coordinates in the IAU system, one has $x_1 = r \cos \hat{\delta}_{IAU} \cos \hat{\alpha}_{IAU}$ and $x_2 = r \cos \hat{\delta}_{IAU} \sin \hat{\alpha}_{IAU}$. Hence,

$$\frac{\partial V}{\partial t} = - \dot{W}\left(x_1\frac{\partial V}{\partial x_2} - x_2\frac{\partial V}{\partial x_1}\right) \tag{D.11}$$

$$= \dot{W}\left(x_1\left[g_2 - \ddot{x}_2\right] - x_2\left[g_1 - \ddot{x}_1\right]\right)$$

Substitution of (D.11) in (D.5) and using $V = U$ finally yields

$$U = \frac{1}{2}\|\dot{x}\|_2^2 - \sum_{k=1}^{3}\int_{t_0}^{t_1} g_k\dot{x}_k dt + \dot{W}\int_{t_0}^{t_1}\left(x_1\left[g_2 - \ddot{x}_2\right] - x_2\left[g_1 - \ddot{x}_1\right]\right)dt - C_0 \tag{D.12}$$

If the non-conservative force $\mathbf{g}$ is also neglected this reduces to

$$U = \frac{1}{2}\|\dot{x}\|_2^2 - \dot{W}\left(x_1\dot{x}_2 - x_2\dot{x}_1\right) - C_1 \tag{D.13}$$

where C_1 is an updated constant of integration, also accounting for the initial value of the velocity components $\dot{x}_1$ and $\dot{x}_2$. The models (D.12) and (D.13) relate the gravitational potential U to satellite position, velocity and, in the case of (D.12), also to the non-conservative forces acting on the spacecraft. It should be kept in mind that this is an approximate model, since the time-dependent geometric transformations are based on linear approximations of the actual non-linear situation.

The term

$$\int_{t_0}^{t_1} \frac{\partial V}{\partial t} dt = -\dot{W}\left(x_1 \dot{x}_2 - x_2 \dot{x}_1\right)$$

accommodates the rotation of the gravitational potential in the inertial frame, in this case the IAU system. It is different from the more commonly seen centrifugal term $\dot{W}^2(x_1^2 + x_2^2)$, which is used to account for the rotation when computing the acceleration due to the selenopotential in the selenocentric, rotating frame. The term has been baptised the "potential rotation" by *Jekeli* [1999].

D.4   The low-low range rate SST signal equation

Using radio tracking between two spacecraft, either the range or the range-rate between two spacecraft is measured. For Earth observation missions, like GRACE, a range-type observable is being implemented [*e.g., Davies et al.*, 1999; *Jekeli*, 1999; *Mazanek et al.*, 2000], while most proposals for lunar missions envisage range rate tracking, mostly because of the technical difficulty in achieving a range measurement precision compared to the range rate measurement in a small, compact and cost-effective relay satellite.

The LOS range ρ between two satellites, denoted by subscripts 1 and 2, is given by

$$\rho = \|\mathbf{r}_2 - \mathbf{r}_1\| = \|\mathbf{r}_{12}\| \tag{D.14}$$

whereas the range-rate $\dot{\rho}$ is simply the time derivative of the range

$$\dot{\rho} = \frac{d\rho}{dt} = \dot{\mathbf{r}}_{12}^T \frac{\mathbf{r}_{12}}{\|\mathbf{r}_{12}\|} = \dot{\mathbf{r}}_{12}^T \mathbf{e}_{12} \tag{D.15}$$

where $\mathbf{e}_{12} = \mathbf{r}_{12}/\rho$ is the unit LOS vector between the two satellites [*e.g., Floberghagen et al.*, 1996]. The range rate measurement is fundamentally the projection of the velocity difference between the two spacecraft onto the line joining them. Notice, therefore, that both the SST range and range-rate measurements are entirely independent of the coordinate system used to describe the motion of the satellites.

In order to relate the selenopotential to the range rate measurement, the potential difference between two locations in inertial space $U_2 - U_1 = U_{12}$ is needed. Similarly, $\dot{\mathbf{x}}_2 - \dot{\mathbf{x}}_1 = \dot{\mathbf{x}}_{12}$, etc. Recalling that

$$\frac{1}{2}\left(\|\dot{\mathbf{x}}_2\|_2^2 - \|\dot{\mathbf{x}}_1\|_2^2\right) = \frac{1}{2}\left(\dot{\mathbf{x}}_2 + \dot{\mathbf{x}}_1\right)^T \dot{\mathbf{x}}_{12} = \frac{1}{2}\|\dot{\mathbf{x}}_{12}\|_2^2 + \dot{\mathbf{x}}_1^T \dot{\mathbf{x}}_{12}$$

yields, after substitution of the two position and velocity vectors in inertial space in (D.13),

$$U_{12} = \frac{1}{2}\|\dot{\mathbf{x}}_{12}\|_2^2 + \dot{\mathbf{x}}_1^T \dot{\mathbf{x}}_{12} - \dot{W}\left(x_{1_{12}}\dot{x}_{2_1} + \dot{x}_{2_{12}}x_{1_1} - x_{2_{12}}\dot{x}_{1_2} - x_{2_1}\dot{x}_{1_{12}}\right) - C_{1_{12}} \tag{D.16}$$

Equation (D.16) relates selenopotential differences to the velocity components of the two spacecraft in inertial space. Although not an exact relationship, it is a first-order approximation of the differenced integral equation (D.5).

A particularly simple relationship can be derived for low-low SST configurations on a circular orbit in a static gravity field (rotation term neglected). If the along-track separation between the two satellites is small (and they hence pass over the same perturbations of the selenopotential), the LOS range rate will be nearly equal to $\|\dot{\mathbf{x}}_{12}\|_2$. Furthermore, the velocity component $\dot{\mathbf{x}}_1$ will be nearly perpendicular to $\dot{\mathbf{x}}_{12}$. Since the component of the acceleration along the track of the satellite is given by $\partial U/\partial \tau$, where τ denotes the along-track position component, cf. Chap. 3, it follows that

$$\dot{\rho} \approx \|\dot{\mathbf{x}}_{12}\|_2 = \int_{t_0}^{t_1} \frac{\partial U}{\partial \tau} dt = \int_{t_0}^{t_1} \frac{\partial U}{\partial \tau} \left(\frac{dt}{d\tau} \right)^{-1} d\tau \qquad (D.17)$$

Given that $d\tau/dt$ equals the velocity of the two satellites $\|\dot{\mathbf{x}}_1\|_2 \approx \|\dot{\mathbf{x}}_2\|_2 \approx v$, it holds that

$$\dot{\rho} \approx \frac{\Delta U(\tau_1, \tau_2)}{v} = \frac{U_{12}}{v} \qquad (D.18)$$

which shows that the low-low range rate signal is approximately proportional to selenopotential difference along the track of the two satellites. The proportionality constant is the orbital velocity.

This low-low range rate SST model is actually the original model used by *Wolff* [1969], and also later by *Wagner* [1983]. The derivation presented here is perhaps more general since it relates the spacecraft velocity differences to potential differences along the orbit, and also takes due account of the rotation of the potential in inertial space.

Bibliography

Akim, E. L. (1966), Determination of the gravitational field of the Moon from the motion of the artificial lunar satellite 'Luna-10', *Doklady Akademii Nauk SSSR*, *170*(4), 799–802, originally in Russian, English translation in Soviet Physics - Doklady, Vol. 11, No. 10, April 1967.

Akim, E. L., and A. Golikov (1997), Combined model of the lunar gravity field, in *12th International Symposium on Space Flight Dynamics*, European Space Agency, ESOC Darmstadt, Germany, 2–6 June 1997, SP-403.

Akim, E. L., and Z. P. Vlasova (1977), Model of the Moon's gravitational field based on the motion of the artificial satellites Luna-10, 12, 14, 19 and 22, *Doklady Akademii Nauk SSSR*, *235*, 38–41, originally in Russian, English translation in Soviet Physics - Doklady, Vol. 22, No. 7, July 1977.

Akim, E. L., and Z. P. Vlasova (1983), Research on the Moon's gravitational from measurement data on the trajectories of Soviet artificial satellites of the Moon, *Kosmich. issled.*, *21*, 499–511, originally in Russian.

Aksnes, K. (1970), A second order artificial satellite theory based on an intermediate orbit, *Astronom. J.*, *75*, 1066–1076.

Albertella, A., F. Sansò, and N. Sneeuw (1999), Band-limited functions on a bounded spherical domain: the Slepian problem on the sphere, *J. Geodesy*, *73*, 436–447.

Allen, R. R. (1967), Satellite resonances with longitude-dependent gravity, II, Effects involving the eccentricity, *Planet. Space Sci.*, *15*, 1829–1845.

Alvarez, W. (1997), *T. Rex and the Crater of Doom*, Princeton University Press, Princeton, NJ.

Ananda, M. P. (1977), Lunar gravity: A mass point model, *J. Geophys. Res.*, *82*(20), 3049–3064.

Ananda, M. P., J. Lorrel, and W. Flury (1976), Lunar farside gravity: An assessment of satellite to satellite tracking techniques and gravity gradiometry, in *Proceedings of the Seventh Lunar Science Conference (Supplement 7, Geochimica et Cosmochimica Acta, Vol. 3)*, pp. 2623–2638, Pergamon Press.

Andolz, F. J., T. A. Dougherty, and A. B. Binder (1998), Lunar Prospector mission handbook, *LMMS/P45841*, Lockheed Martin Missiles & Space Co., Sunnyvale, CA.

Araki, H., M. Ooe, T. Tsubokawa, N. Kawano, H. Hanada, and K. Heki (1999), Lunar laser altimetry in the SELENE project, *Adv. Space Res., 23*(11), 1813–1816.

Arkani-Hamed, J. (1973a), On the formation of the lunar mascons, in *Proceedings of the Fourth Lunar Science Conference (Supplement 4, Geochimica et Cosmochimica Acta, Vol. 3)*, pp. 2673–2684, Pergamon Press.

Arkani-Hamed, J. (1973b), Viscosity of the Moon, I: After mare formation, *The Moon, 6*, 100–111.

Arkani-Hamed, J. (1973c), Viscosity of the Moon, II: During mare formation, *The Moon, 6*, 112–124.

Arkani-Hamed, J. (1998), The lunar mascons revisted, *J. Geophys. Res., 103*(E2), 3709–3739.

Arkani-Hamed, J. (1999a), On the equipotential surface hypothesis of lunar maria floors, *J. Geophys. Res., 104*(E3), 5921–5931.

Arkani-Hamed, J. (1999b), The high-resolution gravity anomaly of the lunar topography, *J. Geophys. Res., 104*(E5), 11865–11874.

Arnold, V. I. (1980), *Mathematical Methods of Classical Mechanics*, vol. 60 of *Graduate Texts in Mathematics*, Springer, Berlin, second edition.

Arnold, V. I., V. V. Kozlov, and A. I. Neishtadt (1997), *Mathematical Aspects of Classical and Celestial Mechanics*, Springer, Berlin, second edition.

Backus, G. E. (1988), Bayesian inference in geomagnetism, *Geophys. J., 92*, 125–142.

Baldwin, R. B. (1949), *The Face of the Moon*, 239 p., University of Chicago Press, Chicago.

Balmino, G. (1993), Orbit choice and the theory of radial orbit error for altimetry, in *Satellite altimetry in geodesy and oceanography*, vol. 50 of *Lecture Notes in Earth Sciences*, edited by F. Sansò and R. Rummel, pp. 243–315, Springer-Verlag, Berlin.

Balmino, G., E. Schrama, and N. Sneeuw (1996), Compatibility of first-order circular orbit perturbations theories; consequences for cross-track inclination functions, *Acta Astronaut., 39*(6), 407–416, based on paper IAF-95-A.5.01 presented at the 46th International Astronautical Congress, Oslo, 2-6 October 1995.

Balmino, G., F. Perosanz, R. Rummel, N. Sneeuw, H. Sünkel, and P. Woodworth (1998), European views on dedicated gravity field missions: GRACE and GOCE, An Earth Sciences Division Consultation Document, *ESD–MAG–REP–CON–001*, European Space Agency, May 1998.

Barbieri, C., and F. Rampazzi (eds.) (2001), *Earth-Moon Relationships*, 575 p., Kluwer Academic Publishers, Dordrecht.

Barriot, J. P. (1994), An inverse problem in planetary geodesy, *Inverse Problems, 10*, 809–816.

Barriot, J. P., and G. Balmino (1992), Estimation of local planetary gravity fields using line-of-sight gravity data and integral operator, *Icarus*, *99*, 202–224.

Barriot, J. P., and G. Balmino (1994), Analysis of the LOS gravity data set from cycle 4 of the Magellan probe around Venus, *Icarus*, *112*, 34–41.

Barriot, J.-P., N. Valès, G. Balmino, and P. Rosenblatt (1998), A 180th degree and order model of the Venus gravity field from Magellan line of sight resiudal Doppler data, *Geophys. Res. Lett.*, *25*(19), 3743–3746.

Beckman, M., and M. Concha (1998), Lunar Prospector orbit determination results, in *Advances in the Astronautical Sciences, Space Flight Dynamics*, AIAA, GSFC, Greenbelt, MD, AIAA-98-4561.

Bedrich, S., F. Flechtner, Ch. Förste, Ch. Reigber, and A. Teubel (1997), PRARE System Performance, in *Proc. 3rd ERS Symposium on Space at the Service of our Environment*, vol. 3, edited by T.-D. Guyenne and D. Danesy, pp. 1637–1642, Florence, Italy, 17-21 March 1997, ESA SP-414.

Benz, W., W. L. Slattery, and A. G. W. Cameron (1986), The origin of the Moon and the single impact hypothesis I, *Icarus*, *66*, 515–535.

Benz, W., W. L. Slattery, and A. G. W. Cameron (1987), The origin of the Moon and the single impact hypothesis II, *Icarus*, *71*, 30–45.

Benz, W., A. G. W. Cameron, and H. J. Melosh (1989), The origin of the Moon and the single impact hypothesis III, *Icarus*, *81*, 113–131.

Berger, J. O. (1985), *Statistical decision theory and Bayesian analysis*, Springer Verlag, New York.

Bertiger, W. I., Y. E. Bar-Sever, E. J. Christensen, E. S. Davies, J. R. Guinn, B. J. Haines, R. W. Ibanez-Meier, J. R. Lee, S. M. Lichten, W. G. Melbourne, R. J. Muellerschoen, T. N. Munson, Y. Vigue, S. C. Wu, T. P. Yunck, B. E. Schutz, P. A. M. Abusali, H. J. Rim, M. M. Watkins, and P. Willis (1994), GPS precise tracking of TOPEX/Poseidon: results and implications, *J. Geophys. Res.*, *99*(C12), 24449–24464.

Bierman, G. J. (1977), *Factorization Methods for Discrete Sequential Estimation*, vol. 128 of *Mathematics in Science and Engineering*, Academic Press, New York, Editor R. Bellman.

Bills, B. G. (1995), Discrepant estimates of moments of inertia of the Moon, *J. Geophys. Res.*, *100*(E12), 26297–26303.

Bills, B. G., and A. J. Ferrari (1980), A harmonic analysis of lunar gravity, *J. Geophys. Res.*, *85*(B2), 1013–1025.

Bills, B. G., and F. G. Lemoine (1995), Gravitational and topographic isotropy of the Earth, Moon, Mars and Venus, *J. Geophys. Res.*, *100*(E12), 26275–26295.

Binder, A. B. (1998), Lunar Prospector: Overview, *Science*, *281*, 1476–1480.

Björk, Å. (1996), *Numerical Methods for Least Squares Problems*, SIAM, Philadelphia.

Blackshear, W. T., and J. P. Gapcynski (1977), An improved value of the lunar moment of inertia, *J. Geophys. Res.*, *82*(11), 1699–1701.

Blakely, R. J. (1995), *Potential Theory in Gravity & Magnetic Applications*, Stanford–Cambridge Program, Cambridge University Press.

Bouman, J. (1998), Quality of regularization methods, *DEOS Report No. 98.2*, Delft Institute for Earth-Oriented Space Research.

Bouman, J. (2000), Quality assessment of satellite-based global gravity field models, *Publications on geodesy. New series No. 48*, Netherlands Geodetic Commission, Delft.

Bouman, J., and R. Koop (1998), Quality differences between Tikhonov regularization and generalized biased estimation in gradiometric analysis, *DEOS Progress Letters*, *98.2*, 69–80.

Bowin, C., B. Simon, and W. R. Wollenhaupt (1975), Mascons: A two-body solution, *J. Geophys. Res.*, *80*(35), 4947–4955.

Brouwer, D. (1959), Solution of the artificial satellite problem without drag, *Astronom. J.*, *64*, 378–3397.

Bryant Jr., W. C., and R. G. Williamson (1974), Lunar gravity analysis results from Explorer 49, in *AIAA Mechanics and Control of Flight Conference, Anaheim, CA*, AIAA, August 5–9 1974, AAS paper 74-810.

Calvetti, D., G. H. Golub, and L. Reichel (1999), Estimation of the L-curve via Lanczos bidiagonalization, *BIT*, *39*(4), 603–619.

Cameron, A. G. W., and W. Benz (1991), The origin of the moon and the single impact hypothesis IV, *Icarus*, *92*, 204–216.

Carranza, E., A. Konopliv, and M. Ryne (1999), Lunar Prospector orbit determination uncertainties using the high resolution lunar gravity models, in *Advances in the Astronautical Sciences, AAS/AIAA Astrodynamics Specialist Conference*, AAS/AIAA, Girdwood, Alaska, AAS paper 99-325.

Coleman Jr., P. J., B. R. Lichtenstein, C. T. Russell, L. R. Sharp, and G. Schubert (1972*a*), Magnetic fields near the Moon, in *Proceedings of the Third Lunar Science Conference (Supplement 3, Geochimica et Cosmochimica Acta, Vol. 3)*, M.I.T. Press, pp. 2271-2286.

Coleman Jr., P. J., C. T. Russel, L. R. Sharp, and G. Schubert (1972*b*), Preliminary mapping of the lunar magnetic field, *Phys. Earth Planet. Interiors*, *6*, 167–174.

Colombo, O. (1981), Global geopotential modelling from satellite-to-satellite tracking, *Report No. 317*, Ohio State University.

Colombo, O. L. (1984), *The Global Mapping of Gravity With Two Satellites*, vol. 7, No. 3, Netherlands Geodetic Commission, Publications on Geodesy, New Series.

Colombo, O. L. (1986), Notes on the mapping of the gravity field using satellite data, in *Mathematical and numerical techniques in physical geodesy*, vol. 7 of *Lecture Notes in Earth Sciences*, edited by H. Sünkel, pp. 263–315, Springer-Verlag.

Colombo, O. L. (1989*a*), Advanced techniques for high-reolution mapping of the gravitational field, in *Theory of Satellite Geodesy and Gravity Field Determination*, vol. 25 of *Lecture Notes in Earth Sciences*, edited by F. Sansò and R. Rummel, pp. 335–369, Springer-Verlag.

Colombo, O. L. (1989*b*), Mapping the Earth's gravity field with orbiting GPS receivers, in *Proc. 1989 IUGG General Assembly, GPS Symposium*, Springer-Verlag.

Comfort, G. C. (1973), Direct mapping of gravity anomalies by using Doppler tracking between a satellite pair, *J. Geophys. Res.*, *78*(29), 6845–6851.

Cook, R. A. (1991), The long-term behavior of near-circular orbits in a zonal field, in *Advances in the Astronautical Sciences, AAS/AIAA Spaceflight Mechanics Meeting*, AAS/AIAA, Durango, Colorado, 19–22 August, 1991, AAS paper 91-453.

Cook, R. A., and Th. H. Sweetser (1992), Orbit maintenance for low altitude near-circular lunar orbits, in *Advances in the Astronautical Sciences, AAS/AIAA Spaceflight Mechanics Meeting*, AAS/AIAA, Colorado Springs, Colorado, AAS paper 92-185.

Cook, R. A., A. B. Sergeyevsky, E. A. Belbruno, and T. H. Sweetser (1990), Return to the Moon: The Lunar Observer mission, in *AIAA/AAS Astrodynamics Conference*, AAS/AIAA, Portland, Oregon, 20–22 August, 1990, AIAA paper 90-2888.

Coradini, A., B. Foing, M. Harrison, H. Hoffmann, P. Janle, Y. Langevin, A. Milani, G. Neukum, G. Picardi, G. Racca, J. Raitala, C. d'Uston, N. Waltman, and H. Wänke (1996), Moro Moon Orbiting Observatory Phase A Study Report, European Space Research and Technology Centre, Noordwijk, The Netherlands, March 1996, ESA Publication SCI(96)1.

Cui, Ch., and D. Lelgemann (2000), On non-linear low–low SST observation equations for the determination of the geopotential based on an analytical solution, *J. Geodesy*, *74*, 431–440.

d'Avanzo, P., P. Teofilatto, and C. Ulivieri (1995), Long-term effects on lunar orbiter, *Acta Astronaut.*, *40*(1), 13–20, October 1995, first presented at the 46th International Astronautical Congress in Oslo, 2–6 October 1995, as paper IAF-95-A.1.08.

Davies, E., C. E. Dunn, R. H. Stanton, and J. B. Thomas (1999), The GRACE mission: Meeting the technical challenge, 50th International Astronautical Congress, Amsterdam, 4–8 October, 1999.

Davies, M. E., V. K. Abalakin, A. Brahic, M. Bursa, B. H. Chovitz, J. H. Lieske, P. K. Seidelmann, A. T. Sinclair, and Y. S. Tjuflin (1992), Report of the IAU/IAG/COSPAR Working Group on Cartographic Coordinates and Rotational Elements of the Planets and Satellites: 1991, *Celest. Mech. & Dynam. Astron.*, *53*, 377–397.

Davies, M. E., V. K. Abalakin, M. Bursa, J. H. Lieske, B. Morando, D. Morrison, P. K. Seidelmann, A. T. Sinclair, and B. Yallop (1996), Report of the

IAU/IAG/COSPAR Working Group on Cartographic Coordinates and Rotational Elements of the Planets and Satellites: 1994, *Celest. Mech. & Dynam. Astron., 63,* 127–148.

Davies, M. E., V. K. Abalakin, M. Bursa, H. Kinoshita, R. L. Kirk, J. H. Lieske, P. K. Seidelmann, and J.-L. Simon (1997), Report of the IAU/IAG/COSPAR Working Group on Cartographic Coordinates and Rotational Elements of the Planets and Satellites: 1997, *Celest. Mech. & Dynam. Astron.,* ??, ??

Dickey, J. O., P. L. Bender, J. E. Faller, X. X. Newhall, R. L. Ricklefs, J. G. Ries, P. J. Shelus, C. Veillet, A. L. Whipple, J. R. Wiant, J. G. Williams, and C. F. Yoder (1994), Lunar laser ranging: A continuing legacy of the Apollo program, *Science, 265,* 482–490, 22 July 1994.

d'Inverno, R. (1992), *Introducing Einstein's Relativity,* Oxford University Press, Oxford.

Dooling, D. (1994), L+25 - a quarter century after the Apollo landing, *IEEE Spectrum, 31*(7), 16–29.

Douglas, B. C., C. C. Goad, and F. F. Morrison (1980), Determination of the geopotential from satellite-to-satellite tracking data, *J. Geophys. Res., 85*(B10), 5471–5480.

Eckstein, M. C., and O. Montenbruck (1995), MORO orbit evolution study using the LOP software, *GSOC TN 95–08,* DLR - GSOC, Wessling, Germany, July 1995.

Eddy, W. F., J. J. McCarthy, D. E. Pavlis, J. A. Marshall, S. B. Luthke, L. S. Tsaoussi, G. Leung, and D. A. Williams (1990), *GEODYN II System Operations Manual, Vol. 1–5,* ST System Corp., Lanham, MD, Contractor Report.

Eldén, L. (1977), Algorithms for the regularization of ill-conditioned least squares problems, *BIT, 17,* 134–145.

Eldén, L. (1982), A weighted pseudoinverse, generalized singular values, and constrained least squares problems, *BIT, 22,* 487–502.

Engl, H. W., M. Hanke, and A. Neubauer (1996), *Regularization of Inverse Problems,* Mathematics and Its Applications, Kluwer Academic Publishers, Dordrecht.

ESA (1979), Polar orbiting lunar observatory (POLO), European Space Agency.

ESA (1996), GOCE: Gravity field and steady-state ocean circulation mission, *ESA SP–1196(1),* ESA, Paris.

ESA (1999), Gravity field and steady-state ocean circulation mission, *ESA SP–1233(1),* ESA, Paris.

ESA (2000), From Eötvös to milligal, *ESA/ESTEC Contract No. 13392/98/NL/GD,* ESA.

Feissel, M., and F. Mignard (1998), The adoption of ICRS on 1 january 1998: meaning and consequences, *Astron. Astrophys., 331,* L33–L36.

Ferrari, A. J. (1977*a*), Lunar gravity: A harmonic analysis, *J. Geophys. Res.*, *82*(20), 3065–3084.

Ferrari, A. J. (1977*b*), Lunar gravity: A long-term Keplerian rate method, *J. Geophys. Res.*, *82*(20), 3085–3097.

Finzi, A. E., and M. Vasile (1997), Numerical solutions for lunar orbits, paper presented at the 48th Int'l. Astronautical Congress, Turin, paper IAF-97-A.5.08, October 1997.

Floberghagen, R. (1995), Global Lunar Gravity Recovery From Satellite-to-Satellite Tracking, Master's thesis, Section Space Research & Technology, Faculty of Aerospace Engineering, Delft University of Technology, May 1995.

Floberghagen, R., and P. N. A. M. Visser (1997), Low lunar orbit analysis, determination and selection for soft landing on the lunar south pole, in *12th International Symposium on Space Flight Dynamics*, European Space Agency, ESOC, Darmstadt, Germany, 2–6 June 1997 1997, SP-403.

Floberghagen, R., R. Noomen, P. N. A. M. Visser, and G. D. Racca (1996), Global lunar gravity field recovery from satellite-to-satellite tracking, *Planet. Space Sci.*, *44*(10), 1081–1097, issue contains a special section on Geodesy of the Moon.

Floberghagen, R., P. N. A. M. Visser, F. Weischede, and M. Vasile (1998), On the analysis of lunar albedo effects on low lunar orbit determination and gravity field estimation, in *Space Flight Dynamics 1998, Advances in the Astronautical Sciences*, edited by T. H. Stengle, AAS, Goddard Space Flight Center, Greenbelt, Maryland, AAS 98-313.

Floberghagen, R., P. Visser, and F. Weischede (1999), Lunar albedo force modeling and its effect on low lunar orbit and gravity field determination, *Adv. Space Res.*, *23*(4), 733–738.

Flury, W. (1981), Lunar gravity field determination, in *Proceedings of an International Symposium on Spacecraft Flight Dynamics*, pp. 37–45, European Space Operations Centre (ESOC), Darmstadt, Germany, 18–22 May 1981, SP-160.

Freeden, W., and M. Schreiner (1998), Regularization wavelets and multiresolution, *Inverse Problems*, *14*, 225–243.

Freeden, W., F. Schneider, and M. Schreiner (1997), Gradiometry - an inverse problem in modern satellite geodesy, in *SIAM Symposium on Inverse Problems: Geophysical Applications*, pp. 179–239.

Frese, R. R. B. von, L. Tan, L. V. Potts, J. Woo Kim, C. J. Merry, and J. D. Bossler (1997), Lunar crustal analysis of Mare Orientale from topographic and gravity correlations, *J. Geophys. Res.*, *102*(E11), 25657–357676.

Gapcynski, J. P., W. T. Blackshear, R. H. Tolson, and H. R.Compton (1975), A determination of the lunar moment of inertia, *Geophys. Res. Lett.*, *2*(8), 353–356.

Garret, H. B., and P. Rustan (1995), Clementine engineering experiments program, *J. Spacecraft and Rockets*, *32*(6), 1045–1048.

Gaudenzi, R. De, E. E. Lijphart, and E. Vassallo (1990), The new ESA multi-purpose tracking system, *ESA J.*, *14*, 23–40.

Geemert, R. van (2000), Development of a timewise parallelized iterative approach for gravity field recovery from satellite gravity gradiometry data, *Internal report*, Faculty of Civil Engineering and Geosciences, Delft University of Technology, Delft, The Netherlands.

Geemert, R. van, R. J. J. Koop, R. Klees, and P. Visser (2000), Application of parallel computing to gravity field recovery from satellite gravity gradiometric data, in *Proc. 6th International Conference on Applications of High-Performance Computers in Engineering*, edited by M. Ingber *et al.*, WIT Press, Southampton, 26–28 January 2000, Maui, Hawaii 2000.

Gleason, D. M. (1991), Obtaining Earth surface gravity disturbances from a GPS-based 'high–low' satellite-to-satellite tracking experiment, *Geophys. J. Int.*, *107*, 13–23.

Gold, K. L. (1994), *GPS Orbit Determination for the Extreme Ultraviolet Explorer*, Ph.D. dissertation, Dept. of Aerospace Engineering Sciences, University of Colorado.

Goldstein, D. B., R. S. Nerem, J. V. Austin, A. B. Binder, and W. C. Feldman (1999), Impacting Lunar Prospector in a cold trap to detect water ice, *Geophys. Res. Lett.*, *26*(12), 1653–1656.

Golub, G. H., and C. F. van Loan (1996), *Matrix computations*, The Johns Hopkins University Press, Baltimore and London, third edition.

Golub, G. H., M. Heath, and G. Wahba (1979), Generalized cross-validation as a method for choosing a good ridge parameter, *Technometrics*, *21*(2), 215–223.

Goossens, S. (1999), Long-term low lunar orbit perturbations due to the selenopotential and solar radiation pressure, Master's thesis, Delft Institute for Earth-Oriented Space Research, Faculty of Aerospace Engineering, Delft University of Technology, April 1999.

Goossens, S., R. Floberghagen, and M. Vasile (1999), Long-term orbit prediction for low-lunar satellites under the influence of gravity and solar radiation pressure, paper presented at the 50th International Astronautical Congress in Amsterdam, paper IAF-99-A.7.09, October 1999.

Green, R. M. (1985), *Spherical Astronomy*, Cambridge University Press, Cambridge.

Groetsch, C. W. (1984), *The theory of Tikhonov regularization for Fredholm integral equations of the first kind*, Pitman, Boston - London - Melbourne.

Groetsch, C. W. (1993), *Inverse problems in the mathematical sciences*, Friedrich Vieweg & Sohn Verlag, Wiesbaden.

Haagmans, R. H. N., and M. van Gelderen (1991), Error variances-covariances of GEM-T1: their characteristics and implications in geoid computation, *J. Geophys. Res.*, *96*(B12), 20011–20022.

Hadamard, J. (1923), *Lectures on Cauchy's Problem in Linear Partial Differential Equations*, Yale University Press, New Haven, CT.

Hajela, D. P. (1979), Tests for the recovery of 5° mean gravity anomalies in local areas from ATS-6/GEOS 3 satellite-to-satellite range-rate observations, *J. Geophys. Res.*, *84*(B12), 6884–6890.

Hanada, H., M. Ooe, N. Kawaguchi, N. Kawano, S. Kuji, T. Sasao, S. Tsuruta, M. Fujishita, and M. Morimoto (1993), Study of the lunar core by VLBI observations of artificial radio sources on the Moon, *J. Geomag. Geoelectr.*, *45*, 1405–1414.

Hanke, M. (1996), Limitations of the L-curve method in ill-posed problems, *BIT*, *36*, 287–301.

Hansen, P. C. (1989), Regularization, GSVD and truncated GSVD, *BIT*, *29*, 491–504.

Hansen, P. C. (1990), The discrete Picard condition for discrete ill-posed problems, *BIT*, *30*, 658–672.

Hansen, P. C. (1992), Analysis of discrete ill-posed problems by means of the L-curve, *SIAM Review*, *34*(4), 561–580.

Hansen, P. C. (1998a), *Rank-Deficient and Ill-Posed Problems: Numerical Aspects of Linear Inversion*, SIAM Monographs on Mathematical Modeling and Computation, SIAM, Philadelphia.

Hansen, P. C. (1998b), *Regularization Tools, A Matlab package for analysis and solution of discrete ill-posed problems, Version 3.0 for Matlab 5.2*, Department of Mathematical Modelling, Technical University of Denmark, http://www.imm.dtu.dk/~pch.

Hansen, P. C. (1999), The L-curve and its use in the numerical treatment of inverse problems, *Adv. Comput. Biomedicine, 3: Inverse Problems In Electrocardiology*, ?, to appear by WIT Press; internal report number IMM-REP-1999-15.

Hansen, P. C., and D. P. O'Leary (1993), The use of the L-curve in the regularization of discrete ill-posed problems, *SIAM J. Sci. Comput.*, *14*(6), 1487–1503.

Hartmann, W. K., and D. R. Davies (1975), Satellite-sized planetesimals and lunar origin, *Icarus*, *24*, 504–515.

Häusler, B. (1998), Gravity field investigations with the Lunarstar-subsatellite GAUSS; complete lunar gravity field investigation, Universität der Bundeswehr München, Institut für Rauhmfahrttechnik, Neubiberg, Germany, Discovery-class mission proposal to NASA and DLR.

Häusler, B., R. Floberghagen, W. Ip, P. Janle, E. Gill, O. Montenbruck, M. Pätzold, S. Asmar, A. Konopliv, A. Stern, and M. Zuber (1998), Complete gravity field determination of the Moon, October 11–14 1998, Third International Conference on Exploration and Utilization of the Moon (abstract).

Haw, R. J., P. G. Antresian, T. P. McElrath, and G. D. Lewis (2000), Galileo prime mission navigation, *J. Spacecraft and Rockets*, *37*(1), 56–63.

Head III, J. W. (1976), Lunar volcanism in space and time, *J. Geophys. Res.*, *14*(2), 265–300.

Heki, K., K. Matsumoto, and R. Floberghagen (1999), Three-dimensional tracking of a lunar satellite with differential very-long-baseline-interferometry, *Adv. Space Res.*, *23*(11), 1821–1824.

Hesper, E. T. (1994), Landsat-5 and TOPEX/POSEIDON Orbit Determination From GPS Observations, in *Proceedings of the 2nd ESA International Conference on GNC*, ESTEC, Noordwijk, The Netherlands, 12–15 April 1994, ESA WPP-071.

Hoerl, A. E., and R. W. Kennard (1970*a*), Ridge regression: biased estimation for nonorthogonal problems, *Technometrics*, *12*(1), 55–67.

Hoerl, A. E., and R. W. Kennard (1970*b*), Ridge regression: applications to nonorthogonal problems, *Technometrics*, *12*(1), 69–82.

Hubbard, G. S., S. A. Cox, M. A. Smith, T. A. Dougherty, and L. Chu-Thielbar (1999), Lunar Prospector: continuing mission results, paper IAF-99-Q.4.02, 50th International Astronautical Congress, Amsterdam, 4–8 October, 1999 1999.

Huntress Jr., W. T. (1999), The Discovery program, *Acta Astronaut.*, *45*(4–9), 207–213.

Ivanov, V. K. (1962), Integral equations of the first kind and an approximate solution for the inverse problem of potential, *Soviet Math. Doklady*, *3*, 210–212.

Jekeli, C. (1999), The determination of gravitational potential differences from satellite-to-satellite tracking, *Celest. Mech. & Dynam. Astron.*, *75*, 85–101.

Jinsong, P., F. Weischede, Y. Kono, H. Hanada, and N. Kawano (2000), High frequency components in Doppler data of Lunar Prospector, *Internal document*, Divison of Earth Rotation, National Astronomical Observatory, Japan, May 2000.

Kahn, W. D., S. M. Klosko, and W. T. Wells (1982), Mean gravity anomalies from a combination of Apollo/ATS 6 and GEOS 3/ATS 6 SST tracking campaigns, *J. Geophys. Res.*, *87*(B4), 2904–2918.

Kaneko, Y., H. Itagaki, Y. Takizawa, and S. Sasaki (1999), The SELENE project and the following lunar mission, paper IAF-99-Q.4.03, 50th International Astronautical Congress, Amsterdam, 4–8 October, 1999.

Kaplan, M. H. (1976), *Modern spacecraft dynamics and control*, John Wiley & Sons, New York.

Kascheev, R. A. (1988), Lunar gravity parameters from line-of-sight acceleration data, *Earth, Moon, and Planets*, *41*, 2904–2918.

Kaula, W. M. (1966), *Theory of satellite geodesy*, Blaisdell Pub. Co., Waltham, Toronto.

Kaula, W. M. (1969), The gravitational field of the Moon, *Science*, *166*(3913), 1581–1588, 26 December 1969.

Kaula, W. M. (1983), Inference of variations in the gravity field from satellite-to-satellite range rate, *J. Geophys. Res.*, *88*(B10), 8345–8349, October 10, 1983.

Kaula, W. M., G. Schubert, and R. E. Lingenfelter (1972), Analysis and interpretation of lunar laser altimetry, in *Proceedings of the Third Lunar Science Conference (Supplement 3, Geochimica et Cosmochimica Acta, Vol. 3)*, M.I.T. Press, pp. 2189-2204.

Kaula, W. M., G. Schubert, R. E. Lingenfelter, W. L. Sjogren, and W. R. Wollenhaupt (1973), Lunar topography from Apollo 15 and 16 laser altimetry, in *Proceedings of the Fourth Lunar Science Conference (Supplement 4, Geochimica et Cosmochimica Acta, Vol. 3)*, Pergamon Press, pp. 2811-2819.

Kaula, W. M., G. Schubert, R. E. Lingenfelter, W. L. Sjogren, and W. R. Wollenhaupt (1974), Apollo laser altimetry and inferences to lunar structure, in *Proceedings of the Fifth Lunar Science Conference (Supplement 5, Geochimica et Cosmochimica Acta, Vol. 3)*, Pergamon Press, pp. 3049-3058.

Kaula, W. M., M. J. Drake, and J. W. Head (1986), The Moon, in *Satellites*, pp. 580-628, The University of Arizon Press, eds. J. A. Burns and M. S. Matthews.

King, R. W., C. C. Counselman III, and I. I. Shapiro (1976), Lunar dynamics and selenodesy: Results from analysis of VLBI and laser data, *J. Geophys. Res.*, *81*(35), 6251-6256.

Kinman, P. W. (1992), Doppler tracking of planetary spacecraft, *IEEE Trans. on Microwave Theory and Techniques*, *40*(6), 1199-1204.

Klees, R. (1993), Gravity field determination using boundary element methods, *Surv. Geophys.*, *14*, 419-432.

Klees, R. (1997), Topics on boundary element methods, in *Geodetic Boundary Value Problems in View of the One Centimeter Geoid*, vol. 65 of *Lecture Notes in Eearth Sciences*, edited by F. Sansò and R. Rummel, pp. 482-531, Springer-Verlag, Berlin.

Klees, R., R. Koop, P. Visser, and J. van den IJssel (2000), Efficient gravity field recovery from GOCE gravity gradient observations, *J. Geodesy*, *74*, 561-571.

Konopliv, A. S. (1991), Simulations of lunar gravity field determination for Lunar Observer, in *AAS/AIAA Astrodynamics Specialist Conference*, American Astronautical Society, Durango, Colorado, 19-22 August, 1991, AAS paper 91-395.

Konopliv, A. S., and W. L. Sjogren (1996), Venus Gravity Handbook, *JPL Publication 96-2*, Jet Propulsion Laboratory, Pasadena, California.

Konopliv, A. S., and D. N. Yuan (1999), Lunar prospector 100th degree gravity model development, 30th Lunar and Planetary Science Conference, Houston, TX, 15-19 March, 1999.

Konopliv, A. S., W. L. Sjogren, R. N. Wimberly, and A. Vijayaraghavan (1993), A High Resolution Lunar Gravity Field and Predicted Orbit Behaviour, in *AAS/AIAA Astrodynamics Specialist Conference*, American Astronautical Society, Victoria, B.C., Canada, 16-19 August, 1993, AAS paper 93-622.

Konopliv, A. S., A. B. Binder, L. L. Hood, A. B. Kucinskas, W. L. Sjogren, and J. G. Williams (1998), Improved gravity field of the Moon from Lunar Prospector, *Science*, *281*, 1476-1480, 4 September 1998.

Konopliv, A. S., W. B. Banerdt, and W. L. Sjogren (1999), Venus gravity: 180th degree and order model, *Icarus, 139*, 3–18.

Konopliv, A. S., S. W. Asmar, E. Carranza, W. L. Sjogren, and D. N. Yuan (2001), Recent gravity models as a result of the Lunar Prospector mission, *Icarus, 150*, 1–18.

Koop, R. J. J. (1993), Global gravity field modelling using satellite gravity gradiometry, *Publications on geodesy. New series No. 38*, Netherlands Geodetic Commission, Delft.

Kozai, Y. (1959), The motion of a close earth satellite, *Astronom. J., 64*, 367–377.

Kozai, Y. (1962), Second-order solution of artificial satellite theory without air drag, *Astronom. J., 67*, 446–461.

Kozai, Y. (1966), The earth gravitational potential derived from satellite motion, *Space Science Reviews, 5*, 818–879.

Kress, R. (1999), *Linear integral equations*, Applied Mathematical Sciences 82, Springer-Verlag, Berlin, second edition.

Kreyszig, E. (1999), *Advanced engineering mathematics*, John Wiley and Sons, New York, eighth edition.

Kubo-oka, T. (1999), Long-term effect of solar radiation pressure on the orbit of an octogonal satellite orbiting around the Moon, *Adv. Space Res., 23*(4), 727–731.

Kubo-oka, T., and A. Sengoku (1999), Solar radiation pressure model for the relay satellite of SELENE, *Earth, Planets and Space, 51*, 979–986.

Kuckes, A. F. (1977), Strength and rigidity of the elastic lunar lithosphere and implications for present-day mantle convection in the Moon, *Phys. Earth Planet. Interiors, 14*, 1–12.

Laskar, J. (1996), Chaos à grande échelle dans le système solaire et implications planétologiques, *Comptes Rendues de l'Academie des Sciences IIA, 322*, 163–180.

Laskar, J., and P. Robutel (1993), The chaotic obliquity of the planets, *Nature, 361*, 608–612.

Laskar, J., F. Joutel, and F. Boudin (1993), Orbital, precessional, and insolation quantities for the Earth from -20 Myr to +10 Myr, *Astron. Astrophys., 270*, 522–533.

Lawson, C. L., and R. J. Hanson (1974), *Solving least squares problems*, Prentice-Hall, Englewood Cliffs, New Jersey.

Lehmann, R., and R. Klees (1999), Numerical solution of geodetic boundary value problems using a global reference field, *J. Geodesy, 73*, 543–554.

Leick, A. (1995), *GPS Satellite Surveying*, John Wiley & Sons, Inc., New York, second edition.

Lemoine, F. G., D. E. Smith, L. Kunz, R. Smith, E. C. Pavlis, N. K. Pavlis, S. M. Klosko, D. S. Chinn, M. H. Torrence, R. G. Williamson, C. M. Cox, K. E. Rachlin, Y. M. Yang, S. C. Kenyon, R. Salman, R. Trimmer, R. H. Rapp, and R. S. Nerem (1997), The development of the NASA GSFC and NIMA joint geopotential model, in *International Association of Geodesy Symposia, Vol. 117, Gravity, Geoid and Marine Geodesy*, pp. 461–469, Springer-Verlag, Heidelberg.

Lemoine, F. G., S. C. Kenyon, J. K. Factor, R. G. Timmer, N. K. Pavlis, D. S. Chinn, C. M. Cox, S. M. Klosko, S. B. Luethcke, M. H. Torrence, Y. M. Yang, R. G. Williamson, E. C. Pavlis, R. H. Rapp, and T. R. Olson (1998), The development of the joint NASA GSFC and the National Imagery and Mapping Agency NIMA geopotential model EGM96, *NASA/TP–1998-206861*, NASA/Goddard Space Flight Center, Greenbelt, Maryland 20771.

Lemoine, F. G. R., D. E. Smith, M. Zuber, G. A. Neumann, and D. D. Rowlands (1997), A 70th degree lunar gravity model (GLGM–2) from Clementine and other tracking data, *J. Geophys. Res.*, 102(E7), 16339–16359.

Lerch, F. J. (1991), Optimum data weighting and error calibration for estimation of gravitational parameters, *Bulletin Géodésique*, 65, 44–52.

Lerch, F. J., J. G. Marsh, S. M. Klosko, G. B. Patel, D. S. Chinn, E. C. Pavlis, and C. A. Wagner (1991), An improved error assessment for the GEM-T1 gravitational model, *J. Geophys. Res.*, 96(B12), 20023–20040.

Lerch, F. J., R. S. Nerem, B. H. Putney, T. L. Felsentreger, B. V. Sanchez, J. A. Marshall, S. M. Klosko, G. B. Patel, R. G. Williamson, D. S. Chinn, J. C. Chan, K. E. Rachlin, N. L. Chandler, J. J. McCarthy, S. B. Luthcke, N. K. Pavlis, D. E. Pavlis, J. W. Robbins, S. Kapoor, and E. C. Pavlis (1994), A geopotential model from satellite tracking, altimeter, and surface gravity data: GEM-T3, *J. Geophys. Res.*, 99(B2), 2815–2839.

Liu, A. S., and P. Laing (1971), Lunar gravity analysis from long-term effects, *Science*, 173, 1017–1020.

Longhi, J. (1977), Magma oceanography: II. Chemical evolution and crustal formation, in *Proceedings of the Eighth Lunar Science Conference (Supplement 8, Geochimica et Cosmochimica Acta, Vol. 1)*, Pergamon Press, pp. 601-621.

Lorell, J., and W. L. Sjogren (1968), Lunar gravity: Preliminary estimates from Lunar Orbiter, *Science*, 159, 625–627.

Louis, A. K. (1989), *Inverse und schlecht gestellte Probleme*, Teubner, Stuttgart.

Lozier, D., K. Galal, D. Folta, and M. Beckman (1998), Lunar Prospector mission design and trajectory support, in *Space Flight Dynamics 1998, Advances in the Astronautical Sciences*, edited by T. H. Stengle, AAS, Goddard Space Flight Center, Greenbelt, Maryland, AAS 98-323.

Lucey, P. G., G. J. Taylor, and E. Malaret (1995), Abundande and dsitribution of iron on the Moon, *Science*, 268, 1150–1153.

Luthke, S. B., J. A. Marshall, S. C. Rowton, K. E. Rachlin, C. M. Cox, and R. G. Williamson (1997), Enhanced radiative force modeling of the tracking and data relay satellites, *Journal of the Astronautical Sciences*, 45(1), 349–370, January–March 1997.

Manabe, S., T. Hara, O. Kameya, N. Kawano, S. Kuji, T. Sasao, K. Sato, F. Glasik, W. Köhnlein, H. Kynast, C. Lehner, H. Wenderoth, K. Wiedemann, S. Yasuda, M. Fujishita, S. Hama, and Y. Takahashi (1993), First VLBI observations between Weilheim, Kashima and Nobeyama, *J. Geodetic Soc. Japan*, 39(4), 353–361.

Marini, J-W. (1970), The effect of satellite spin on two-way Doppler range-rate measurements, *IEEE Trans. Aerospace and Electronic Systems*, 7, 316–320.

Marsh, J. G., F. J. Lerch, B. H. Putney, D. C. Christodoulidis, D. E. Smith, T. L. Felsentreger, B. V. Sanchez, S. M. Klosko, E. C. Pavlis, T. V. Martin, J. W. Robbins R. G. Williamson, O. L. Colombo, D. D. Rowlands, W. F. Eddy, N. L. Chandler, K. E. Rachlin, G. B. Patel, S. Bhati, and D. S. Chinn (1988), A new gravitational model for the earth from satellite tracking data: GEM-T1, *J. Geophys. Res.*, 93(B6), 6169–6215.

Marsh, J. G., F. J. Lerch, B. H. Putney, T. L. Felsentreger, B. V. Sanchez, S. M. Klosko, G. B. Patel, J. W. Robbins, R. G. Williamson, T. L. Engelis, W. F. Eddy, N. L. Chandler, D. S. Chinn, S. Kapoor, K. E. Rachlin, L. E. Braatz, and E. C. Pavlis (1990), The GEM-T2 gravitational model, *J. Geophys. Res.*, 95(B13), 22043–22071.

Marshall, J. A., F. G. Lemoine, S. B. Luthcke, J. C. Chan, C. M. Cox, S. C. Rowton, and R. G. Williamson (1995), Precision orbit determination and gravity field improvement derived from TDRSS, October 1995, paper IAF-95-A.1.05 presented at the 46th Int'l. Astronautical Congress in Oslo.

Marshall, J. A., F. J. Lerch, S. B. Luthcke, R. G. Williamson, and J. C. Chan (1996), An assessment of TDRSS for precision orbit determination, *J. Astronaut. Sci.*, 44(1), 115–127.

Martensen, E., and S. Ritter (1997), Potential theory, in *Geodetic Boundary Value Problems in View of the One Centimeter Geoid*, vol. 65 of *Lecture Notes in Earth Sciences*, edited by F. Sansò and R. Rummel, pp. 19–66, Springer-Verlag, Berlin.

Matsumoto, K., K. Heki, and D. Rowlands (1999), Impact of far-side satellite tracking on gravity estimation in the SELENE project, *Adv. Space Res.*, 23(11), 1809–1812.

Matsumoto, K., K. Heki, and D. Rowlands (2000), Lunar gravity recovery with 4-way Doppler and ΔVLBI in the SELENE project, presented at the 33rd COSPAR Scientific Assembly, Warsaw, Poland, 16-23 July, 2000.

Mazanek, D. D., R. R. Kumar, H. Seywald, and M. Qu (2000), GRACE mission design: Impact of uncertainties in disturbance environment and satellite force models, in *AAS/AIAA Spaceflight Mechanics Meeting*, American Astronautical Society, Clearwater, Florida, 23–26 January, 2000, AAS paper 00-163.

Meissl, P. (1971), A study of covariance functions related to the earth's disturbing potential, *Report No. 151*, Ohio State University.

Meissner, R. (1977), Lunar viscosity models, *Phil. Trans. R. Soc. Lond. A.*, *285*, 463–467.

Melbourne, W. G., E. S. Davies, T. P. Yunck, and B. D. Tapley (1994), The GPS flight experiment of TOPEX/Poseidon, *Geophys. Res. Lett.*, *21*(19), 2171–2174.

Merson, R. H., and D. G. King-Hele (1958), The uses of artificial satellites to explore the Earth's gravitational field: Results from Sputnik 2, *Nature*, *182*, 640–641.

Meyer, K. W., J. J. Buglia, and P. N. Desai (1994), Lifetimes of Lunar Satellite Orbits, *NASA Technical Paper 3394*, National Aeronautics and Space Administration, Langley Research Center, Mapton, Virginia, March 1994.

Michael, W. H., and T. Blackshear (1972), Recent results on the mass, gravitational field and moments of inertia of the moon, *The Moon*, *3*, 388–402.

Milani, A., and Z. Knežević (1995), Selenocentric Proper Elements – A Tool for Lunar Satellite Mission Analysis, Consorzio Pisa Ricerche, Pisa, Italy, May 1995, Final Report of a study conducted for ESA under purchase order 144506 of 19 December 1994.

Milani, A., M. Luise, and F. Scortecci (1996), The lunar sub-satellite experiment of the ESA MORO mission: goals and performances, *Planet. Space Sci.*, *44*(10), 1065–1076, issue contains a special section on Geodesy of the Moon.

Minter, C. F., and D. A. Cicci (1998), Improvements in Mars gravity field determination using ridge-type estimation methods, *J. Astronautical Sciences*, *46*(4), 411–424.

Montenbruck, O. (1992), Numerical integration methods for orbital motion, *Celest. Mech. & Dynam. Astron.*, *53*, 59–69.

Montenbruck, O., and E. Gill (2000), *Satellite Orbits*, Springer, Heidelberg.

Moreaux, G., J. P. Barriot, and L. Amodei (1999), A harmonic spline model for local estimation of planetary gravity fields from line-of-sight acceleration data, *J. Geodesy*, *73*, 130–137.

Moritz, H. (1989), *Advanced physical geodesy*, Wichmann, Karlsruhe, second edition.

Morozov, V. A. (1984), *Methods for solving incorrectly posed problems*, Springer-Verlag, New York, Berlin, Heidelberg.

Muller, P. M., and W. L. Sjogren (1968), Mascons: Lunar Mass Concentrations, *Science*, *161*, 680–684.

Muller, P. M., W. L. Sjogren, and W. R. Wollenhaupt (1974), Lunar gravity: Apollo 15 Doppler radio tracking, *The Moon*, *10*, 195–205.

Nagae, Y., Y. Takizawa, and S. Sasaki (1999), The system concept of SELENE, *Acta Astronaut.*, *45*(4–9), 197–205.

Nakamura, Y. (1983), Seismic velocity structure of the lunar mantle, *J. Geophys. Res.*, *88*(B1), 677–686.

Nakamura, Y., G. V. Latham, H. J. Dorman, P. Horvath, and A-B. K. Ibrahim (1977), Seismic indication of broad-scale lateral heterogeneities in the lunar interior: A preliminary report, in *Proceedings of the Eighth Lunar Science Conference (Supplement 8, Geochimica et Cosmochimica Acta, Vol. 1)*, Pergamon Press, pp. 487-498.

Namiki, N., H. Hanada, T. Tsubokawa, N. Kawano, M. Ooe, K. Heki, T. Iwata, M. Ogawa, T. Takano, and RSAT/VRAD/LALT mission groups (1999), Selenodetic experiments of SELENE: relay satellite, differential VLBI, and laser altimeter, *Adv. Space Res.*, *23*(11), 1817–1820.

Nau, H., C. Lehner, R. Manetsberger, J. Hahn, A. Bauch, and N. B. Koshelyaevsky (1994), The hydrogen maser clocks of DLR: Relative stability and time comparisons to clocks operated at PTB Braunschweig and IMVP, NPO, VNIFTRI near Moscow, in *Proceedings of the 8th European Frequency and Time Forum EFTF*, pp. 429–436.

Nerem, R. S., B. G. Bills, and J. B. McNamee (1993), A high resolution gravity model for Venus: GVM-1, *Geophys. Res. Lett.*, *20*(7), 599–602.

Nerem, R. S., F. J. Lerch, J. A. Marshall, E. C. Pavlis, B. H. Putney, B. D. Tapley, R. J. Eanes, J. C. Ries, B. E. Schutz, C. K. Shum, M. M. Watkins, S. M. Klosko, J. C. Chan, S. B. Luthcke, G. B. Patel, N. K. Pavlis, R. G. Williamson, R. H. Rapp, R. Biancale, and F. Nouel (1994), Gravity model development for Topex/Poseidon: Joint Gravity Models 1 and 2, *J. Geophys. Res.*, *99*(C12), 24421–24447.

Nerem, R. S., C. Jekeli, and W. M. Kaula (1995), Gravity field determination and characteristics: Retrospective and prospective, *J. Geophys. Res.*, *100*(B8), 15053–15074.

Néron de Surgy, O., and J. Laskar (1997), On the long term evolution of the spin of the Earth, *Astron. Astrophys.*, *318*, 975–989.

Neumann, G. A., M. T. Zuber, D.E. Smith, and F. G. Lemoine (1996), The lunar crust: global structure and signature of major basins, *J. Geophys. Res.*, *101*(E7), 16841–16863.

Newhall, X. X., and J. G. Williams (1997), Estimation of the lunar physical librations, *Celest. Mech. & Dynam. Astron.*, *66*, 21–30, also published as part of *Dynamics and Astronomy of Natural and Artificial Celestial Bodies*, Proc. of IAU Colloquium 165, Poznań, Poland, 1-5 July, 1996.

NIMA (1997), Department of Defence World Geodetic System 1984 – Its Definition and Relationships with Local Geodetic Systems, *TR 8350.2*, National Imagery and Mapping Agency (NIMA), Bethesda, MD.

Nozette, S., P. Rustan, L. P. Pleasance, D. M. Horan, P. Regeon, E. M. Shoemaker, P. D. Spudis, C. H. Acton, D. N. Baker, J. E. Blamont, B. J. Buratti, M. P. Corson, M. E. Davies, T. C. Duxbury, E. M. Eliason, B. M. Jakosky, J. F. Kordas,

I. T. Lewis, C. L. Lichtenberg, P. G. Lucey, E. Malaret, M. A. Massie, J. H. Resnick, C. J. Rollins, H. S. Park, A. S. McEwen, R. E. Priest, C. M. Pieters, R. A. Reisse, M. S. Robinson, R. A. Simpson, D. E. Smith, T. C. Sorenson, R. W. Vorder Breugge, and M. T. Zuber (1994), The Clementine mission to the Moon: Scientific overview, *Science, 266*, 1835–1862, 16 December 1994.

O'Keefe, J. A. (1959), Zonal harmonics of the Earth's gravitational field and the basic hypothesis of geodesy, *J. Geophys. Res., 64*(12), 2389–2392.

Olson, T. R. (1996), *Geopotential improvement from Explorer platform single-frequency GPS tracking*, Ph.D. dissertation, University of Colorado at Boulder.

Park, S-Y., and J. L. Junkins (1995), Orbital mission analysis for a lunar mapping satellite, *J. Astron. Sci., 43*(2), 207–217.

Pavlis, D. E., S. Luo, P. Dahiroc, J. J. McCarthy, and S. B. Luthcke (1998), *GEODYN II Operations Manual*, STX Systems Corp., Greenbelt, MD.

Pereverzev, S., and E. Schock (1999), Error estimates for band-limited spherical regularization wavelets in an inverse problem of satellite geodesy, *Inverse Problems, 15*, 881–890.

Phillips, D. L. (1962), A technique for the numerical solution of certain integral equations of the first kind, *J. Assoc. for Comput. Mach., 9*, 84–97.

Press, W. H., S.A. Teukolsky, W. T. Vetterling, and B. P. Flannery (1992), *Numerical Recipes in FORTRAN – The Art of Scientific Computing*, 653 p., Cambridge University Press, second edition.

Racca, G. D., B. H. Foing, and C. P. Chaloner (1996), Smallsat version of the European Moon Orbiting Observatory, *Acta Astronaut., 39*(1–4), 121–131.

Regińska, T. (1996), A regularization parameter in discrete ill-posed problems, *SIAM J. Sci. Comput., 17*(3), 740–749.

Reigber, C. H., P. Schwintzer, Ph. Hartl, K. H. Ilk, R. Rummel, M. van Gelderen, E. J. O. Schrama, K. F. Wakker, B. A. C. Ambrosius, and H. Leenman (1987), Study of a Satellite-to-Satellite Tracking Gravity Mission – Final Report, Deutsches Geodätisches Forschungsinstitut – ABT I and Technische Universität München–Institut für Astronomische und Physikalische Geodäsie and Technische Universiteit Delft – Faculteit der Geodesie and Faculteit der Luchtvaart- en Ruimtevaarttechniek, March 1987, ESTEC/Contract No. 6557/85/NL/PP(SC).

Reigber, Ch., R. Bock, Ch. Förste, L. Grunwaldt, N. Jakowski, H. Lühr, and P. Schwintzer (1996), CHAMP Phase B – Executive Summary, *G.F.Z. STR96/13*, GeoForschungsZentrum Potsdam.

Ridenoure R., (Editor) (1991), Lunar Observer: Back to the Moon, 1991 mission and system definition summary, *JPL Internal Document D–8607*, Jet Propulsion Laboratory, Pasadena, California.

Rosborough, G. W. (1986), *Satellite Orbit perturbations Due to the Geopotential*, Ph.D. dissertation, Center for Space Research, The University of Texas at Austin, CSR 86-1.

Rosborough, G. W. (1987), Orbit Error due to Gravity Model Error, in *AAS/AIAA Astrodynamics Specialist Conference, Kalispell, Montana, August 10-13*, American Astronautical Society, AAS paper 87-534.

Rosborough, G. W., and B. D. Tapley (1987), Radial, transverse and normal satellite position perturbations due to the geopotential, *Celestial Mechanics, 40*, 409–421.

Rossi, A., F. Marzari, and P. Farinella (1999), Orbital evolution around irregular bodies, *Earth, Planets and Space, 51*(11), 1173–1180.

Rowlands, D. D., S. B. Luthke, J. A. Marshall, C. M. Cox, R. G. Williamson, and S. C. Rowton (1997), Space shuttle precision orbit determination in support of SLA–1 using TDRSS and GPS tracking data, *J. Astronaut. Sci., 45*(1), 113–129, January–March 1997.

Rummel, R. (1979), Determination of short-wavelength components of the gravity field from satellite-to-satellite tracking or satellite gradiometry; an attempt to an identification of problem areas, *Manuscripta Geodetica, 4*, 107–148.

Rummel, R. (1997), Spherical spectral properties of the Earth's gravitational potential and its first and second derivatives, in *Geodetic Boundary Value Problems in View of the One Centimeter Geoid*, vol. 65 of *Lecture Notes in Earth Sciences*, edited by F. Sansò and R. Rummel, pp. 359–404, Springer-Verlag, Berlin.

Rummel, R., and O. L. Colombo (1985), Gravity field determination from satellite gradiometry, *Bulletin Géodésique, 59*, 233–246.

Rummel, R., and M. van Gelderen (1995), Meissl scheme - spectral characteristics of physical geodesy, *Manuscripta Geodetica, 20*, 379–385.

Rummel, R., Ch. Reigber, and K. H. Ilk (1978), The use of satellite to satellite tracking for gravity parameter recovery, in *Proceedings of the European Workshop on Space Oceanography, Navigation and Geodynamics (SONG) at Schloss Elmau*, pp. 16–21, ESA SP–137.

Rummel, R., K. P. Schwarz, and M. Gerstl (1979), Least squares collocation and regularization, *Bulletin Géodésique, 53*, 343–361.

Rummel, R., M. van Gelderen, R. Koop, E. Schrama, F. Sansò, M. Brovelli, F. Miggliaccio, and F. Sacerdote (1993), Spherical harmonic analysis of satellite gradiometry, *Publications on geodesy, New series No. 39*, Netherlands Geodetic Commission, Delft.

Runcorn, S. K. (1974), On the origin of mascons and moonquakes, in *Proceedings of the Fifth Lunar Science Conference (Supplement 5, Geochimica et Cosmochimica Acta, Vol. 3)*, Pergamon Press, pp. 3115-3126.

Runcorn, S. K. (1977), Early melting of the Moon, in *Proceedings of the Eighth Lunar Science Conference (Supplement 8, Geochimica et Cosmochimica Acta, Vol. 1)*, Pergamon Press, pp. 463-469.

Russell, C. T., P. J. Coleman Jr., B. R. Lichtenstein, G. Schubert, and L. R. Sharp (1973), Subsatellite measurements of the lunar magnetic field, in *Proceedings of the Fourth Lunar Science Conference (Supplement 4, Geochimica et Cosmochimica Acta, Vol. 3)*, Pergamon Press, pp. 2833-2845.

Sagitov, M. U., B. Bodri, V. S. Nazarenko, and Kh. G. Tadzhidinov (1986), *Lunar Gravimetry*, vol. 35 of *International Geophysics Series*, Academic Press.

Sasaki, S., Y. Iijima, K. Tanaka, M. Kato, M. Hasjimoto, H. Mitzutani, K. Tsuruda, and Y. Takizawa (1999), Scientific research in the SELENE mission, paper IAF-99-Q.4.04, 50th International Astronautical Congress, Amsterdam, 4–8 October, 1999.

Saunders, M. (1999), The Discovery experiment – the selction process; results to date, *Acta Astronaut.*, *45*(4–9), 215–225.

Scharroo, R., and P. Visser (1998), Precise orbit determination and gravity field improvement for the ERS satellites, *J. Geophys. Res.*, *103*(C4), 8113–8127, April 15 1998.

Schmitt, C., and H. Bauer (2000), CHAMP attitude and orbit control system, *Acta Astronaut.*, *46*(2–6), 327–333.

Schneider, F. (1997), *Inverse problems in satellite geodesy and their approximate solution by splines and wavelets*, Ph.D. dissertation, Universität Kaiserslauteren.

Schrama, E. J. O. (1986), A Study of a Satellite-to-Satellite Tracking Configuration by Application of Linear Perturbation Theory, Delft University of Technology, Department of Geodesy, November 1986.

Schrama, E. J. O. (1989), The role of orbit errors in processing of satellite altimeter data, *Publications on geodesy. New series No. 33*, Netherlands Geodetic Commission.

Schrama, E. J. O. (1991), Gravity field error analysis: Applications of Global Positioning System receivers and gradiometers on low orbiting platforms, *J. Geophys. Res.*, *96*(B12), 20041–20051.

Schröder, P., and W. Sweldens (2000), Spherical wavelets: Efficiently representing functions on a sphere, in *Wavelets in the Geosciences*, vol. 90 of *Lecture Notes in Eearth Sciences*, edited by R. Klees and R. Haagmans, pp. 158–188, Springer-Verlag, Berlin.

Schuh, W. D. (1996), Tailored numerical solution strategies for the global determination of the earth's gravity field, *Folge 81*, Mitteilungen der geodätischen Institute der Technischen Universität Graz, Graz.

Schuh, W. D., H. Sünkel, W. Hausleitner, and E. Höck (1996), Refinement of iterative procedures for the reduction of spaceborne gravimetry data, in *Study of advanced reduction methods for spaceborne gravimetry data, and of data combination with geophysical parameters - CIGAR IV, final report*, edited by M. Schuyer and H. Sünkel, ESA.

Schwintzer, P. (1990), Sensitivity analysis in least squares gravity field modelling by means of redundancy decomposition of stochastic a priori information, Deutsches Geodätiches Forschungs-Zentrum, Internal Report.

Schwintzer, P., Ch. Reigber, W. Barth, F.-H. Massmann, J. C. Raimondo, M. Gerstl, A. Bode, H. Li, R. Biancale, G. Balmino, B. Moyonot, J.M. Lemoine, J. C. Marty, F. Barlier, and Y. Boudon (1992), GRIM4 - Globale Erdschwerefeldmodelle, *Zeitschrift für Vermessungswesen*, 4, 227–247.

Schwintzer, P., C. Reigber, A. Bode, Z. Kang, S. Y. Zhu, F. H. Massmann, J. C. Raimondo, R. Biancale, G. Balmino, J. M. Lemoine, B. Moynot, J. C. Marty, F. Barlier, and Y. Boudon (1997), Long-wavelength global gravity field models: GRIM4-S4, GRIM4-C4, *J. Geodesy*, 71, 189–208.

Seeber, G. (1993), *Satellite Geodesy*, Walter de Gruyter, Berlin.

Sehnal, L., D. Vokrouhlicky, L. Pospisilova, and A. Kohlhase (1997), Nongravitational forces acting on the CHAMP satellite, *Acta Astronaut.*, 19(11), 1695–1698.

Seidelmann, P. K. (ed.) (1992), *Explanatory Supplement to the Astronomical Almanac*, University Science Books, Mill Valley, CA.

SELENE (1996), SELENE PROJECT (SELenological and ENgineering Explorer), ISAS/NASDA Joint Moon Orbiting Satellite Project, ISAS/NASDA, Tokyo, Japan, project folder.

Shampine, L. F., and M. K. Gordon (1975), *Computer Solution of Ordinary Differential Equations*, W. H. Freeman and Company.

Shewchuk, J. R. (1994), An introduction to the conjugate gradient method without the agonizing pain, edition $1\frac{1}{4}$, School of Computer Science, Carnegie Mellon University, Pittsburgh, PA, USA.

SID (2000), GOCE end to end performance analysis, *ESA/ESTEC Contract No. 12735/98/NL/GD*, SRON, IAPG and DEOS.

Simons, M., S. C. Solomon, and B. H. Hager (1997), Localization of gravity and topography: constraints on the tectonics and mantle dynamics of Venus, *Geophys. J. Int.*, 131, 24–44.

Sjogren, W. L., P. M. Muller, and W. R. Wollenhaupt (1972a), Apollo 15 gravity analysis from the S-band transponder experiment, *Moon*, 4, 411–418.

Sjogren, W. L., P. M. Muller, and P. Gottlieb (1972b), Lunar gravity via Apollo 14 Doppler radio tracking, *Science*, 175, 165–168.

Sjogren, W. L., R. N. Wimberly, and W. R. Wollenhaupt (1974a), Lunar gravity via the Apollo 15 and 16 subsatellites, *The Moon*, 9, 115–128.

Sjogren, W. L., R. N. Wimberly, and W. R. Wollenhaupt (1974b), Lunar gravity: Apollo 16, *The Moon*, 11, 35–40.

Sjogren, W. L., R. N. Wimberly, and W. R. Wollenhaupt (1974c), Lunar gravity: Apollo 17, *The Moon*, 11, 41–52.

Smit, J. (1994), Extinctions at the cretatceous-tertiary boundary; the link to the Chixulub impact, in *Hazards due to comets and asteroids*, edited by T. Gehrels, pp. 859–878, Arizona University Press.

Smith, A. J. E., E. T. Hesper, D. C. Kuijper, G. J.Mets, P. N. A. M. Visser, B. A. C. Ambrosius, and K. F. Wakker (1994), TOPEX/POSEIDON data analysis study, Section Space Research & Technology, Faculty of Aerospace Engineering, Delft University of Technology, Delft.

Smith, D. E., F. J. Lerch, R. S. Nerem, G. B. Patel, S. K. Fricke, and F. G. Lemoine (1993), An improved gravity model for Mars - Goddard Mars Model–1, *J. Geophys. Res.*, *98*(E11), 20871–20889.

Smith, D. E., M. T. Zuber, G. A. Neumann, and F. G. Lemoine (1997), Topography of the Moon from the Clementine lidar, *J. Geophys. Res.*, *102*(E1), 1591–1611.

Smith, D. E., M. T. Zuber, S. C. Solomon, R. J. Phillips, J. W. Head, J. B. Garwin, W. B. Banerdt, D. O. Muhlemann, G. H. Pettengill, G. A. Neumann, F. G. Lemoine, J. B. Abshire, O. Aharonson, C. D. Brown, S. A. Hauck, A. B. Ivanov, P. J. McGovern, H. J. Zwally, and T. C. Duxbury (1999*a*), The global topography of Mars and implications for surface evolution, *Science*, *284*, 1495–1503.

Smith, D. E., W. L. Sjogren, G. L. Tyler, G. Balmino, F. G. Lemoine, and A. S. Konopliv (1999*b*), The gravity field of Mars: results from Mars Global Surveyor, *Science*, *286*, 94–97.

Sneeuw, N. J. (1991), Inclination functions - group theoretical background and a recursive algorithm, *91.2*, Reports of the Faculty of Geodetic Engineering, Delft.

Snieder, R. (1998), The role of nonlinearity in inverse problems, *Inverse Problems*, *14*, 387–404.

Solomon, S. C., and J. W. Head (1979), Vertical movement in mare basins: Relation to mare emplacement, basin tectonics, and lunar thermal history, *J. Geophys. Res.*, *84*(B4), 1667–1682.

Solomon, S. C., and J. W. Head (1980), Lunar mascon basins: Lava filling, tectonics and evolution of the lithosphere, *Rev. Geophys. Spa. Phys.*, *18*(1), 107–141.

Solomon, S. C., and J. Longhi (1977), Magma oceanography: I. Thermal evolution, in *Proceedings of the Eighth Lunar Science Conference (Supplement 8, Geochimica et Cosmochimica Acta, Vol. 1)*, Pergamon Press, pp. 583-599.

Spudis, P. D. (1992), To the Moon: Faster, cheaper - and better, *The Planetary Report*, *12*(4), 8–9, July/August 1992.

Spudis, P. D. (1996), *The Once and Future Moon*, Smithsonian Library of the Solar System, Smithsonian Institution Press.

Spudis, P. D., R. A. Reisse, and J. J. Gillis (1994), Ancient mulitring basins on the Moon revealed by Clementine laser altimetry, *Science*, *266*, 1848–1851, 16 December 1994.

Standish, E. M. (1998), JPL Planetary and Lunar Ephemeris, DE405/LE405, *IOM 312.F–98–048*, Jet Propulsion Laboratory.

Standish, E. M., X. X. Newhall, J. G. Williams, and W. M. Folkner (1995), JPL Planetary and Lunar Ephemeris, DE403/LE403, *IOM 314.10–127*, Jet Propulsion Laboratory.

Stanley, H. R. (1975), The GEOS-3 project, *J. Geophys. Res.*, *84*(B8), 3779–3783.

Tapley, B. D., and G. W. Rosborough (1985), Geographically correlated orbit error and its effect on satellite altimetry missions, *J. Geophys. Res.*, *90*(C6), 11817–11831.

Tapley, B. D., M. M. Watkins, J. C. Ries, G. W. Davis, R. J. Eanes, S. R. Poole, H. J. Rim, B. E. Schutz, C. K. Shum, R. S. Nerem, F. J. Lerch, J. A. Marshall, S. M. Klosko, N. K. Pavlis, and R. G. Williamson (1996), The Joint Gravity Model 3, *J. Geophys. Res.*, *101*(B12), 28029–28049.

Teunissen, P. J. G. (2000), *Adjustment theory - an introduction*, Series on Mathematical Geodesy and Positioning, Delft University Press, Delft, The Netherlands.

Thomas, P. C. (1991), Planetary geodesy, *Rev. Geophys.*, *29*, 182–187.

Thomas, P. C. (1993), Gravity, tides, and topography on small satellites and asteroids - application to surface-features of the martian satellites, *Icarus*, *105*(2), 326–344.

Tikhonov, A. N. (1963*a*), Solution of incorrectly formulated problems and the regularization method, *Soviet Math. Dokl.*, *4*, 1035–1038, originally published in Russian in *Dokl. Akad. Nauk. SSSR*, Vol. 151, pp. 501-504, (1963).

Tikhonov, A. N. (1963*b*), Regularization of incorrectly posed problems, *Soviet Math. Dokl.*, *4*, 1624–1627.

Toksöz, M. N., S. C. Solomon, J. W. Minear, and D. H. Johnston (1972), Thermal evolution of the Moon, *The Moon*, *4*, 190–213.

Toksöz, M. N., A. M. Dainty, S. C. Solomon, and K. R. Anderson (1974), Structure of the Moon, *Rev. Geophys. Spa. Phys.*, *12*(4), 539–567.

Tranquilla, J. M., and B. G. Colpitts (1988), GPS antenna design characteristics for high-precision applications, in *Proceedings of the ASCE Specialty Conference GPS88: Engineering Applications of GPS Satellite Surveying Technology*, Nashville, TN, pp. 2-14.

Tscherning, C. C. (2001), Computation of spherical harmonic coefficients and their error estimates using least-squares collocation, *J. Geodesy*, *75*, 12–18.

Tscherning, C. C., and R. H. Rapp (1974), Closed covariance expressions for gravity anomalies, geoid undulations, and deflections of the vertical implied by anomaly degree variance models, *Report No. 208*, Ohio State University.

Tuckness, D. G. (1995*a*), Influence of suboptimal navigation filter design on lunar landing navigation accuracy, *J. Spacecraft and Rockets*, *32*(2).

Tuckness, D. G. (1995*b*), Selenopotential field effects on lunar landing accuracy, *J. Spacecraft and Rockets*, *32*(6).

Ullman, R. E. (1993), *SOLVE Program - Mathematical Formulation*, March 1993, doc. no. HSTX/G&G-9201.

Vasile, M. (1996), Approccio numerico per la determinazione di orbite seleniche frozen e periodiche, Master's thesis, Dipartimento di Ingegneria Aerospaziale, Politecnico di Milano, in Italian.

Verger, F. (ed.) (1992), *Atlas de Géographie de l'Espace*, Sides-Reclus, Antony and Montpellier, France.

Vinod, H.D., and A. Ullah (1981), *Recent advances in regression methods*, Marcel Dekker Inc., New York, Basel.

Visser, P. N. A. M. (1992), *The Use of Satellites in Gravity Field Determination and Model Adjustment*, Ph.D. dissertation, Delft University of Technology, Delft University Press.

Visser, P. N. A. M. (1999), Gravity field determination with GOCE and GRACE, *Adv. Space Res.*, *23*(4), 741–746.

Visser, P. N. A. M., and B. A. C. Ambrosius (1996), TDRSS Inter-Satellite Tracking Evaluation Using TOPEX/POSEIDON - Final Report, Delft Institute for Earth-Oriented Space Research, Delft University of Technology, Delft, ESOC contract 11179/94/D/IM.

Visser, P. N. A. M., B. A. C. Ambrosius, and K. F. Wakker (1995*a*), Gravity field model adjustment from ERS-1 and TOPEX altimetry using an analytical orbit perturbation theory, *Adv. Space Res.*, *16*(12), 143–147.

Visser, P. N. A. M., K. F. Wakker, and B. A. C. Ambrosius (1995*b*), Global gravity field recovery from the ARISTOTELES mission, *J. Geophys. Res.*, *99*(B2), 2841–2851.

Visser, P. N. A. M., J. van den IJssel, R. Koop, and R. Klees (2000), Exploring gravity field determination from orbit perturbations of the European gravity mission GOCE, *J. Geodesy, submitted.*

Vogel, C. R. (1996), Non-convergence of the L-curve regularization parameter selection method, *Inverse Problems*, *12*, 535–547.

Vonbun, F. O., W. D. Kahn, W. T. Wells, and T. D. Conrad (1980), Determination of 5° × 5° gravity anomalies using satelliet-to-satellite tracking between ATS-6 and Apollo, *Geophys. J. R. astr. Soc.*, *61*, 645–657.

Wagner, C. A. (1983), Direct determination of gravitational harmonics from low–low GRAVSAT data, *J. Geophys. Res.*, *88*(B12), 10309–10321, December 10, 1983.

Wagner, C. A. (1985), Radial variations of a satellite orbit due to gravitational errors: Implications for satellite altimetry, *J. Geophys. Res.*, *90*(B4), 3027–3036, March 10, 1985.

Wagner, C. A. (1987*a*), Geopotential orbit variations: Applications to error analysis, *J. Geophys. Res.*, *92*(B8), 8136–8146, July 10, 1987.

Wagner, C. A. (1987*b*), Improved gravitational recovery from a geopotential research mission satellite pair flying en echelon, *J. Geophys. Res.*, *92*(B8), 8147–8155, July 10, 1987.

Wahba, G. (1990), *Spline models for observational data*, CMBMS-NSF Conference Series in Applied Mathematics 59, SIAM, Philadelphia, Pennsylvania.

Wahr, J., M. Molenaar, and F. Bryan (1998), Time variability of the Earth's gravity field: Hydrological and oceanic effects and their possible detection using GRACE, *J. Geophys. Res.*, *103*(B12), 30205–30229.

Wänke, H., and G. Dreibus (1986), Geochemical evidence for the formation of the Moon by impact-induced fission of the proto-Earth, in *Origin of the Moon*, edited by W. K. Hartmann *et al.*, pp. 649–672, Lunar and Planetary Institute, Houston.

Warren, P. H. (1985), The magma ocean concept and lunar evolution, *Ann. Rev. Earth Planet. Sci.*, *13*, 201–240.

Watkins, M. M., E. S. Davies, W. G. Melbourne, T. P. Yunck, J. Sharma, and B. D. Tapley (1995), GRACE: A new mission concept for high resolution gravity field mapping, EGS, Hamburg, Germany (abstract), April 3–7 1995.

Weischede, F. (2000), Description of the DEEPEST deep space trajectory propagation, estimation and tracking data simulation software, *FDS–GEN–xxxx*, DLR/GSOC, Wessling, Germany.

Weischede, F., E. Gill, O. Montenbruck, and R. Floberghagen (1999), Lunar Prospector orbit determination and gravity field modeling based on Weilheim 3-way Doppler measurements, *J. Brazilian Soc. Mechanical Sci.*, *21*, 280–286, Proceedings 14th International Symposium on Space Flight Dynamics, Foz do Iguacu, Brazil, 8-12 February.

Wieczorek, M. A., and R. J. Phillips (1997), The structure and composition of the lunar highland crust, *J. Geophys. Res.*, *102*(E5), 10933–10943, May 25, 1997.

Wieczorek, M. A., and R. J. Phillips (1998), Potential anomalies on a sphere: Applications to the thickness of the lunar crust, *J. Geophys. Res.*, *103*(E1), 1715–1724, January 25, 1998.

Wiejak, W., E. J. O. Schrama, and R. Rummel (1991), Spectral representation of the satellite-to-satellite tracking observable, *Adv. Space Res.*, *11*(6), 6197–6224.

Williams, J. G., M. A. Slade, D. E. Eckhardt, and W. M. Kaula (1973), Lunar physical librations and laser ranging, *The Moon*, *8*, 469–483.

Wise, D. U., and M. T. Yates (1970), Mascons as structural relief on a lunar 'moho', *J. Geophys. Res.*, *75*(2), 261–268.

Wnuk, E. (1999), Recent progress in analytical orbit theories, *Adv. Space Res.*, *23*(4), 677–687.

Wolff, M. (1969), Direct measurement of the earth's gravitational potential using a satellite pair, *J. Geophys. Res.*, *74*, 5295–5300.

Wong, L., W. Downs, W. Sjogren, P. Muller, and P. Gottlieb (1971), A surface-layer representation of the lunar gravitational field, *J. Geophys. Res.*, *76*(26), 6220–6236.

Wood, J. A. (1986), Moon over Mauna Kea: a review of hypotheses of formation of Earth's Moon, in *Origin of the Moon*, edited by W. K. Hartmann *et al.*, pp. 17–55, Lunar and Planetary Institute, Houston.

Woodworth, P. L., J. Johanessen, P. Le Grand, C. Le Provost, G. Balmino, R. Rummel, R. Sabadini, H. Sünkel, C. C. Tscherning, and P. Visser (1998), Towards the definitive space gravity mission, *International WOCE Newsletter*, *33*, December 1998, ISSN 1029-1725.

Xu, P. (1992*a*), Determination of surface gravity anomalies using gradiometric observables, *Geophys. J. Int.*, *110*, 321–332.

Xu, P. (1992*b*), The value of minimum norm estimation of geopotential fields, *Geophys. J. Int.*, *111*, 170–178.

Xu, P. (1998), Truncated SVD methods for discrete linear ill-posed problems, *Geophys. J. Int.*, *135*, 505–514.

Xu, P., and R. Rummel (1994*a*), A simulation study of smoothness methods in recovery of regional gravity fields, *Geophys. J. Int.*, *117*, 472–486.

Xu, P., and R. Rummel (1994*b*), Generalized ridge regression with applications in determination of potential fields, *Manuscripta Geodetica*, *20*, 8–20.

Yeomans, D. K., J.-P. Barriot, D. W. Dunham, R. W. Farquhar, J. D. Giorgini, C. E. Helfrich, A. S. Konopliv, J. V. McAdams, J. K. Miller, W. Owen Jr., D. J. Scheeres, S. P. Synnot, and B. G. Williams (1997), Estimating the mass of asteroid 253 Mathilde from tracking data during the NEAR flyby, *Science*, *278*, 2106–2109.

Yoder, C. F., J. G. Williams, J. O. Dickey, B. E. Schutz, R. J. Eanes, and B. D. Tapley (1983), Secular variation of Earth's gravitational harmonic J_2 from Lageos and non-tidal acceleration of Earth rotation, *Nature*, *303*, 757–762.

Yunck, T. P., and W. G. Melbourne (1996), Spaceborne GPS for Earth Science, in *GPS Trends in Precise Terrestrial, Airborne and Spaceborne Applications*, vol. 115 of *International Association of Geodesy Symposium*, pp. 113–122, Springer-Verlag, Berlin.

Yunck, T. P., W. I. Bertiger, S. M. Lichten, and S. C. Wu (1986), Tracking Landsat-5 by a Differential GPS Technique, in *AIAA/AAS Astrodynamics Conference*, American Institute of Aeronautics and Astronautics, August 18–22, 1986, AIAA 86-2215-CP.

Yunck, T. P., S. C. Wu, J. T. Wu, and C. L. Thornton (1990), Precise tracking of remote sensing satellites with the Global Positioning System, *IEEE Trans. Geosci. and Remote Sens*, *28*(1), 108–116.

Yunck, T. P., W. I. Bertiger, S. C. Wu, Y. Bar-Sever, E. J. Christensen, B. J. Haines, S. M. Lichten, R. J. Muellerschoen, Y. Vigue, and P. Willis (1994), First assessment of GPS-based reduced dynamic orbit determination on TOPEX/Poseidon, *Geophys. Res. Lett.*, *21*(7), 541–544.

Zuber, M. T., D. E. Smith, F. G. Lemoine, and G. A. Neumann (1994), The shape
and internal structure of the Moon from the Clementine mission, *Science, 266,*
1839–1843, 16 December 1994.

Index

Previously published in Astrophysics and Space Science Library book series:

- Volume 269:**Mechanics of Turbulence of Multicomponent Gases**
  Authors: Mikhail Ya. Marov, Aleksander V. Kolesnichenko
  Hardbound, ISBN 1-4020-0103-7, December 2001
- Volume 268:**Multielement System Design in Astronomy and Radio Science**
  Authors: Lazarus E. Kopilovich, Leonid G. Sodin
  Hardbound, ISBN 1-4020-0069-3, November 2001
- Volume 267: **The Nature of Unidentified Galactic High-Energy Gamma-Ray Sources**
  Editors: Alberto Carramiñana, Olaf Reimer, David J. Thompson
  Hardbound, ISBN 1-4020-0010-3, October 2001
- Volume 266: **Organizations and Strategies in Astronomy II**
  Editor: André Heck
  Hardbound, ISBN 0-7923-7172-0, October 2001
- Volume 265: **Post-AGB Objects as a Phase of Stellar Evolution**
  Editors: R. Szczerba, S.K. Górny
  Hardbound, ISBN 0-7923-7145-3, July 2001
- Volume 264: **The Influence of Binaries on Stellar Population Studies**
  Editor: Dany Vanbeveren
  Hardbound, ISBN 0-7923-7104-6, July 2001
- Volume 262: **Whistler Phenomena**
  Short Impulse Propagation
  Authors: Csaba Ferencz, Orsolya E. Ferencz, Dániel Hamar, János Lichtenberger
  Hardbound, ISBN 0-7923-6995-5, June 2001
- Volume 261: **Collisional Processes in the Solar System**
  Editors: Mikhail Ya. Marov, Hans Rickman
  Hardbound, ISBN 0-7923-6946-7, May 2001
- Volume 260: **Solar Cosmic Rays**
  Author: Leonty I. Miroshnichenko
  Hardbound, ISBN 0-7923-6928-9, May 2001
- Volume 259: **The Dynamic Sun**
  Editors: Arnold Hanslmeier, Mauro Messerotti, Astrid Veronig
  Hardbound, ISBN 0-7923-6915-7, May 2001
- Volume 258: **Electrohydrodynamics in Dusty and Dirty Plasmas**
  Gravito-Electrodynamics and EHD
  Author: Hiroshi Kikuchi
  Hardbound, ISBN 0-7923-6822-3, June 2001
- Volume 257: **Stellar Pulsation - Nonlinear Studies**
  Editors: Mine Takeuti, Dimitar D. Sasselov
  Hardbound, ISBN 0-7923-6818-5, March 2001
- Volume 256: **Organizations and Strategies in Astronomy**
  Editor: André Heck
  Hardbound, ISBN 0-7923-6671-9, November 2000
- Volume 255: **The Evolution of the Milky Way**
  Stars versus Clusters
  Editors: Francesca Matteucci, Franco Giovannelli
  Hardbound, ISBN 0-7923-6679-4, January 2001